ABANDONED TRAIN STATIONS

ABANDONED TRAIN STATIONS

RAIL STATIONS, YARDS, SIGNAL BOXES AND TRACK THAT THE WORLD LEFT BEHIND

DAVID ROSS

amber
BOOKS

Published by Amber Books Ltd
United House
North Road
London
N7 9DP
United Kingdom
www.amberbooks.co.uk
Instagram: amberbooksltd
Facebook: amberbooks
Twitter: @amberbooks
Pinterest: amberbooksltd

ISBN: 978-1-83886-199-5

Project Editor: Michael Spilling
Designer: Mark Batley
Picture Research: Justin Willsdon

Printed in China

Contents

Introduction

An abandoned station is one that is no longer staffed, and no longer served by trains. It may be on a working line, or on a disused branch, or stand in rail-less isolation. There are many thousands of them, all over the world: reminders of the hundred-year period when railways had a virtual monopoly of inland transport. Across the American prairies, the African savannah, and among city streets, the whistle of the steam locomotive announced the departure to far-off places.

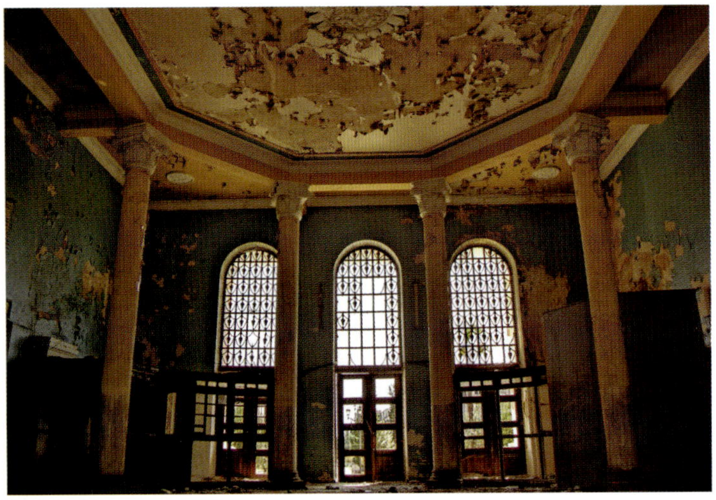

But even in the heyday of railways, stations might come and go. The world's first train station, at Stockton in England, opened in 1824 and was replaced by a relocated, purpose-built, better-equipped depot within a couple of years. Growth of towns and cities, the development and decay of industries like mining and quarrying, the movement of population, all might mean the closure of railways or resiting of stations, leaving abandoned buildings and fixtures behind. Every one of them offers an intriguing glimpse into a past era.

ABOVE:
Gudauta Station, Abkhazia
Abkhazia's disputed secession from Georgia in 1993 left the town of Gudauta with a fine station but no railway connection. With little hope of a resumption of services, the building was taken in hand by local residents in 2018, in the hope of finding a new use for it.

OPPOSITE:
16th Street Station, Oakland, California, USA
Once a terminal for the Southern Pacific Railroad's transcontinental trains, the station has been out of use since 1994.

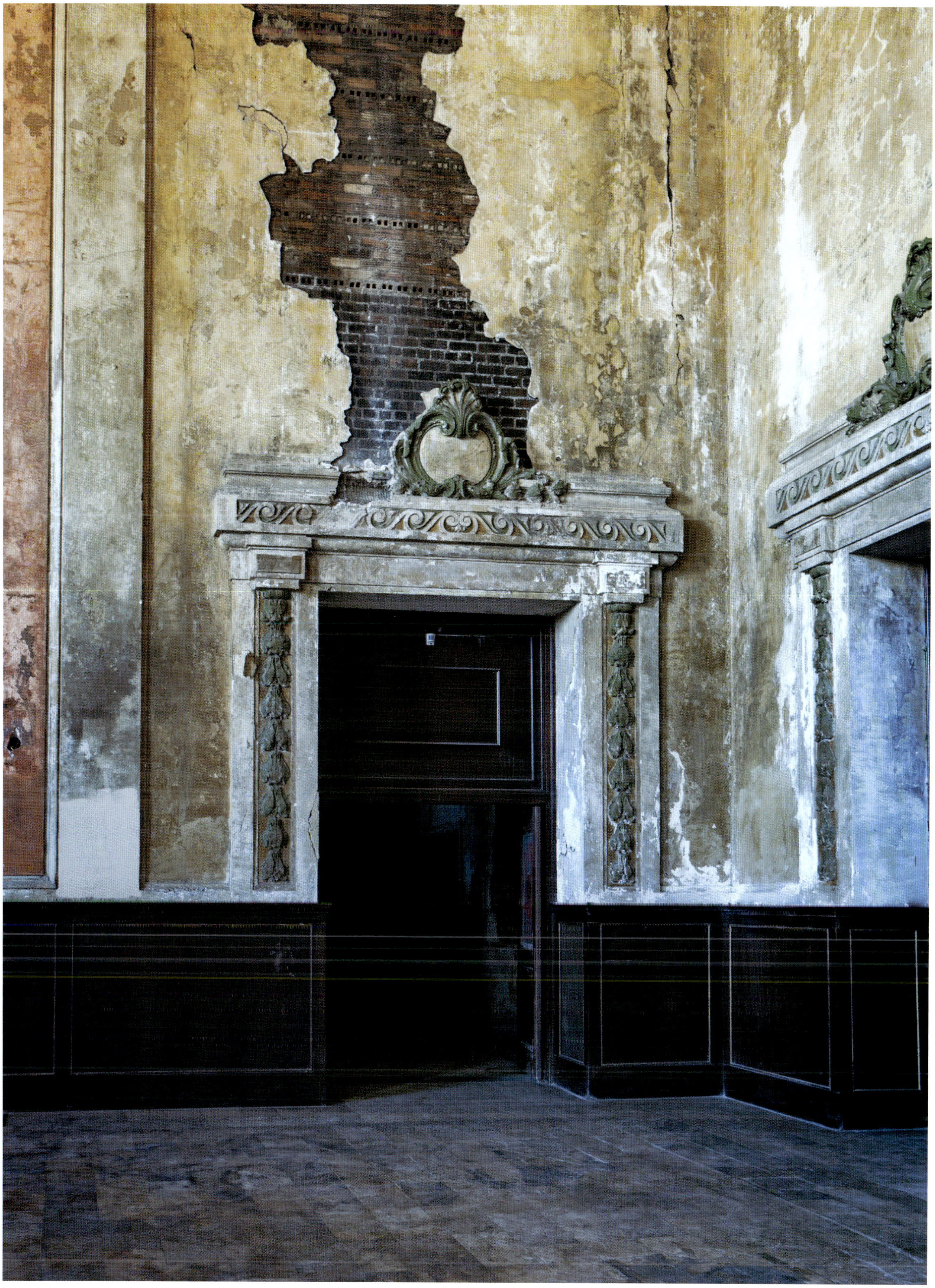

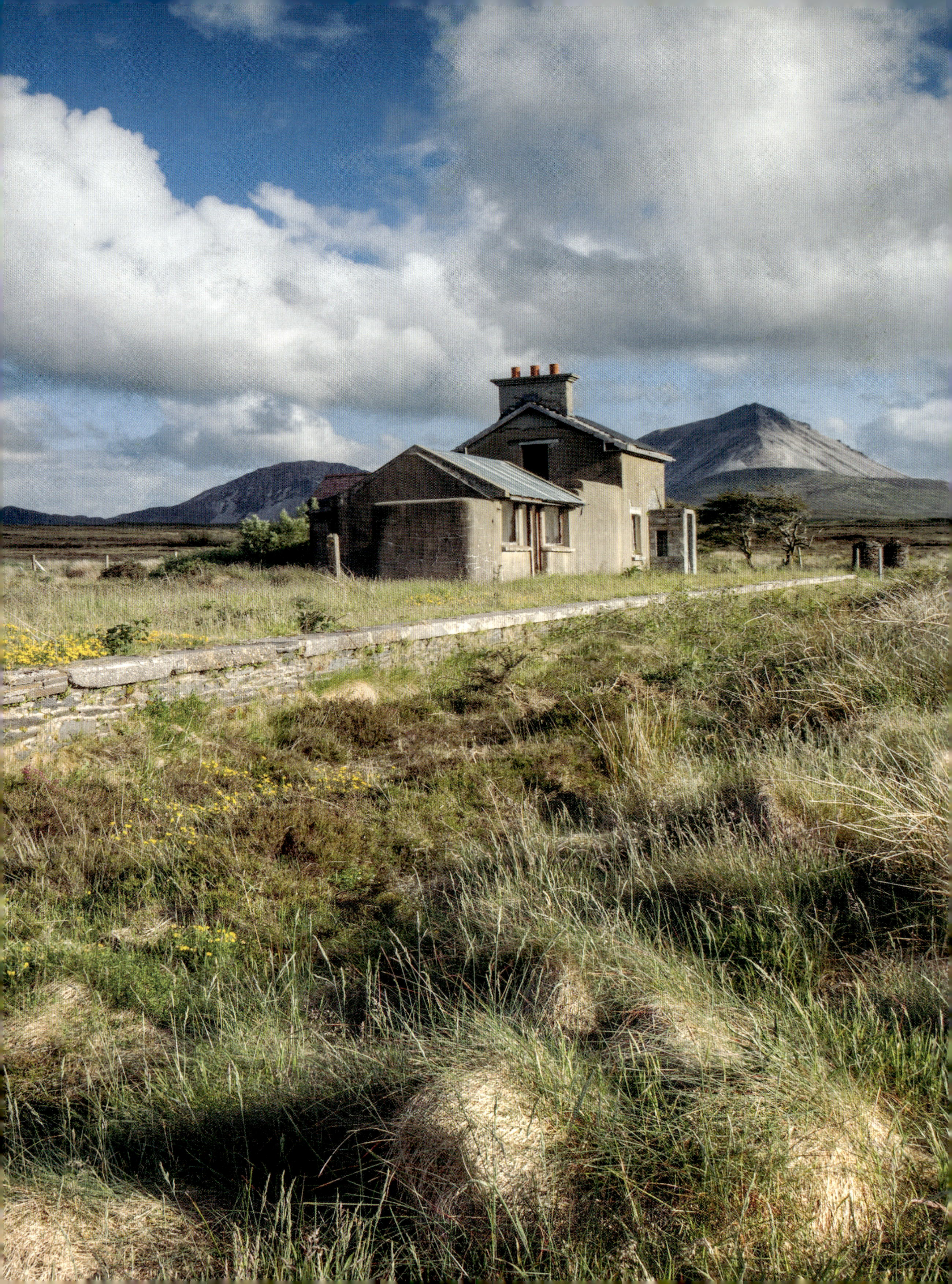

Europe

Europe has the densest railway network of any continent, though many lines and stations have been closed. This is especially true of rural areas, where thousands of small stations have been abandoned. In farming districts, it was once thought that two railways within twenty miles of each other could each make a profit.

Steam road engines debunked this notion even before the internal combustion engine killed it off. The range of abandoned stations is enormous, from wooden sheds to buildings intended to resemble grand country houses, and to the nondescript utilitarian architecture of the 1960s. Perhaps one extreme was reached early on, when the builders of England's Hereford & Shrewsbury Railway, in 1853, installed the Moreton on Lugg booking office inside a large hollow tree at the lineside (no photograph survives); and another in the palatial Canfranc (1928), in the Spanish Pyrenees.

Abandoned stations reflect many changes in commercial and social life, like the rise and fall of iron mining in the Glens of Antrim (Northern Ireland), and the changes in cross-Channel traffic between England and France. Berlin's vast abandoned railway yards echo the impact of post-war division and reunification in Germany. Lines of dumped modern locomotives in Albania remain from the former communist government's ambitious expansion schemes. Like cities in North America, Prague and Athens struggle to find a role for disused central stations: much-loved buildings that were once part of the essence of the modern city but no longer have trains.

OPPOSITE:

Cashelnagor, County Donegal, Ireland

A remote station on a remote line, the 914mm (3ft) gauge Letterkenny and Burtonport Extension in Ireland's extreme north-west, it opened in 1903 and closed in 1940. In 1992 it was partially restored for the film *The Railway Station Man*. Seen here in its abandoned state, in 2017 the station building was fully restored and converted into two holiday cottages.

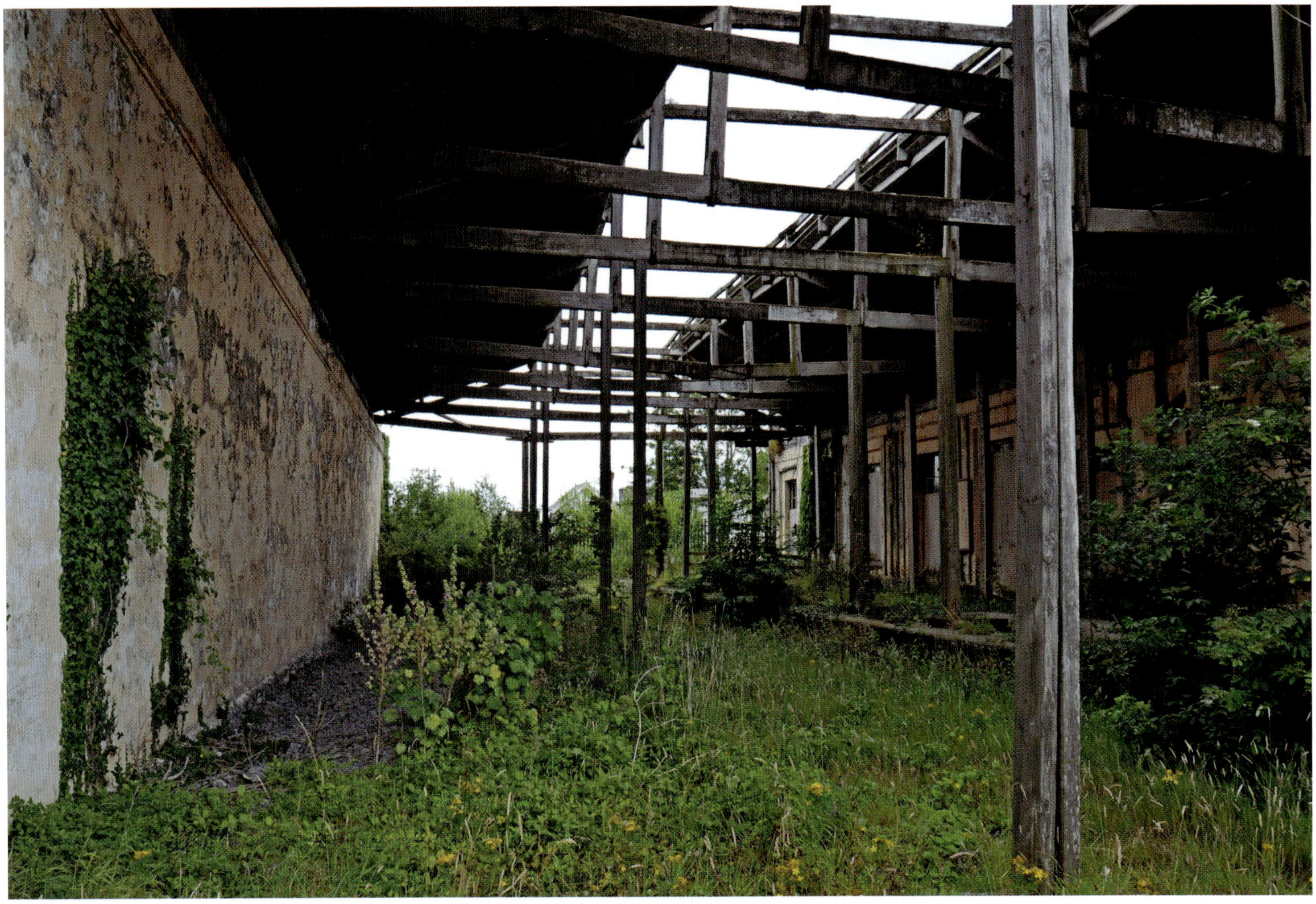

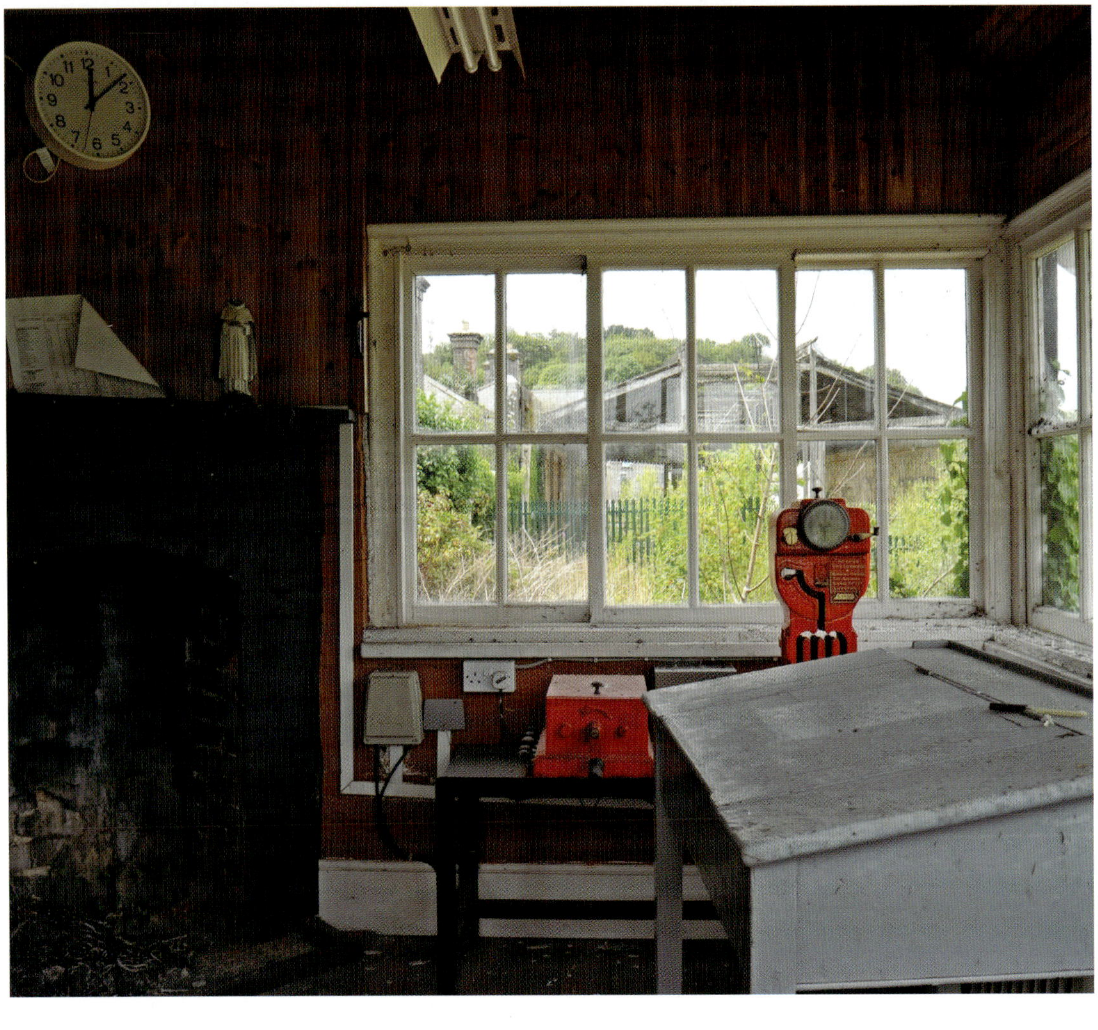

ALL PHOTOGRAPHS:
Foynes, County Limerick, Ireland
From 1858, Foynes was the terminus of a branch line from Limerick. Passenger services ended in 1963; occasional goods trains ran until 2000. The current status of the line is mothballed but with no maintenance work, its condition is steadily deteriorating. The passenger station, once with an overall roof, is a ruin. As with other disused lines in Ireland and elsewhere, there are calls and proposals for reinstatement, but the economic difficulties are great.

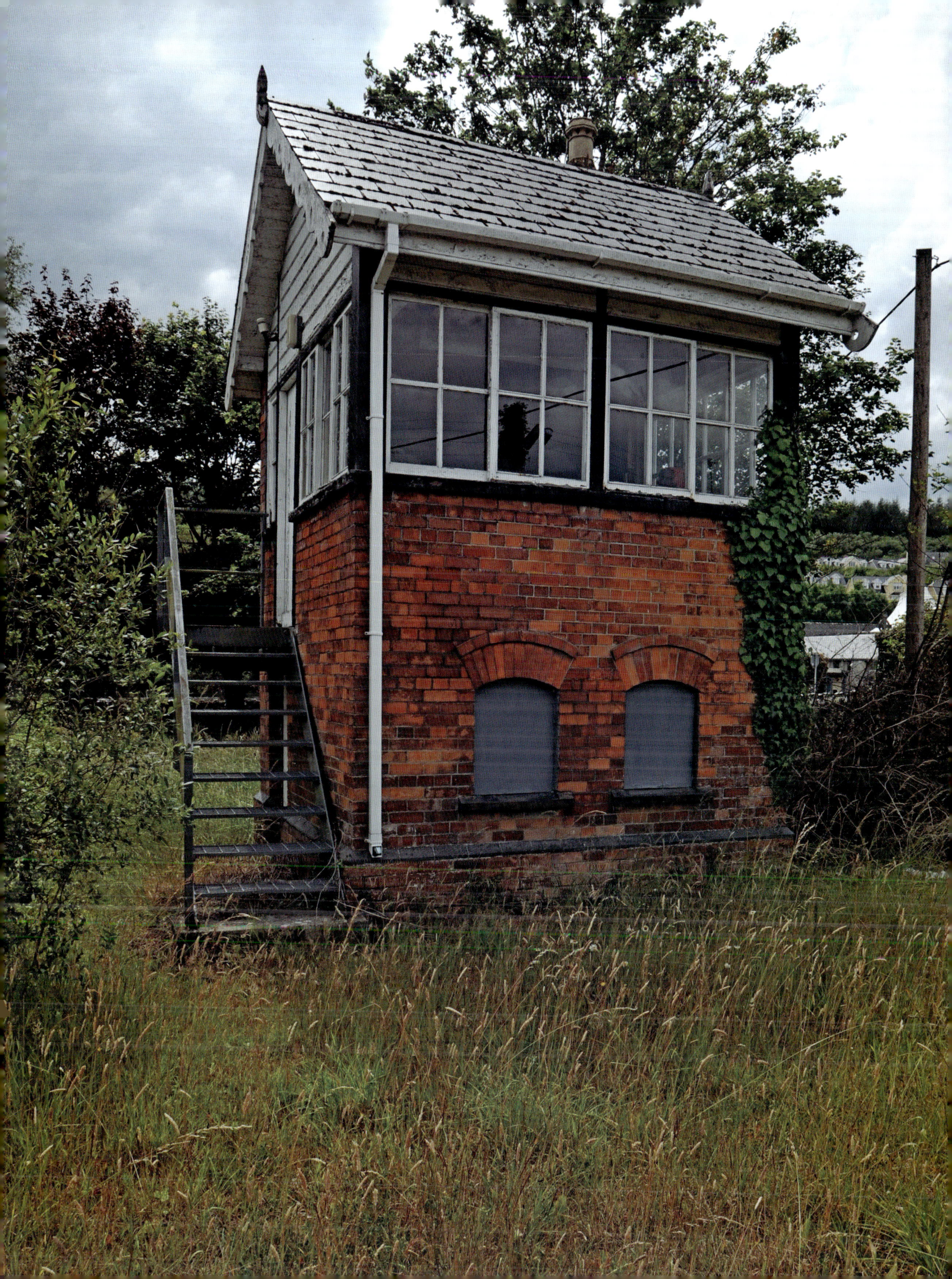

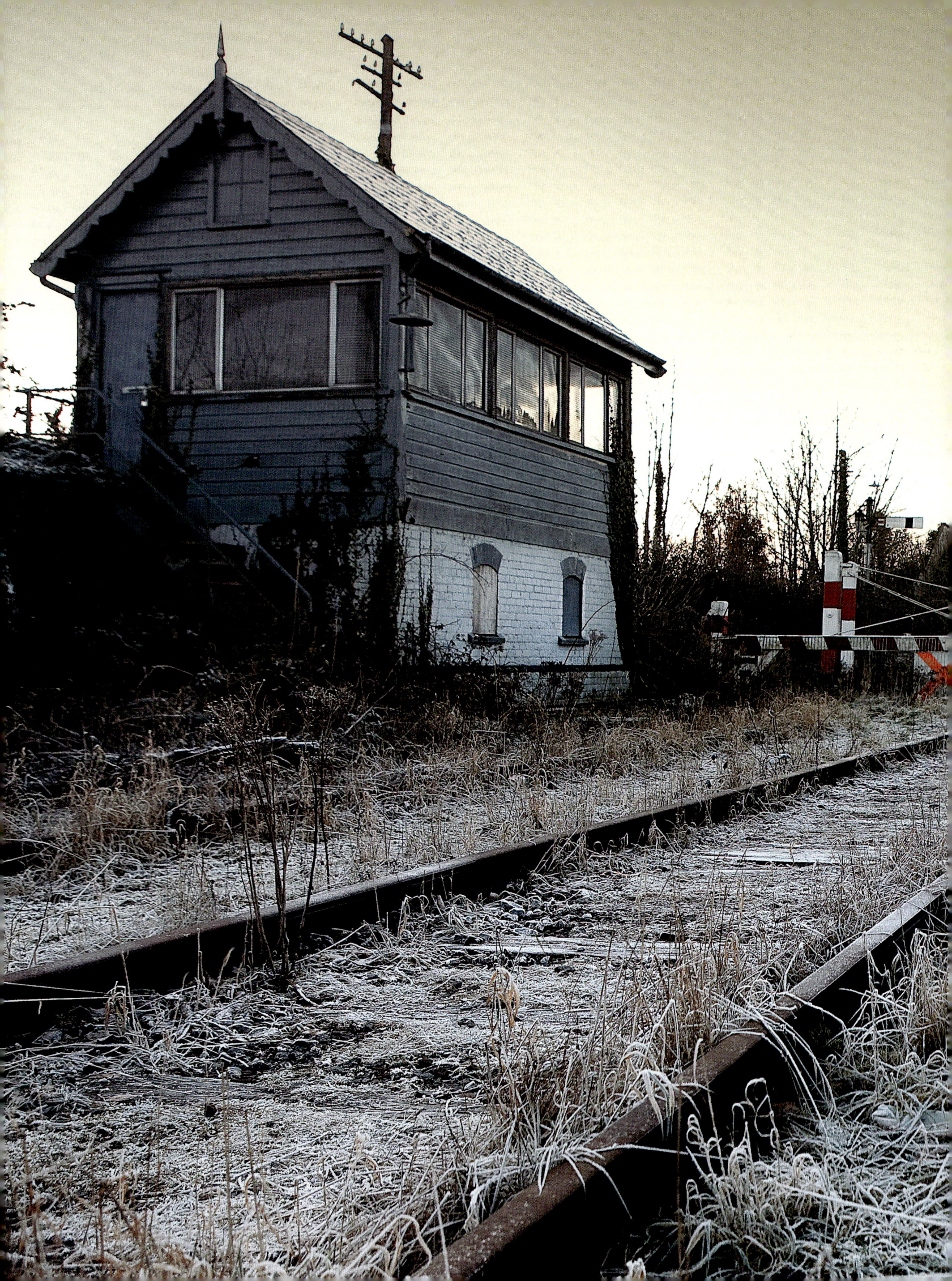

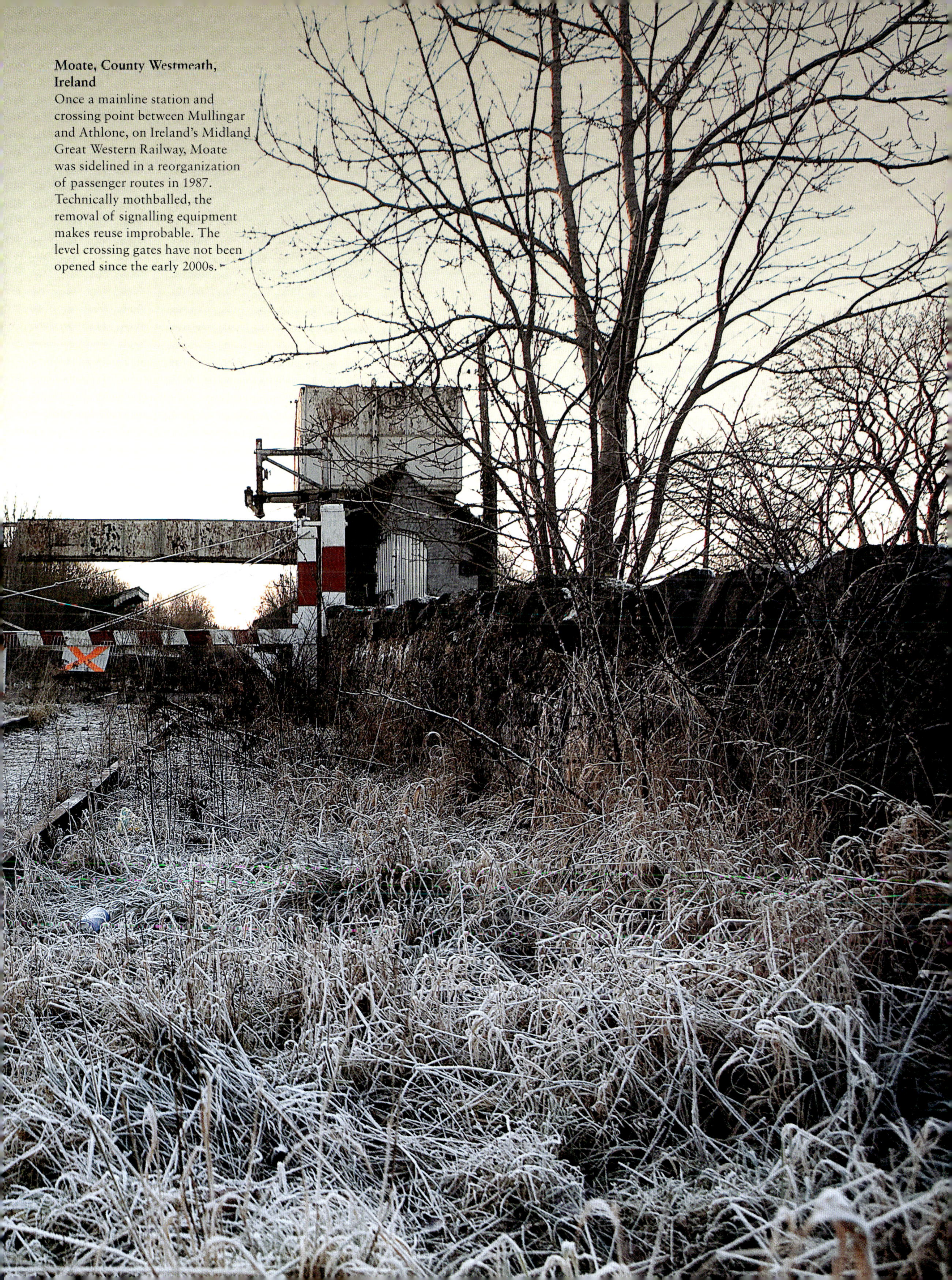

Moate, County Westmeath, Ireland
Once a mainline station and crossing point between Mullingar and Athlone, on Ireland's Midland Great Western Railway, Moate was sidelined in a reorganization of passenger routes in 1987. Technically mothballed, the removal of signalling equipment makes reuse improbable. The level crossing gates have not been opened since the early 2000s.

Blacksboat, Moray, Scotland
Stations on the Strathspey branch of the Great North of Scotland Railway were built in local cottage style, using local stone. The line opened in 1863 and was closed in 1965. Often known as 'the Whisky Line', it served numerous distilleries, but was always lightly used. At Blacksboat the wooden goods shed still survives. This part of the line is now a walkers' and cyclists' path.

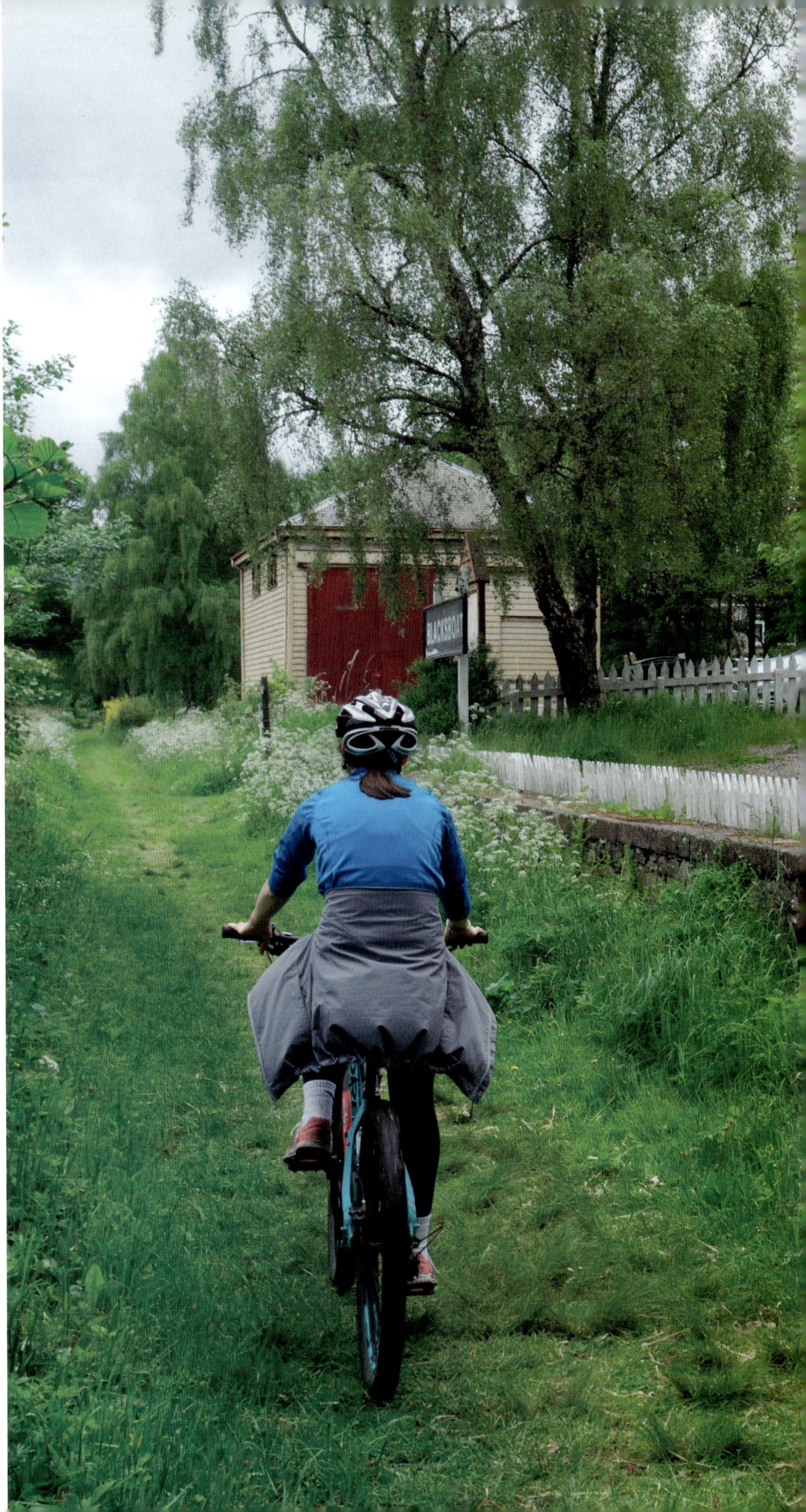

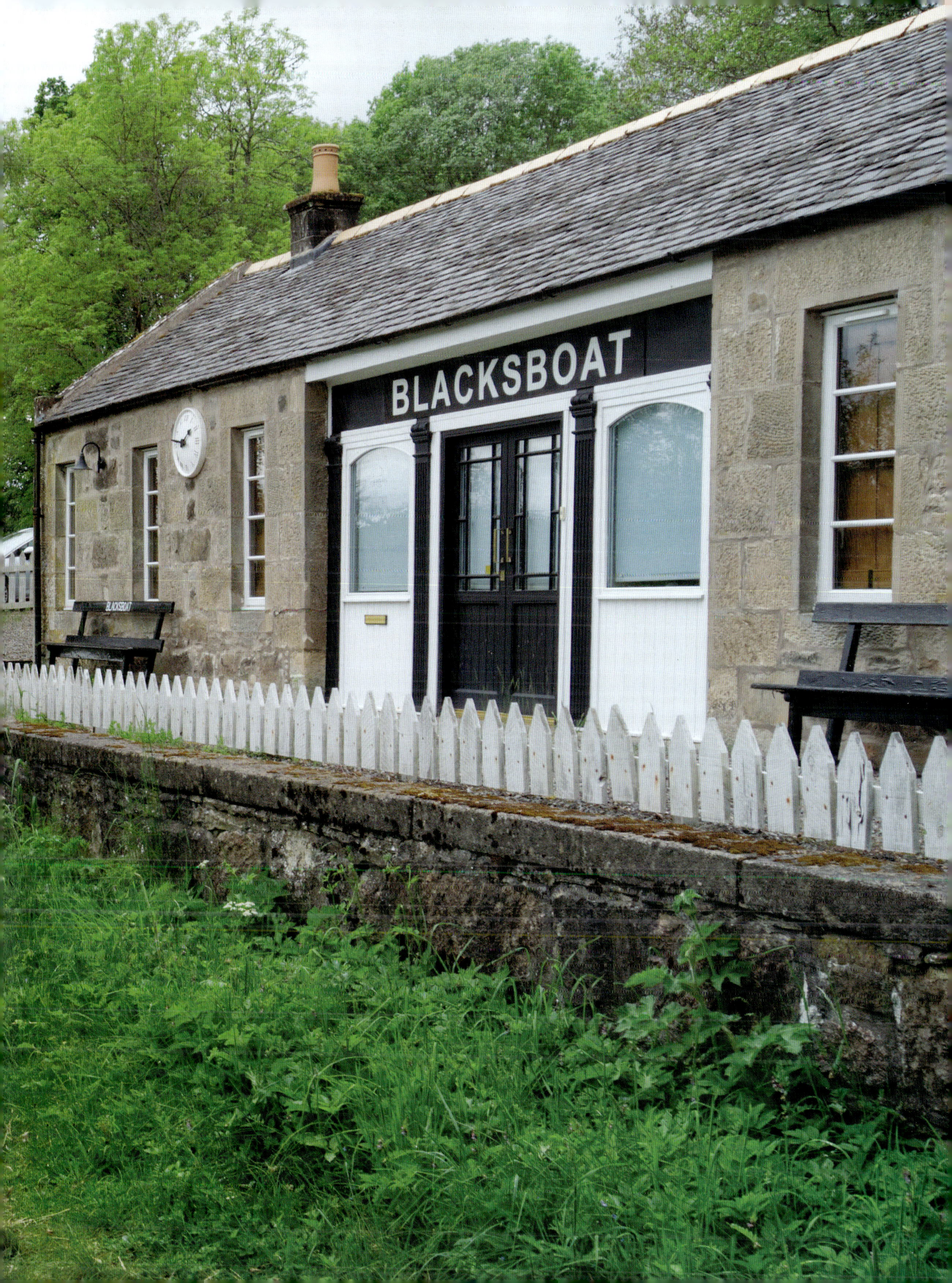
BLACKSBOAT

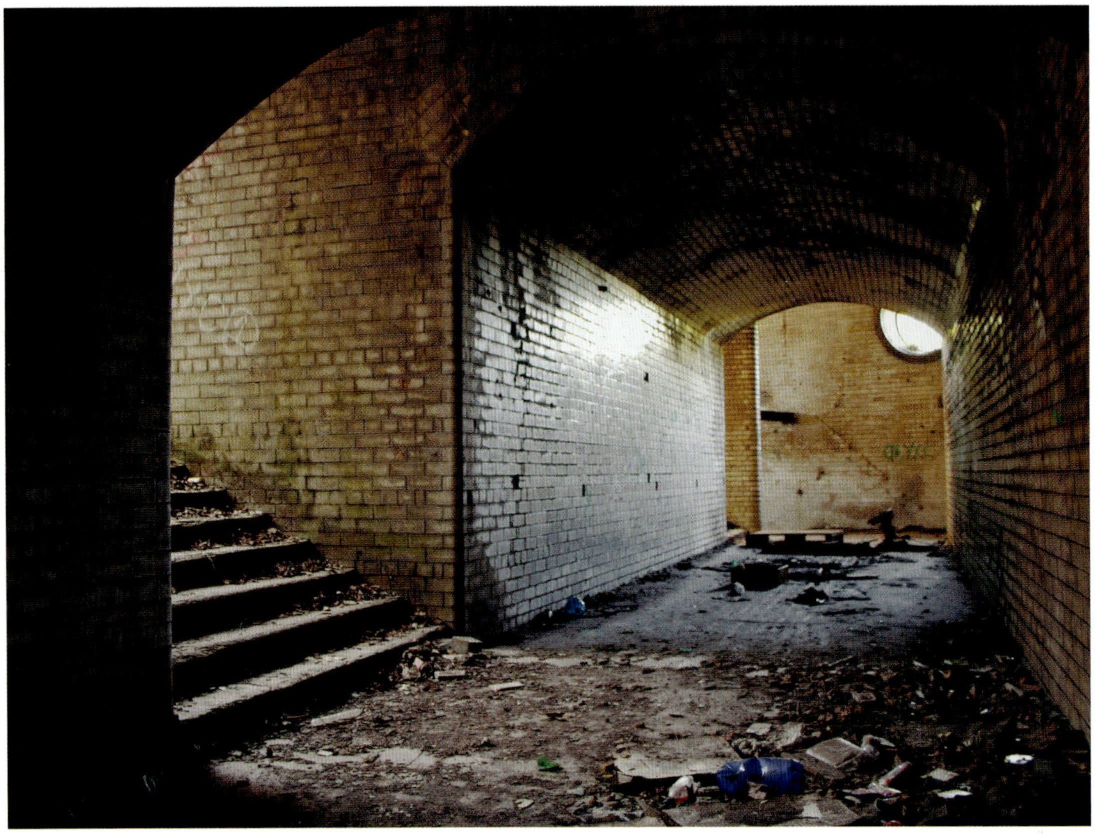

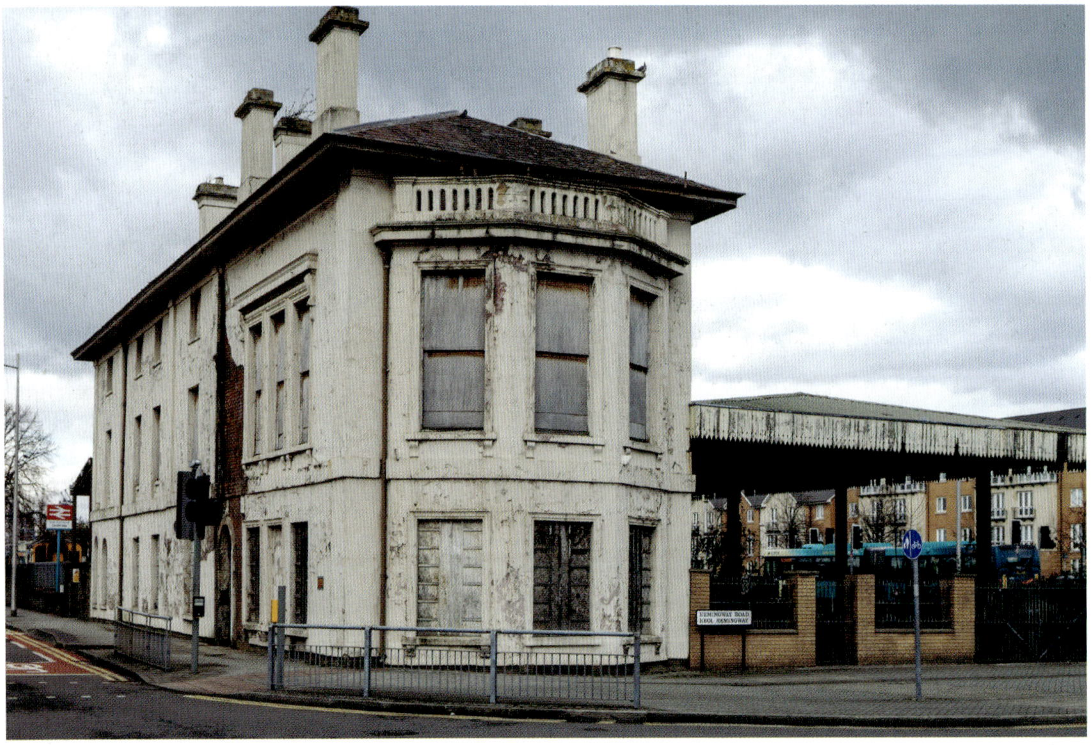

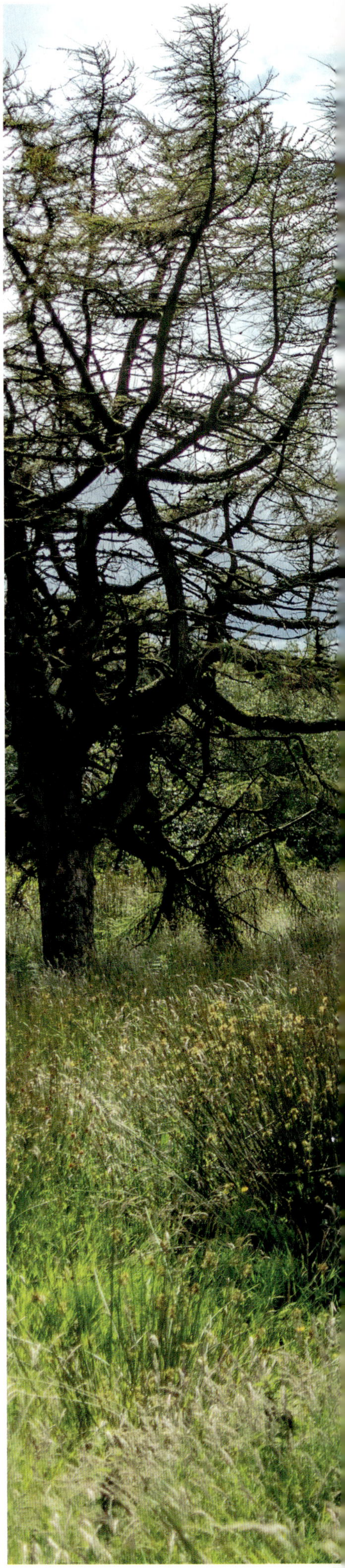

ABOVE TOP:
Cumberland Street Station, Glasgow, Scotland
Typically in inner-city stations, a tiled subway would link the platforms, though this one is at ground level, as the platforms were elevated. Originally called Eglinton Street, the station was opened by the Glasgow & South Western Railway in 1900, and closed in 1966, since when it has been left to deteriorate.

ABOVE BELOW:
Cardiff Bay, Wales
Terminus of a short dockland branch from the city, this station has had various names in its 170-year-plus history, starting as Cardiff (Bute Dock). Its handsome building, scruffy and boarded up in the photograph, was completely restored in 2020, though it no longer functions as a station. A new Cardiff Bay station is scheduled to open in 2023.

RIGHT:
Parkmore, Glenariff, County Antrim, Northern Ireland
Relics of a railway amidst the verdure at Parkmore reveal an industrial past – the Ballymena, Cushendall & Red Bay Railway was built in the early 1870s to carry iron ore from local mines. From 1888 the 994mm (3ft) gauge line also carried passengers as far as Parkmore. Passenger services ended in 1930 and the whole line closed in 1940.

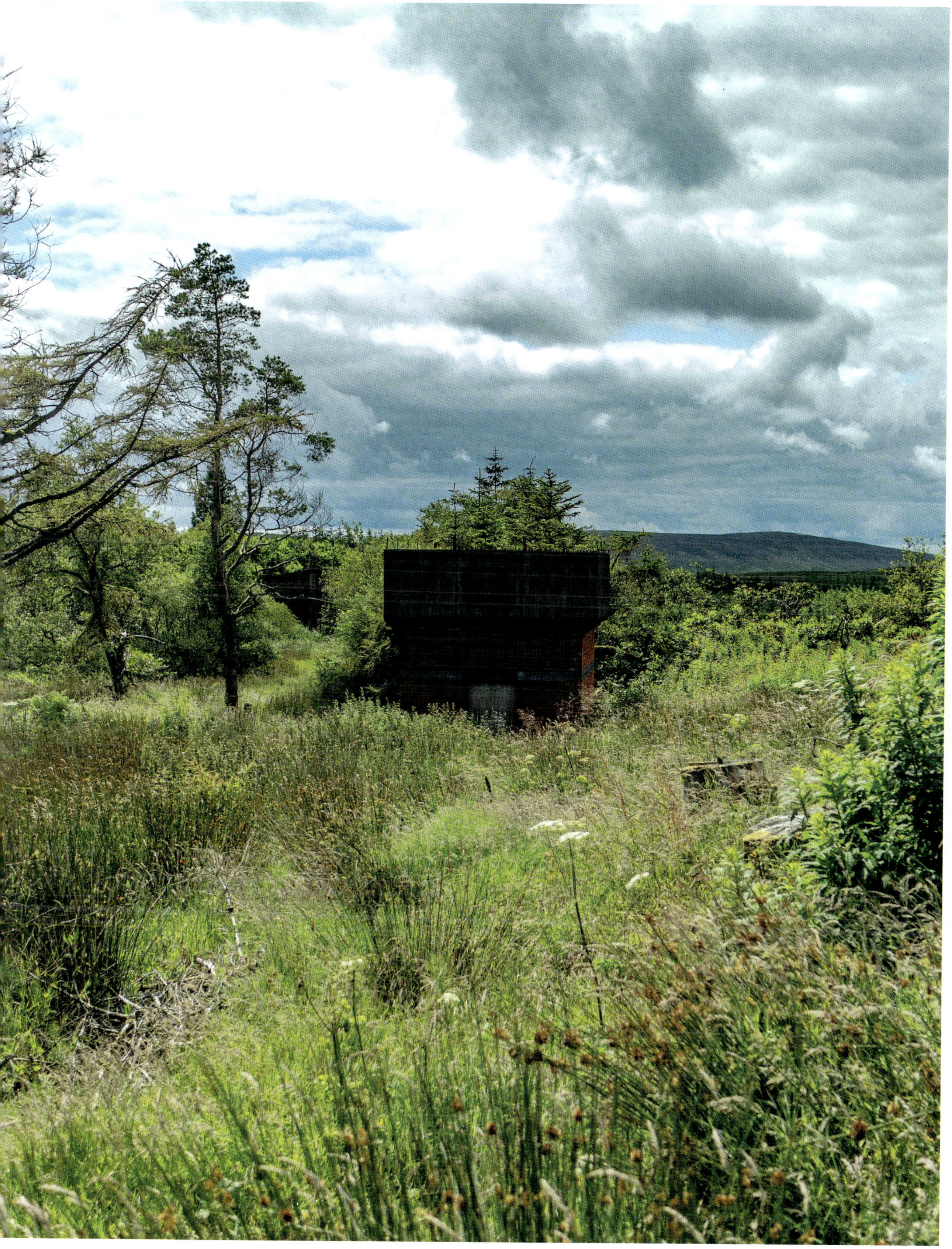

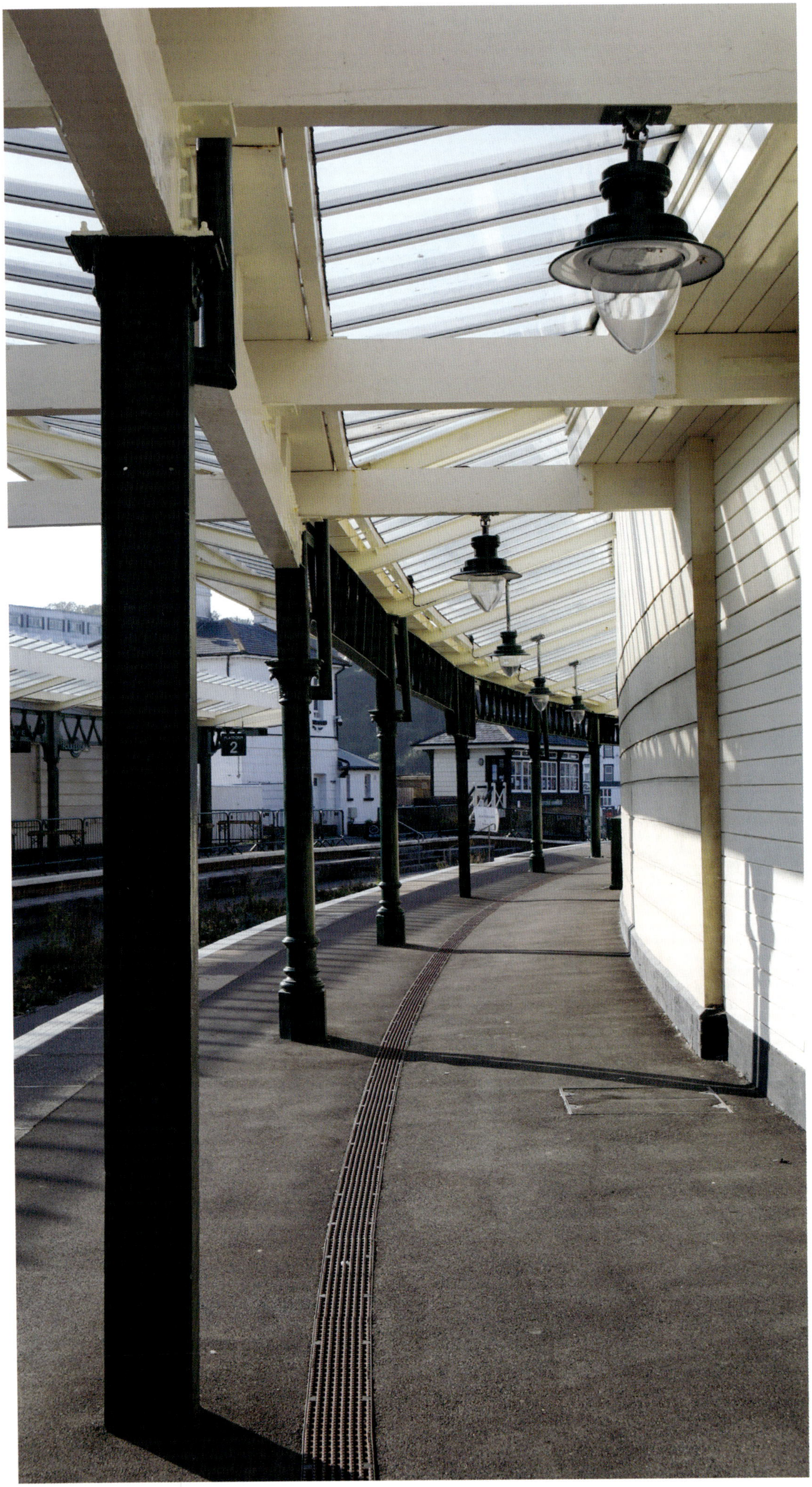

LEFT:
Folkestone Harbour Station, Kent, England
For over a century, railway and steamboat monopolized transport between England and France. In January 1849, Folkestone and Boulogne became exchange points in the world's first international rail-sea-rail service, though the harbour station was not opened until 1850. Car ferries and the Channel Tunnel (1994) led to the station being declared permanently out of use in 2014, but it survives as a heritage site.

OPPOSITE ABOVE:
Felmingham, Norfolk, England
The Midland & Great Northern Joint Railway was a cross-country line linking the Midlands to the sea at Great Yarmouth, by way of numerous small country towns. It was closed in its entirety in 1959. Felmingham, a typical small brick-built station opened in 1883, is now an access point on a walkers' trail, the Weavers' Way, using the old track-bed.

OPPOSITE BELOW:
Honing, Norfolk, England
Here as at Felmingham, the old Midland & Great Northern line is now a walkers' path. The two gently curving platforms have been partly restored but little else remains of the buildings, erected in 1882 and abandoned in 1959. Its former signal box has been re-erected at a miniature railway and museum at nearby Wroxham.

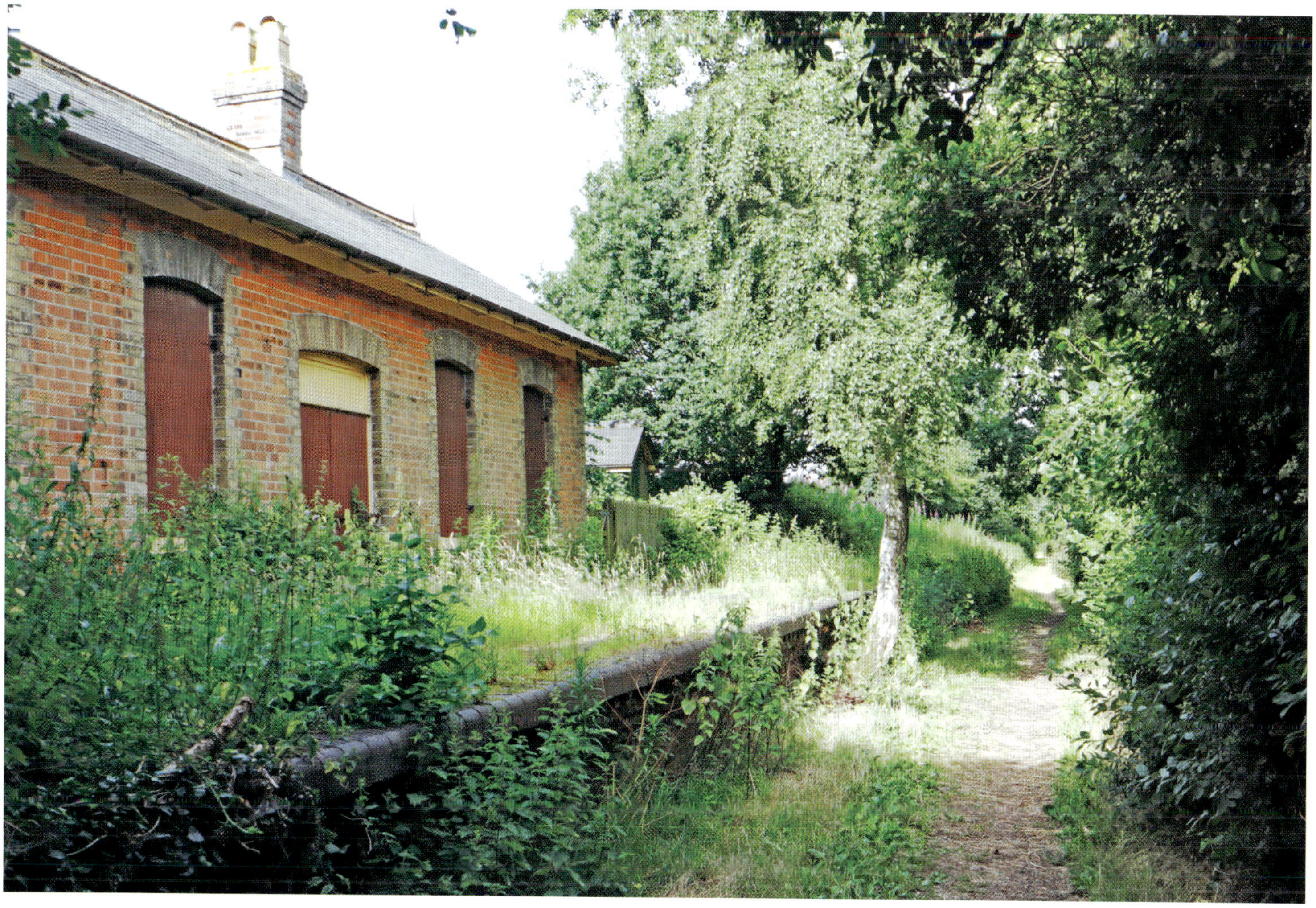

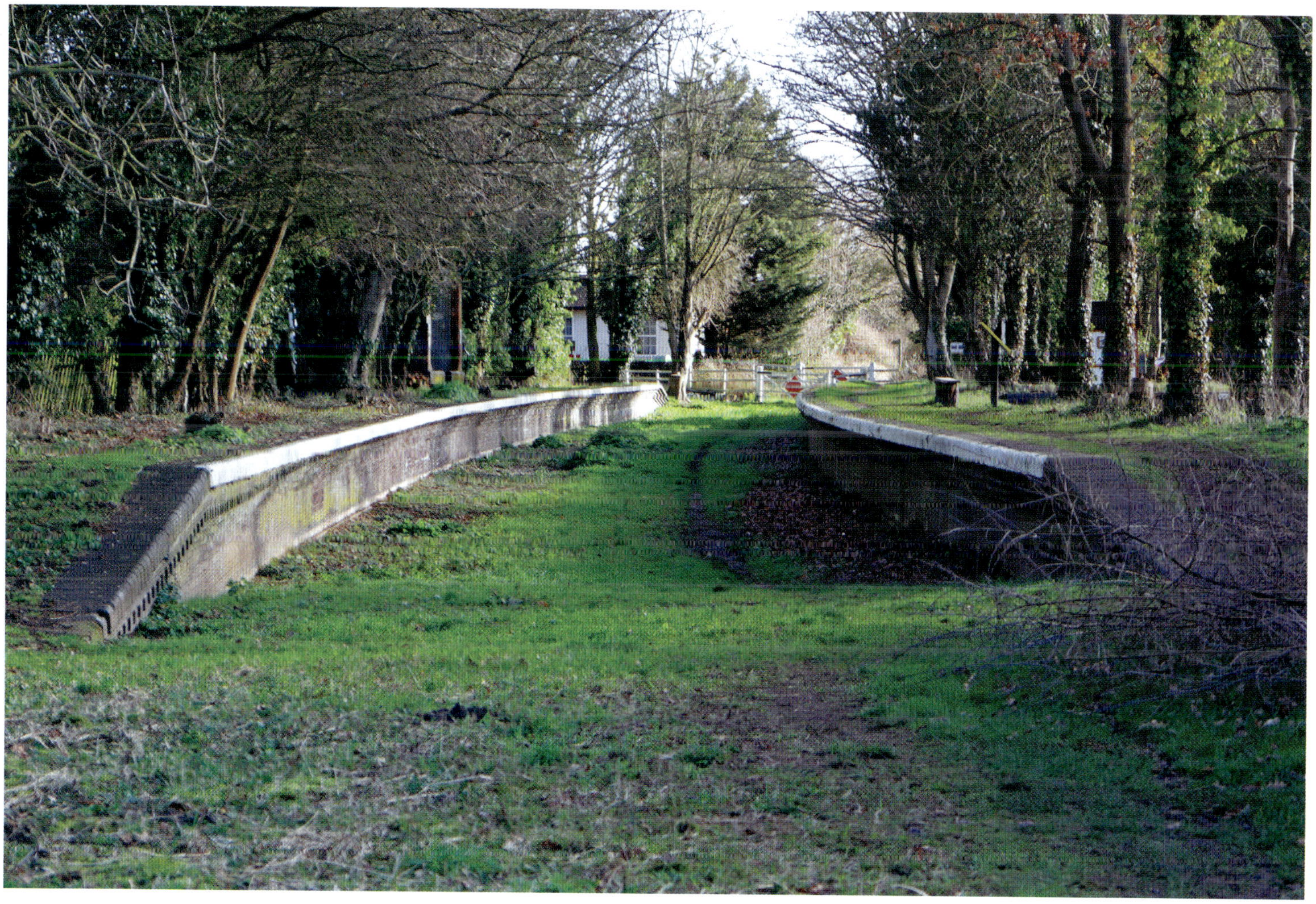

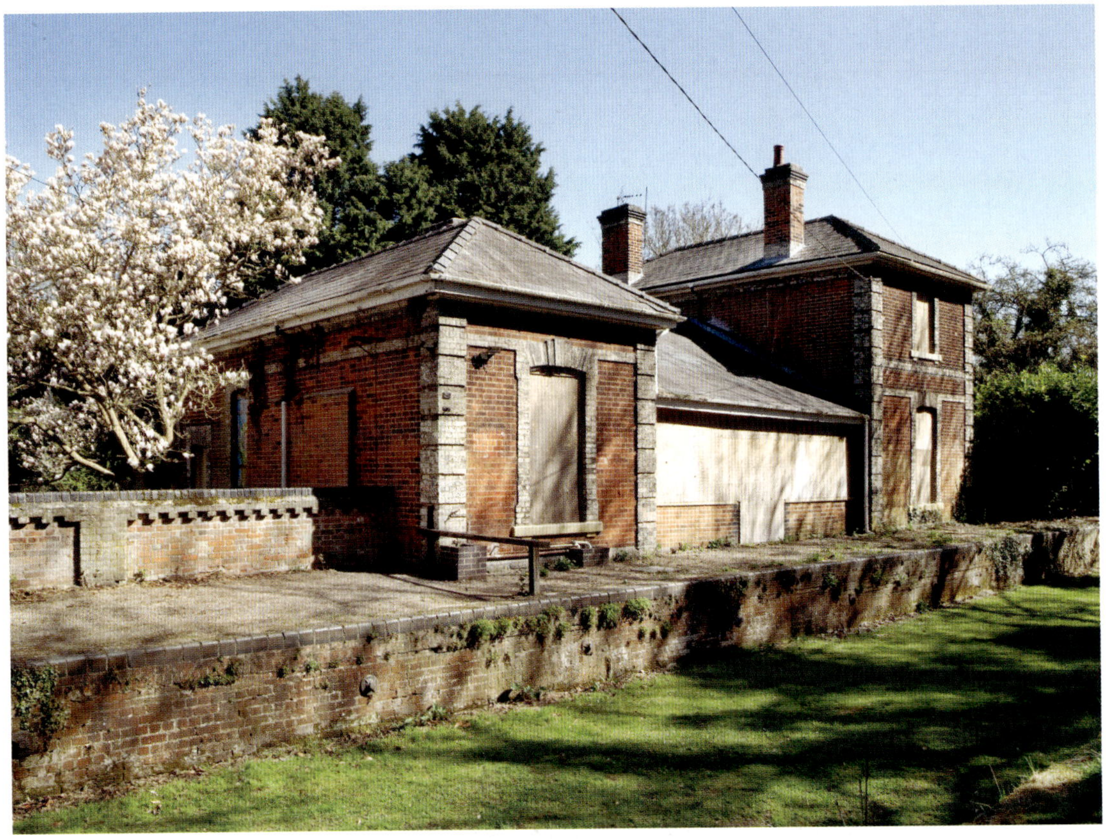

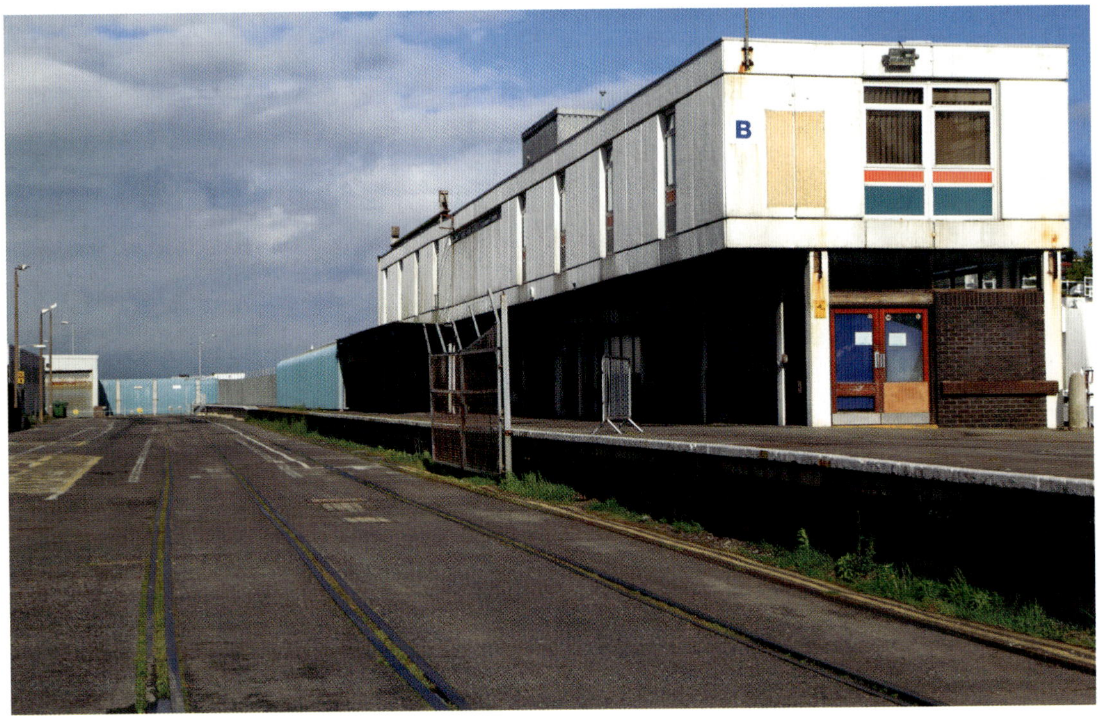

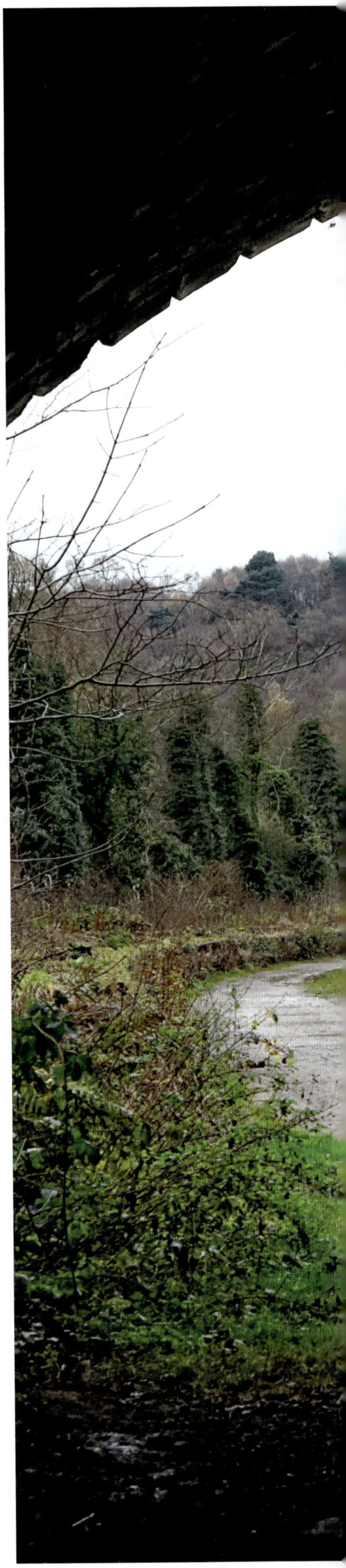

ABOVE TOP:
Clare, Suffolk, England
Preserved as part of Clare Country Park, the station was on the former Great Eastern line between Sudbury and Cambridge, closed in 1967. It became a visitor centre and museum, though these facilities are currently closed. Opened in 1865, its pavilion-style design is put together from modular designs (the GER was a frugal company) with the stationmaster's house as the larger part.

ABOVE BELOW:
Weymouth Quay Station, Dorset, England
Trains ran to here from London to connect with Channel Island steamers. Opened in 1865, the short branch to the quay passed through the streets of the town and horses were the original motive power until 1880, when small steam locomotives took over. Though regular services ended in 1987, special services used it until May 1999. The unlovely buildings date from the 1960s.

OPPOSITE:
Alton Towers, Staffordshire, England
On the North Staffordshire Railway's Churnet Valley line, opened in 1849, this station served the village of Alton and the nearby stately home, Alton Towers (now an amusement park). The Italianate style was unusual for this railway. The line was closed in 1965, and in 1979 the buildings were converted to provide holiday accommodation.

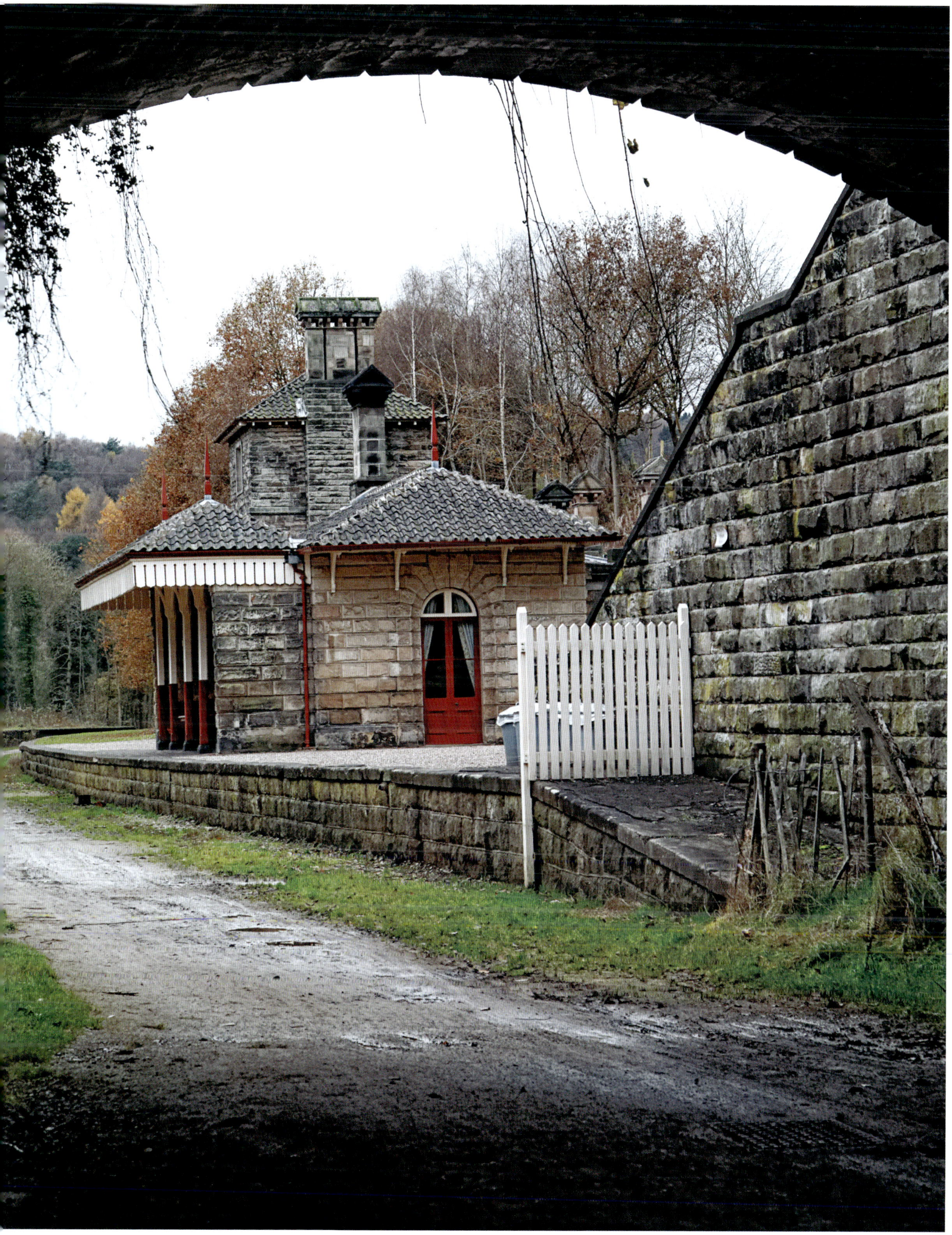

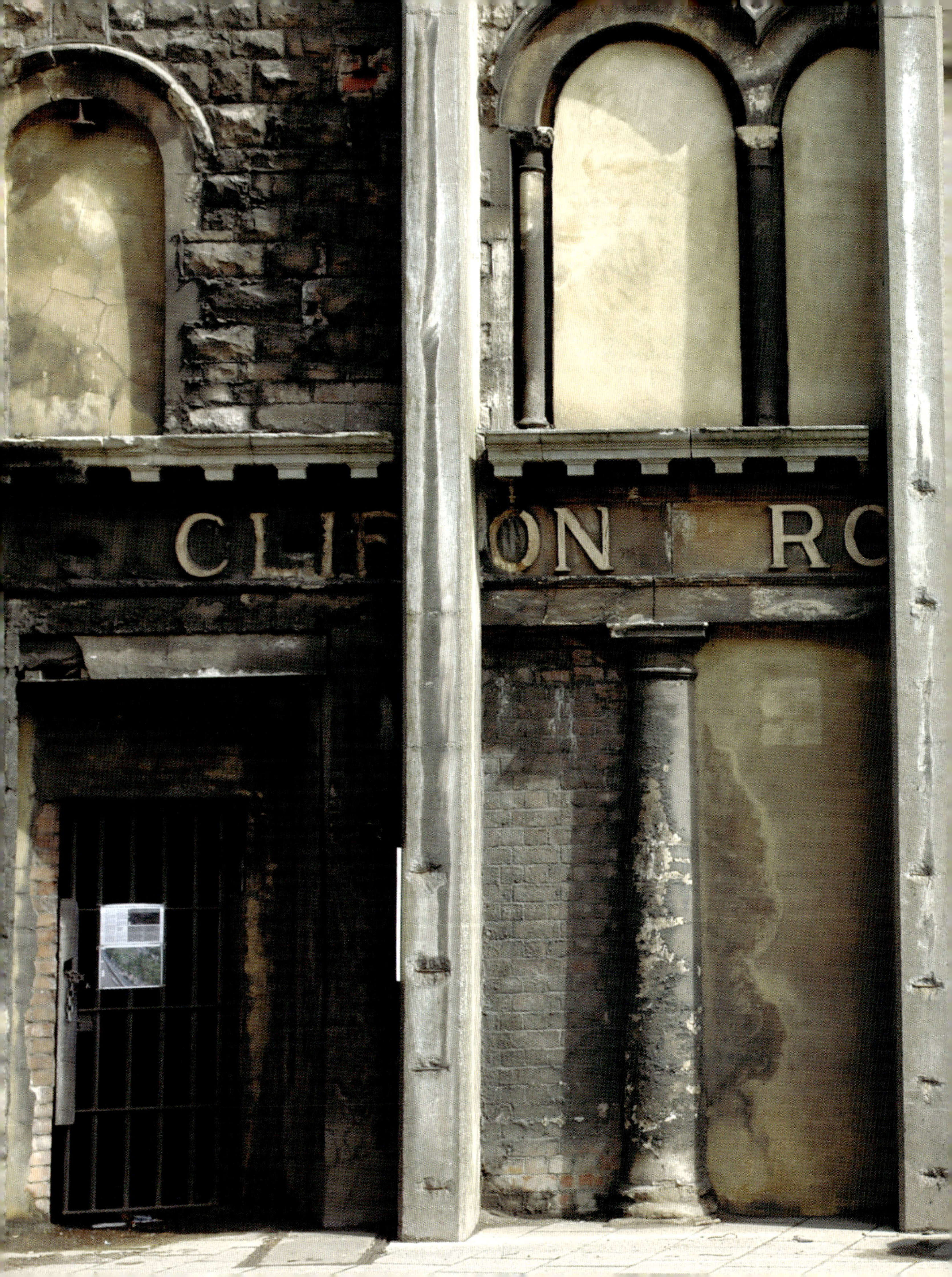
CLIF
ON
RC

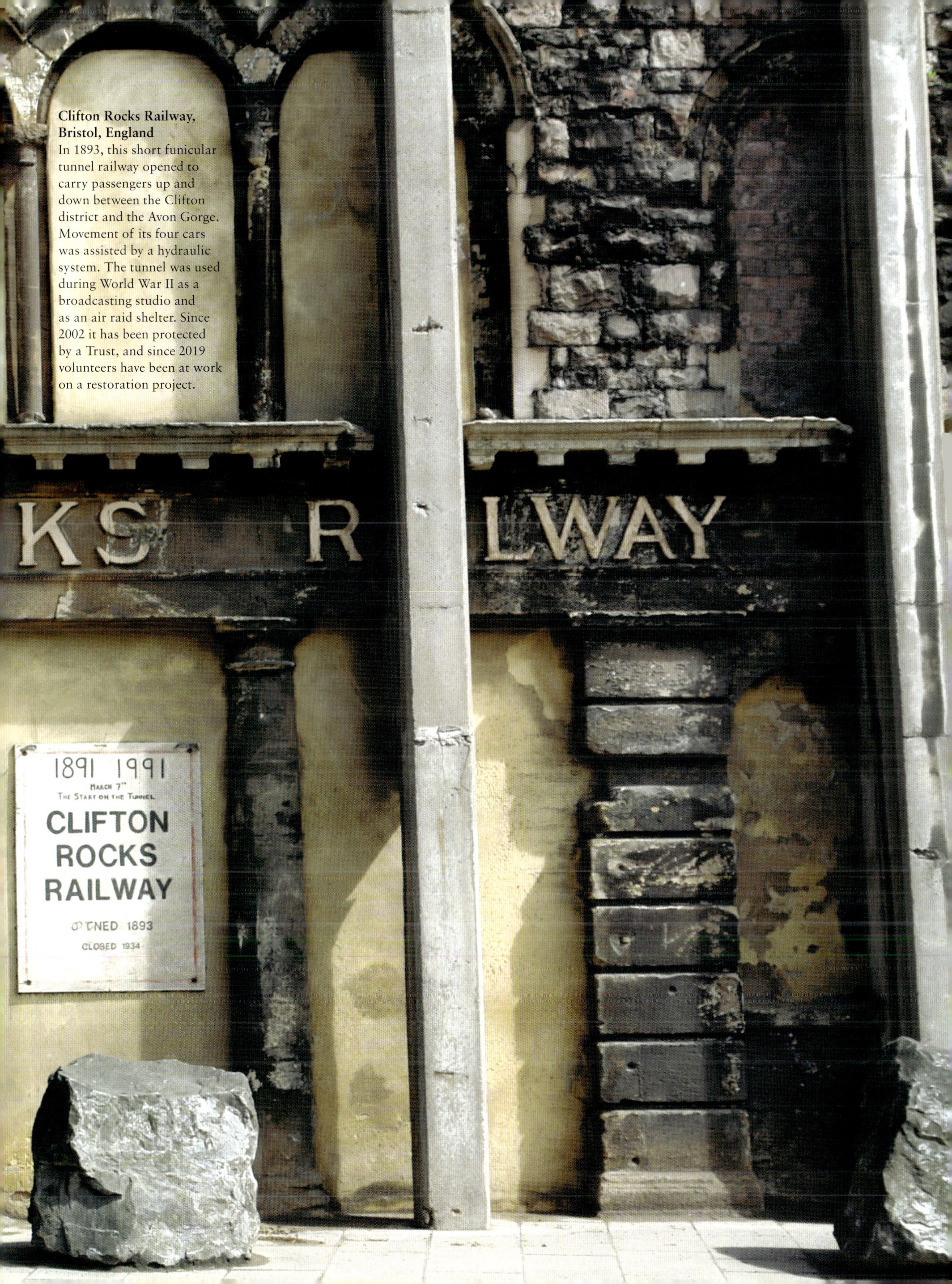

Clifton Rocks Railway, Bristol, England
In 1893, this short funicular tunnel railway opened to carry passengers up and down between the Clifton district and the Avon Gorge. Movement of its four cars was assisted by a hydraulic system. The tunnel was used during World War II as a broadcasting studio and as an air raid shelter. Since 2002 it has been protected by a Trust, and since 2019 volunteers have been at work on a restoration project.

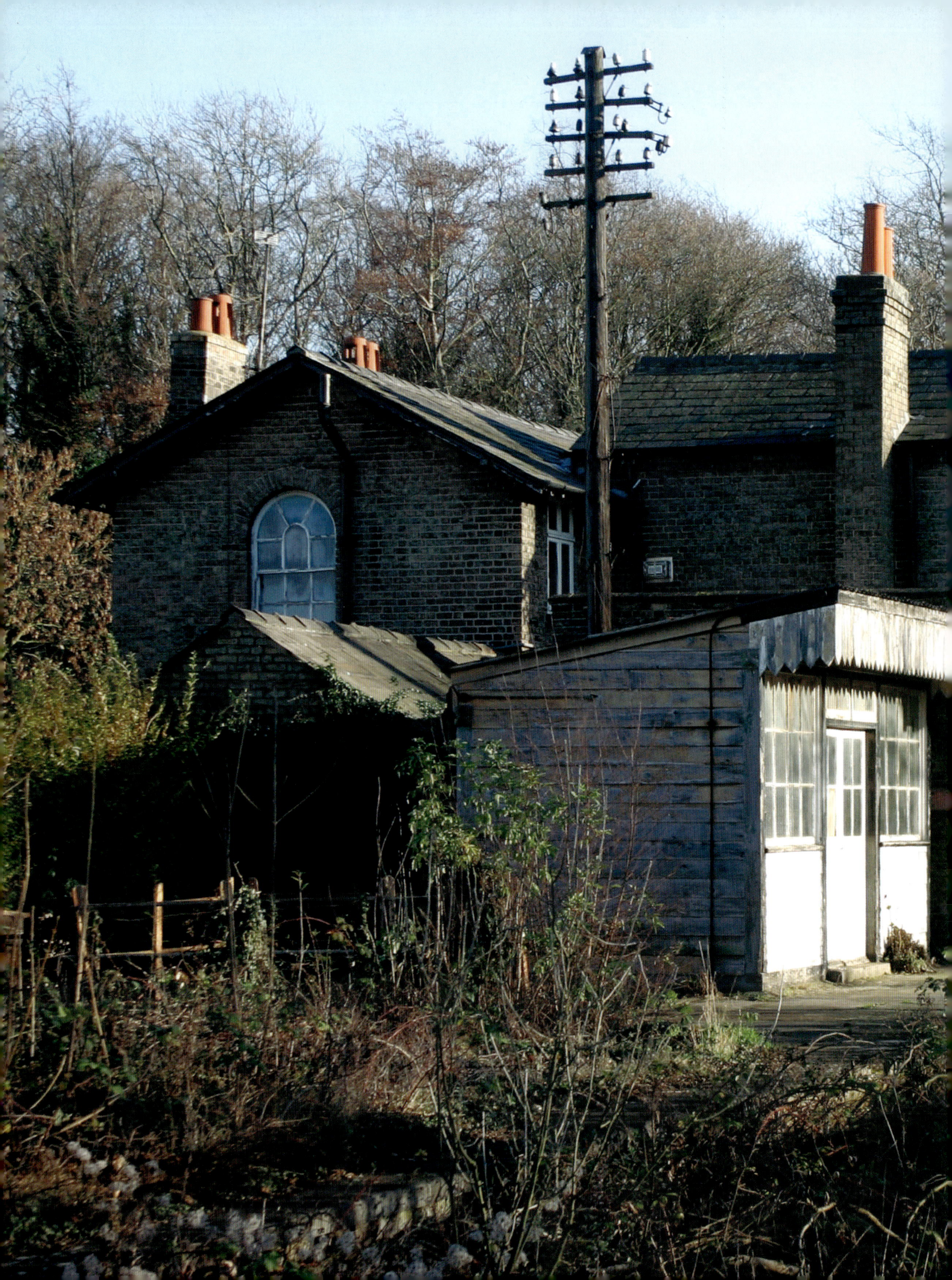

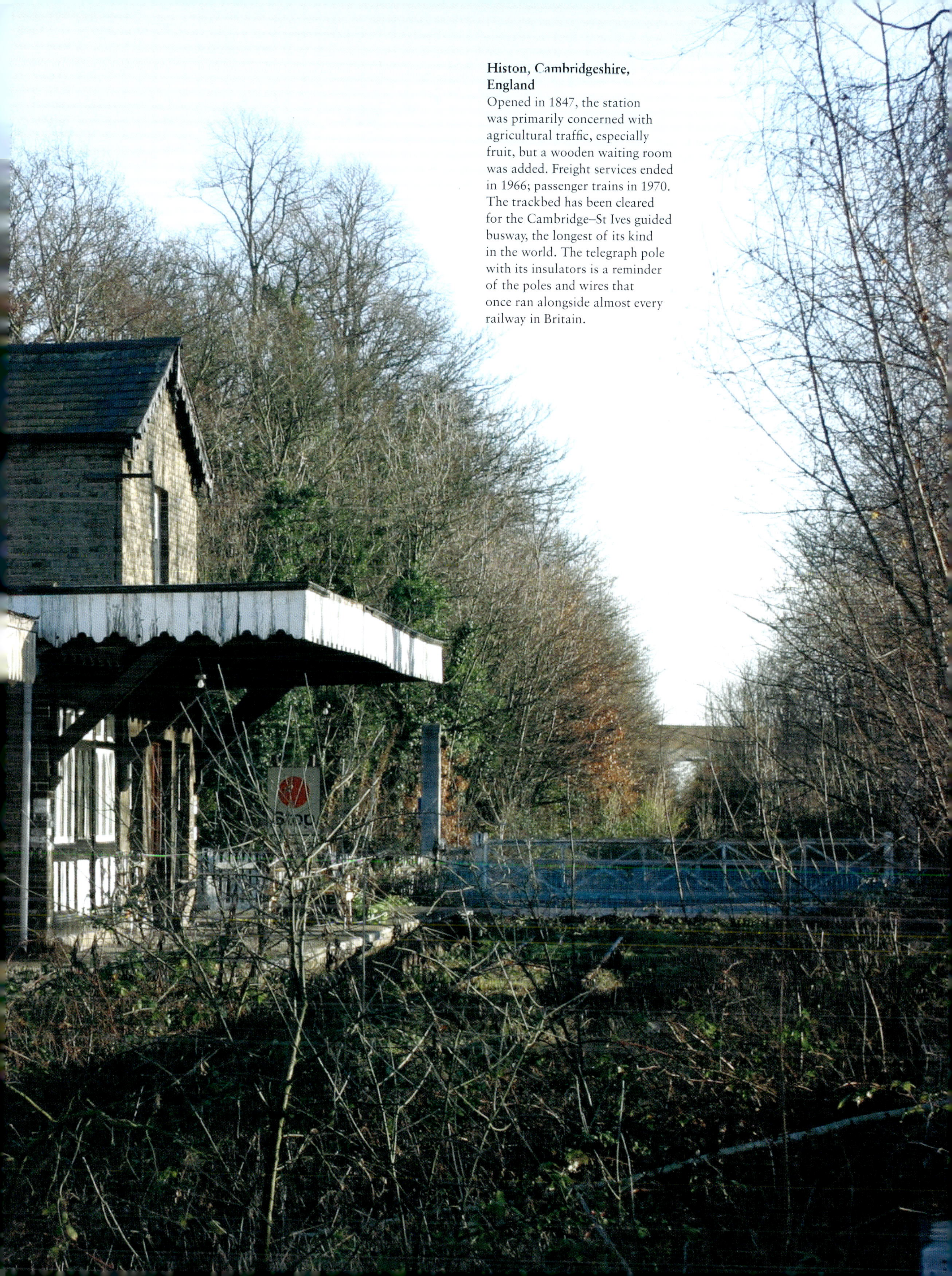

Histon, Cambridgeshire, England
Opened in 1847, the station was primarily concerned with agricultural traffic, especially fruit, but a wooden waiting room was added. Freight services ended in 1966; passenger trains in 1970. The trackbed has been cleared for the Cambridge–St Ives guided busway, the longest of its kind in the world. The telegraph pole with its insulators is a reminder of the poles and wires that once ran alongside almost every railway in Britain.

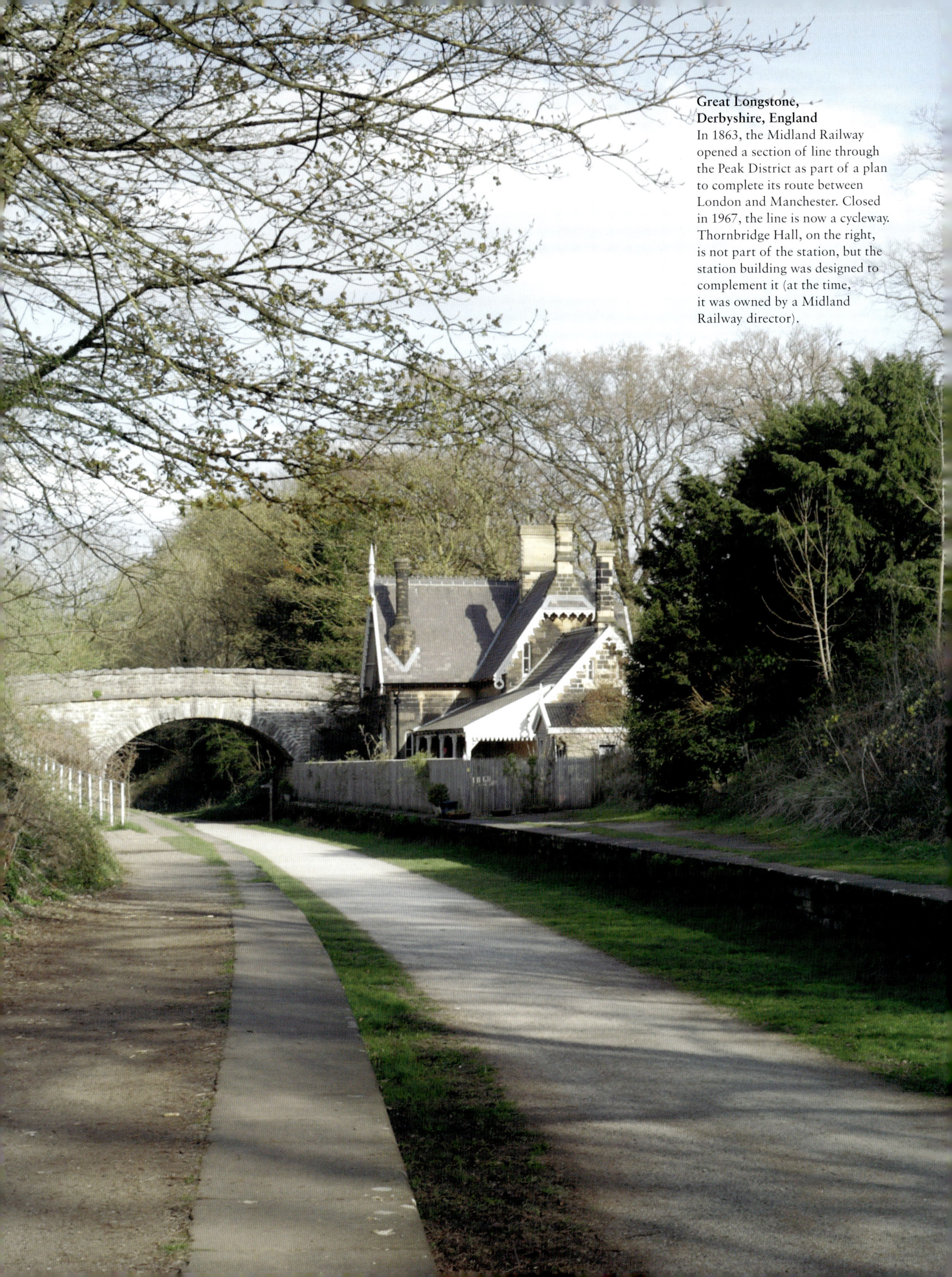

Great Longstone, Derbyshire, England
In 1863, the Midland Railway opened a section of line through the Peak District as part of a plan to complete its route between London and Manchester. Closed in 1967, the line is now a cycleway. Thornbridge Hall, on the right, is not part of the station, but the station building was designed to complement it (at the time, it was owned by a Midland Railway director).

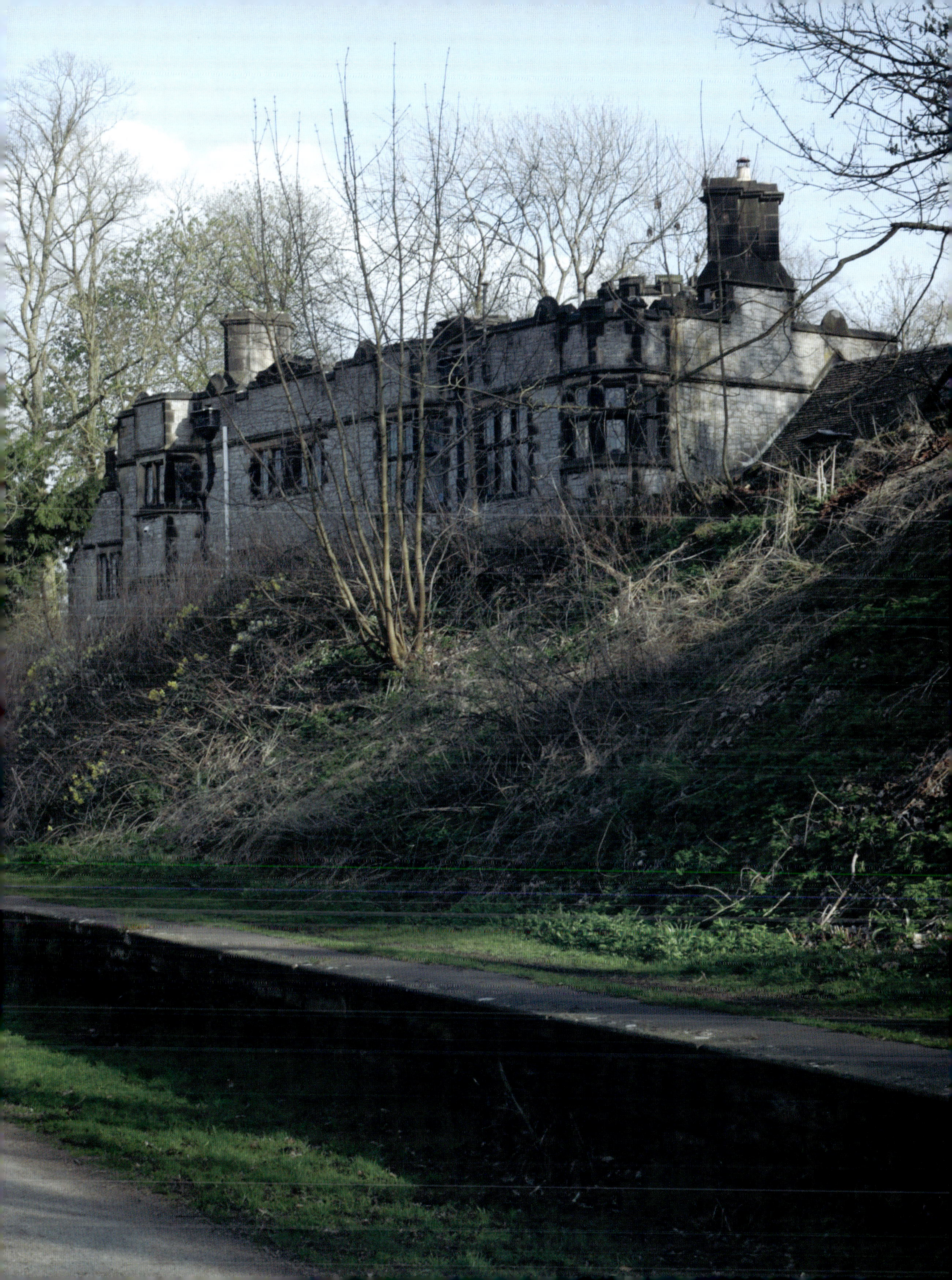

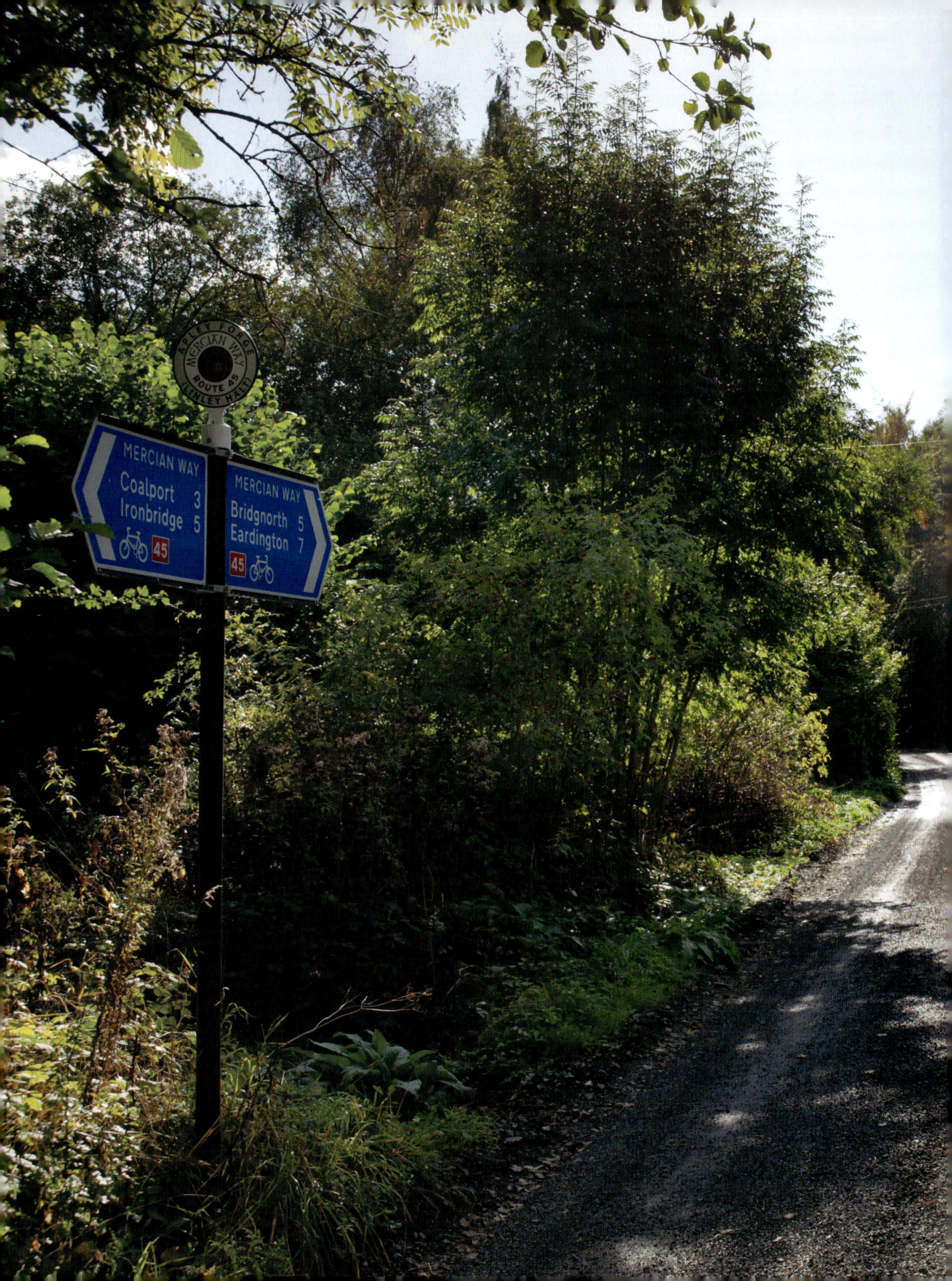
APLEY FORGE
MERCIAN WAY
ROUTE 45
(LINLEY HALT)
MERCIAN WAY
Coalport 3
Ironbridge 5
45
MERCIAN WAY
Bridgnorth 5
Eardington 7
45

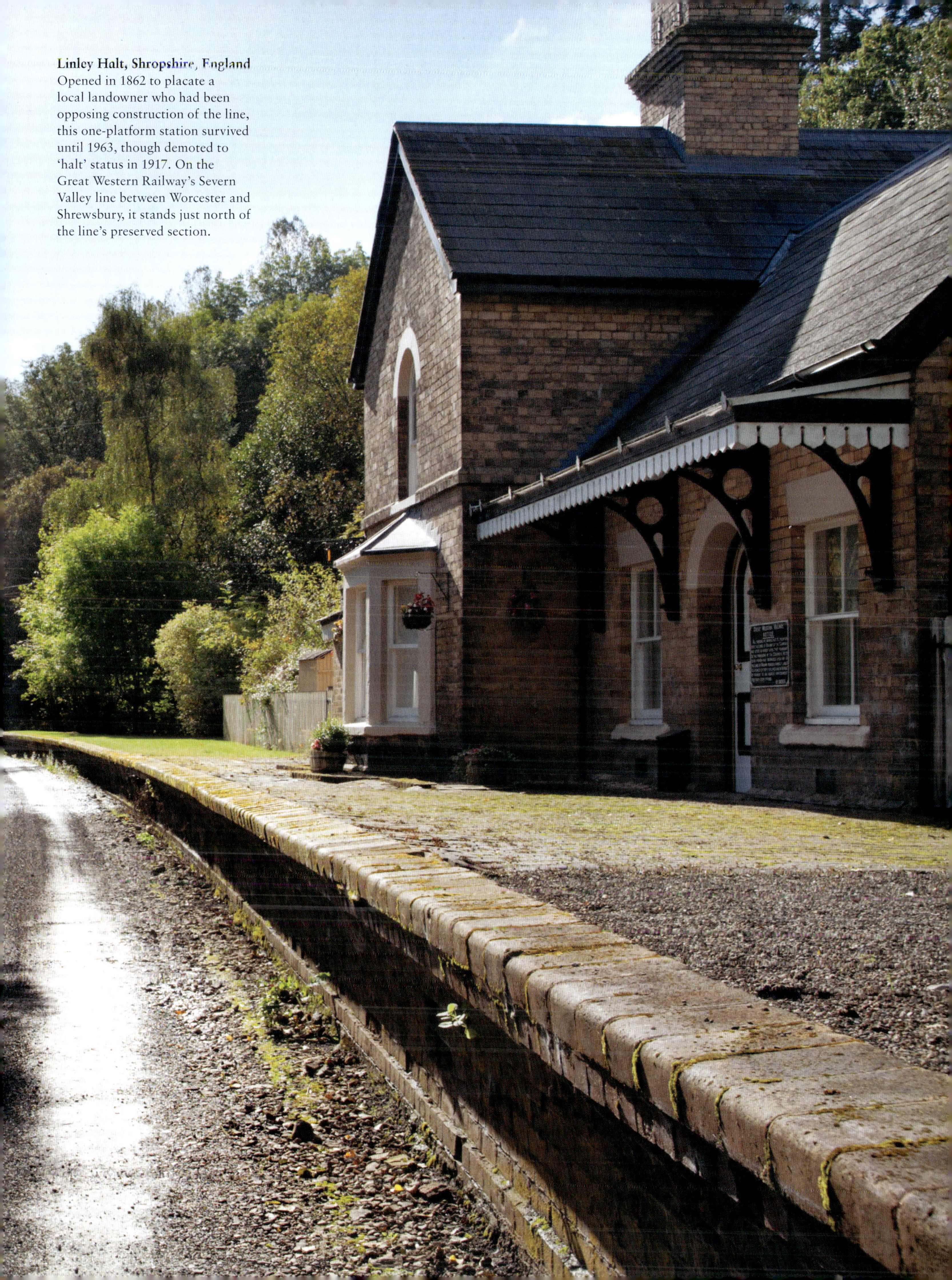

Linley Halt, Shropshire, England
Opened in 1862 to placate a local landowner who had been opposing construction of the line, this one-platform station survived until 1963, though demoted to 'halt' status in 1917. On the Great Western Railway's Severn Valley line between Worcester and Shrewsbury, it stands just north of the line's preserved section.

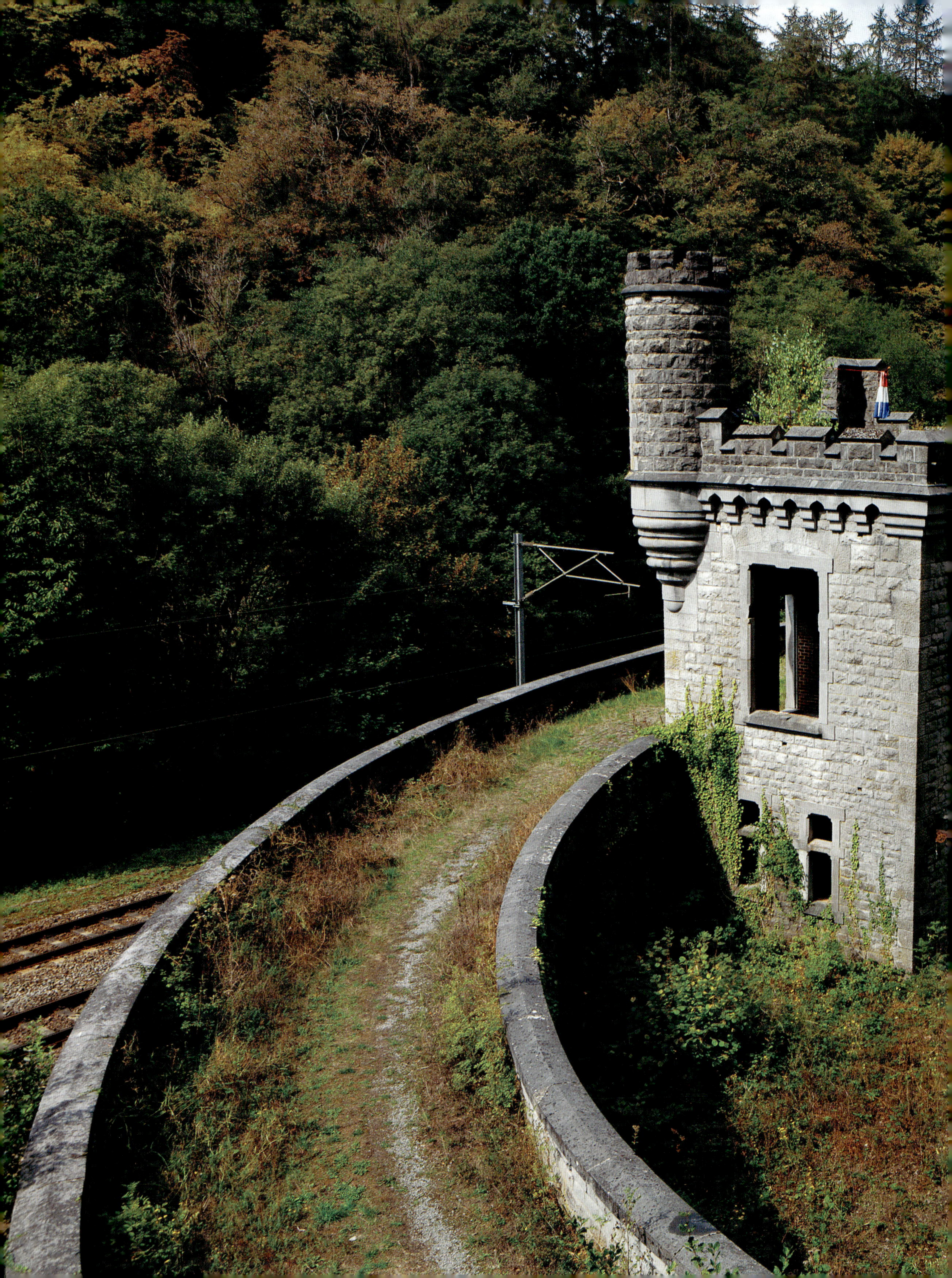

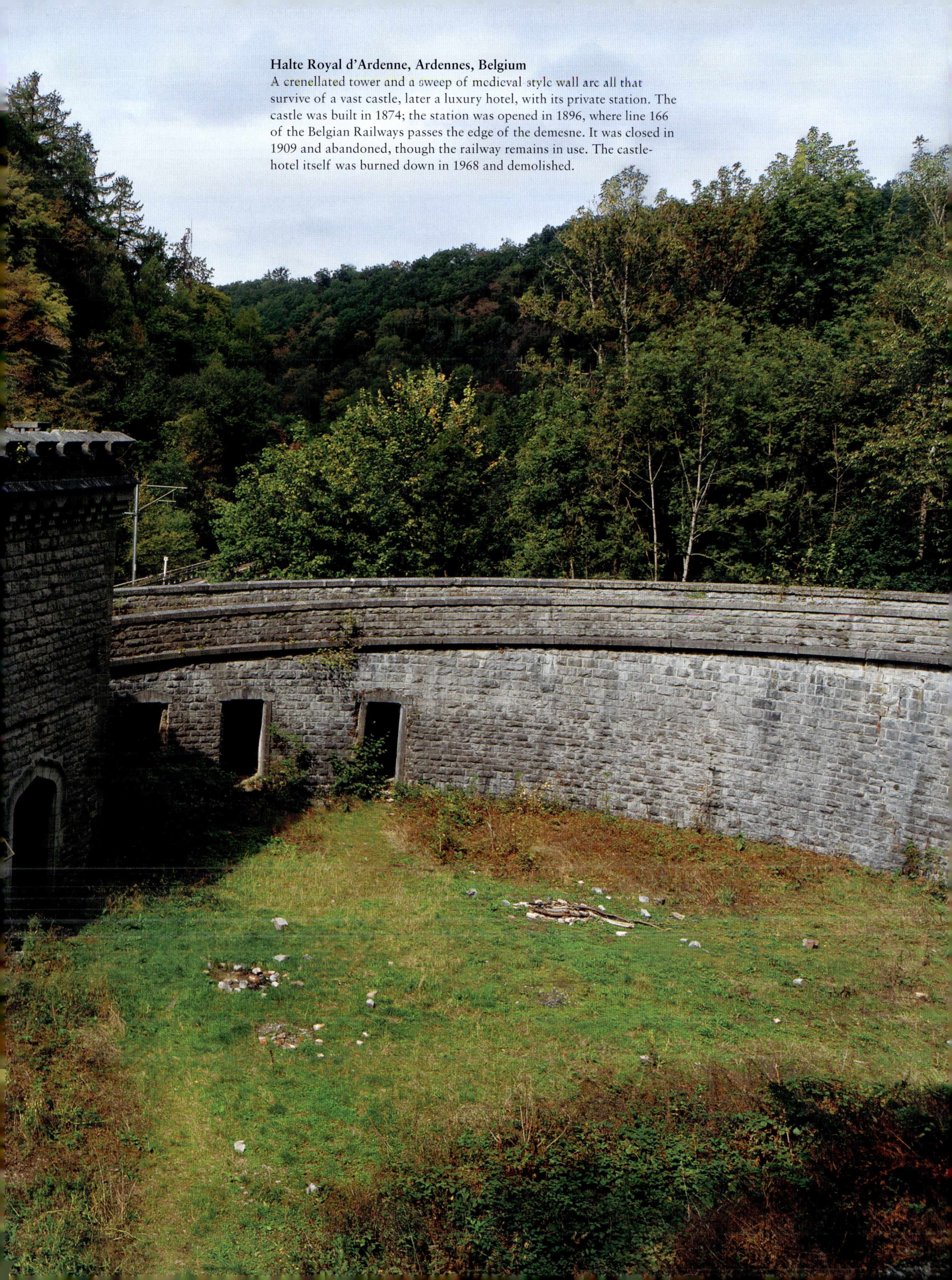

Halte Royal d'Ardenne, Ardennes, Belgium
A crenellated tower and a sweep of medieval style wall are all that survive of a vast castle, later a luxury hotel, with its private station. The castle was built in 1874; the station was opened in 1896, where line 166 of the Belgian Railways passes the edge of the demesne. It was closed in 1909 and abandoned, though the railway remains in use. The castle-hotel itself was burned down in 1968 and demolished.

ALL PHOTOGRAPHS:

Güterbahnhof Pankow, Berlin, Germany

Begun in 1893 and abandoned in 1997, this railway complex on the north-east side of Berlin adjoins the Pankow-Heinersdorf S-Bahn station. It encompassed a large goods station and locomotive depot. The image (right) shows a detail of the current decrepit state of the huge site. The roundhouse-type engine shed (below), with its cupola-ventilator, is one of only two remaining in Germany. Some buildings have been demolished; others have been set on fire, and much equipment like the locomotive turntable (opposite) remains in a derelict state. In 2009 the site was acquired by a developer, but the buildings have advanced further towards collapse while wrangles continue between preservationists, the city council and the development company.

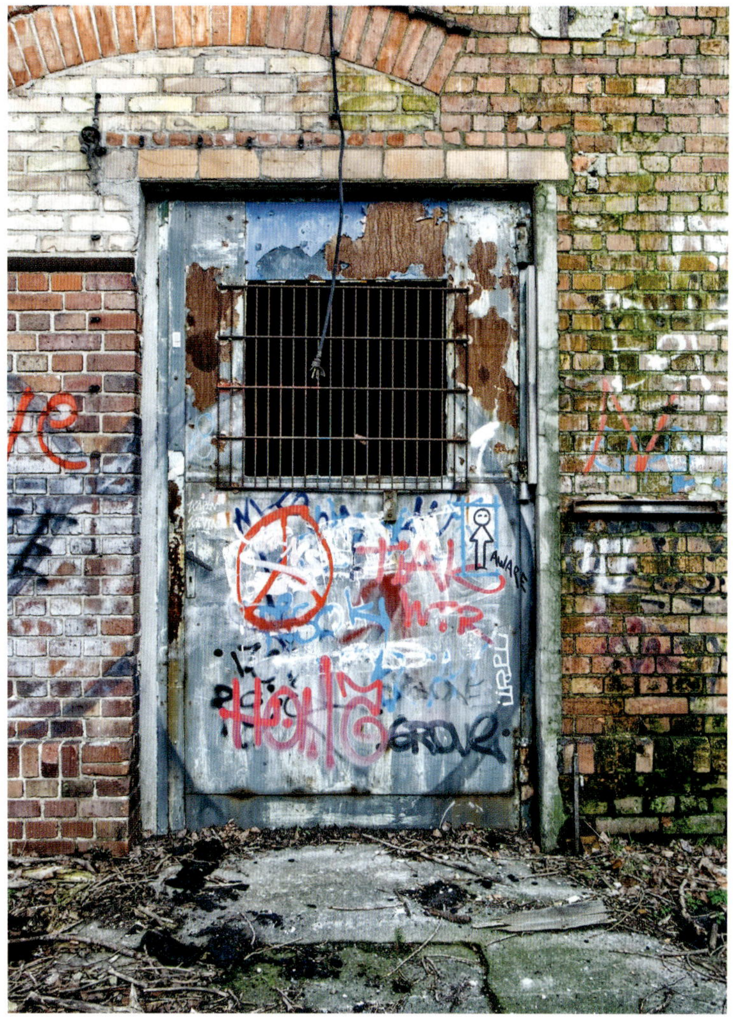

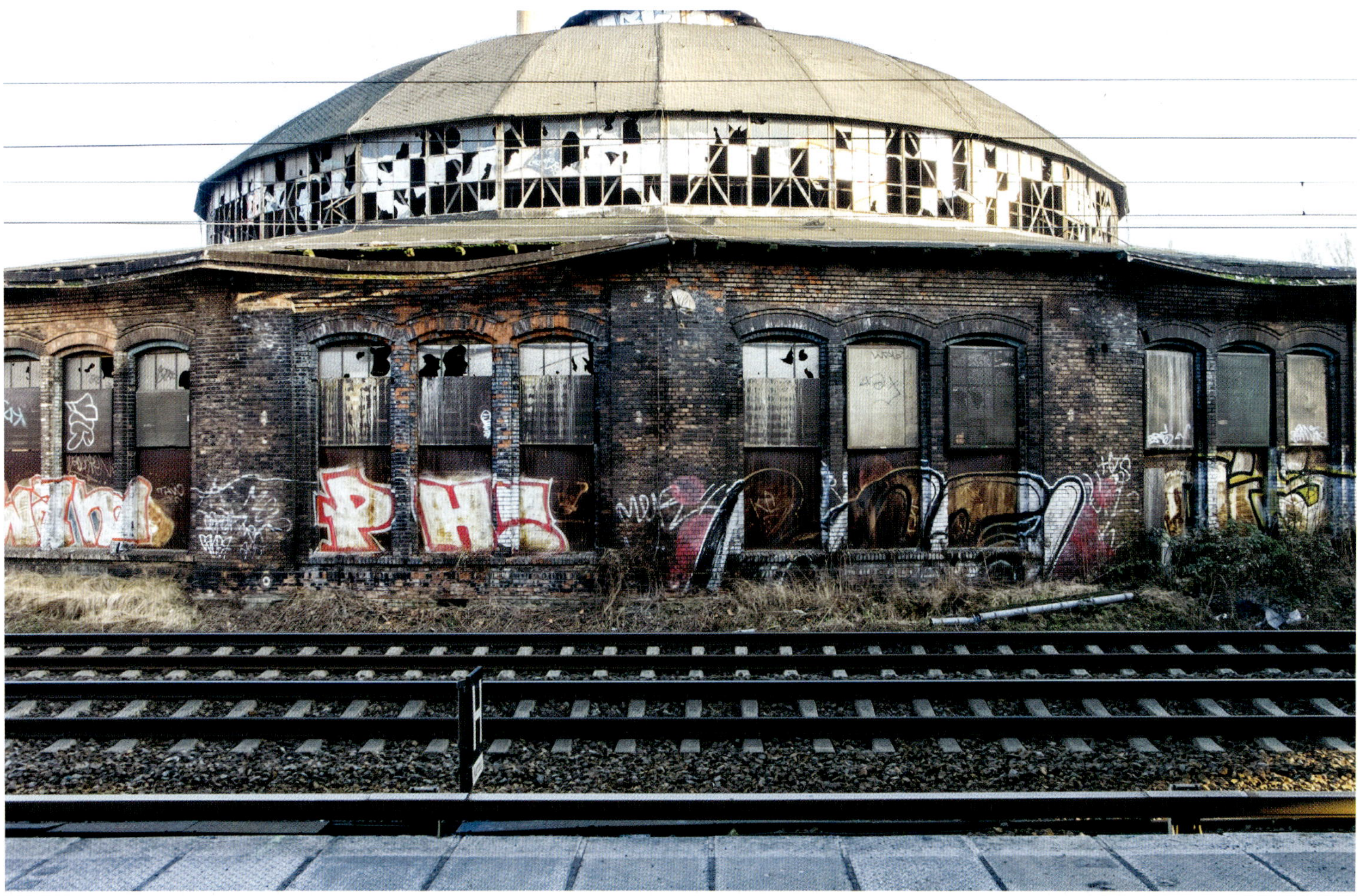

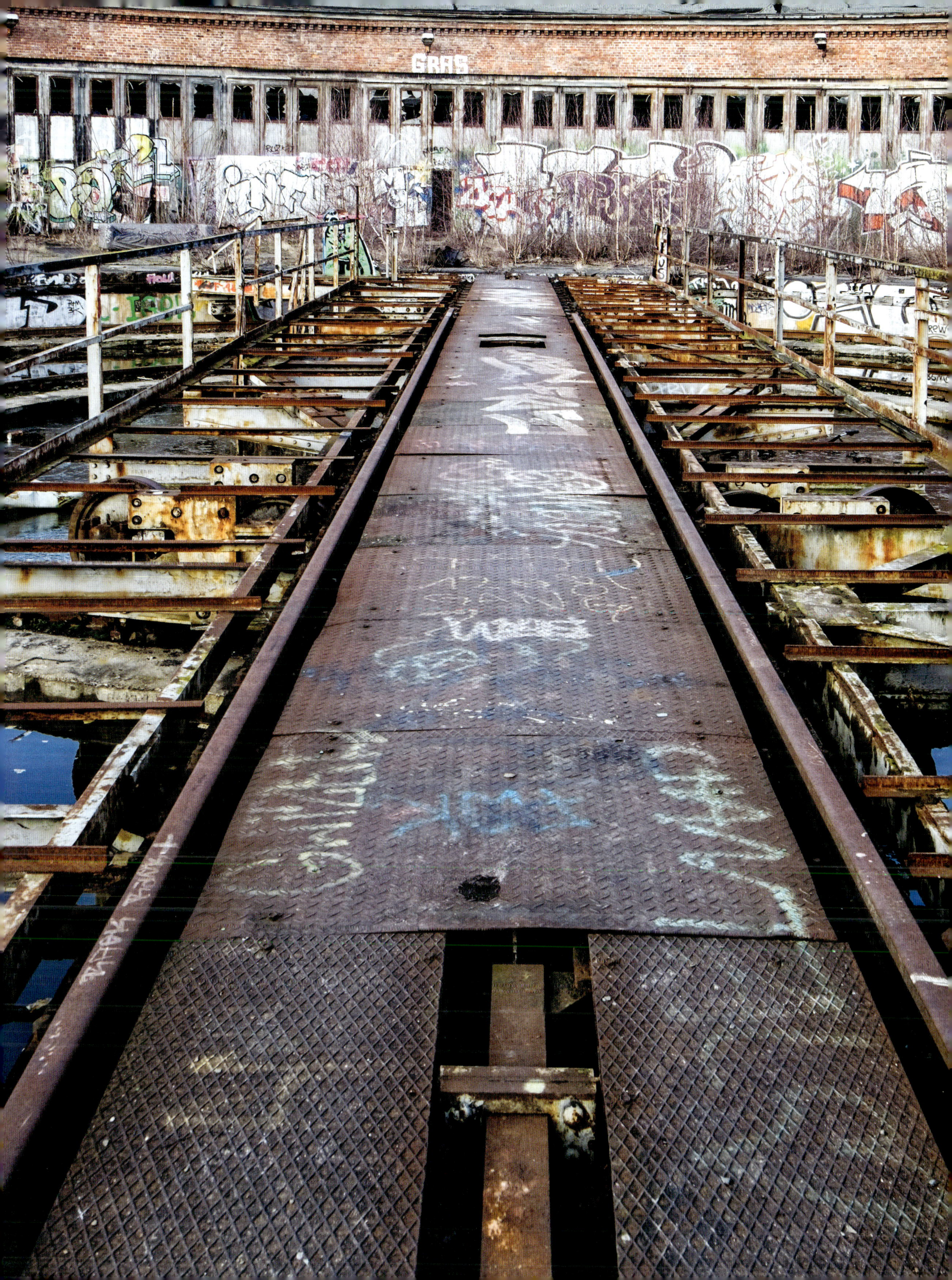
GRAS

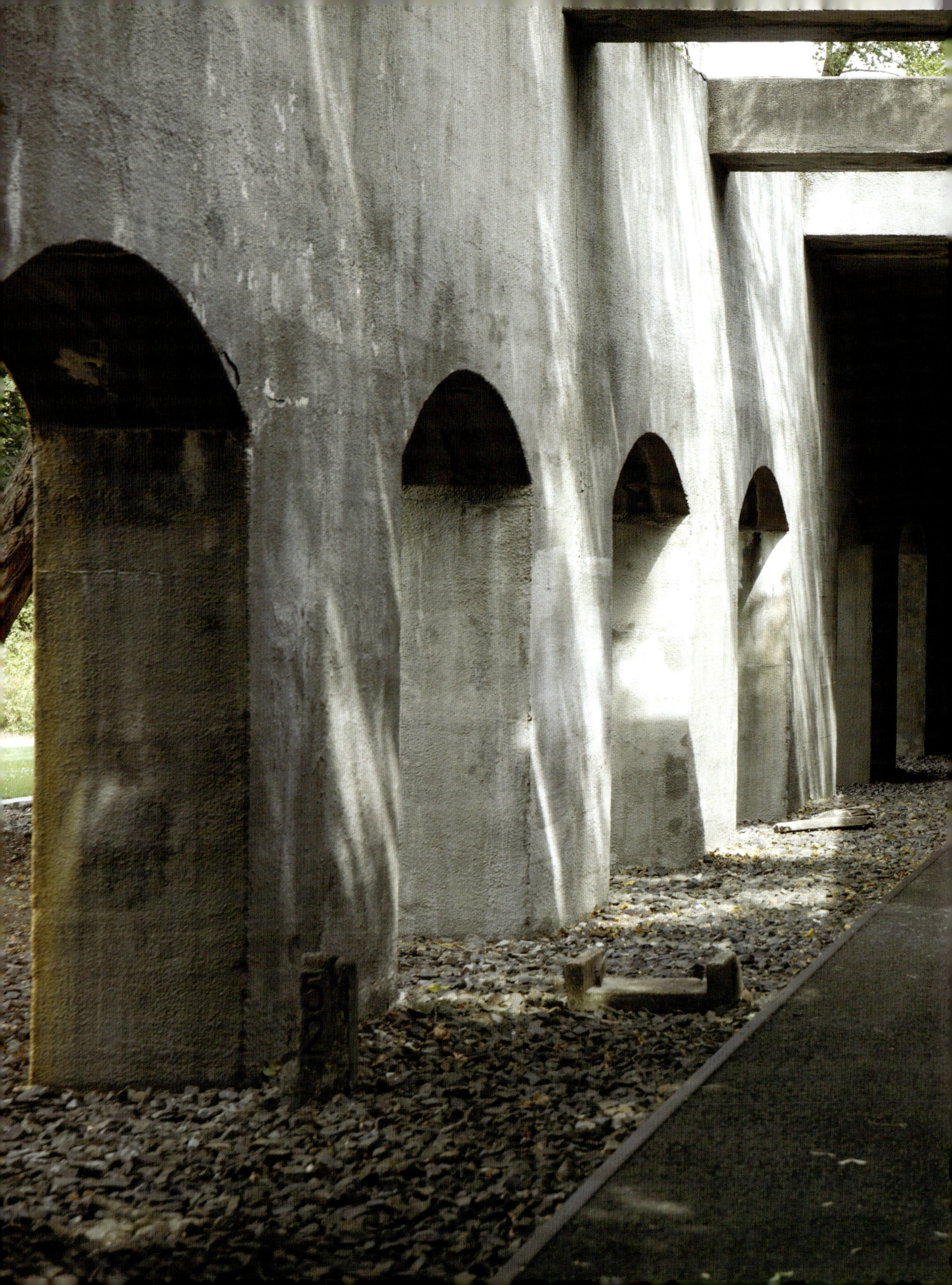

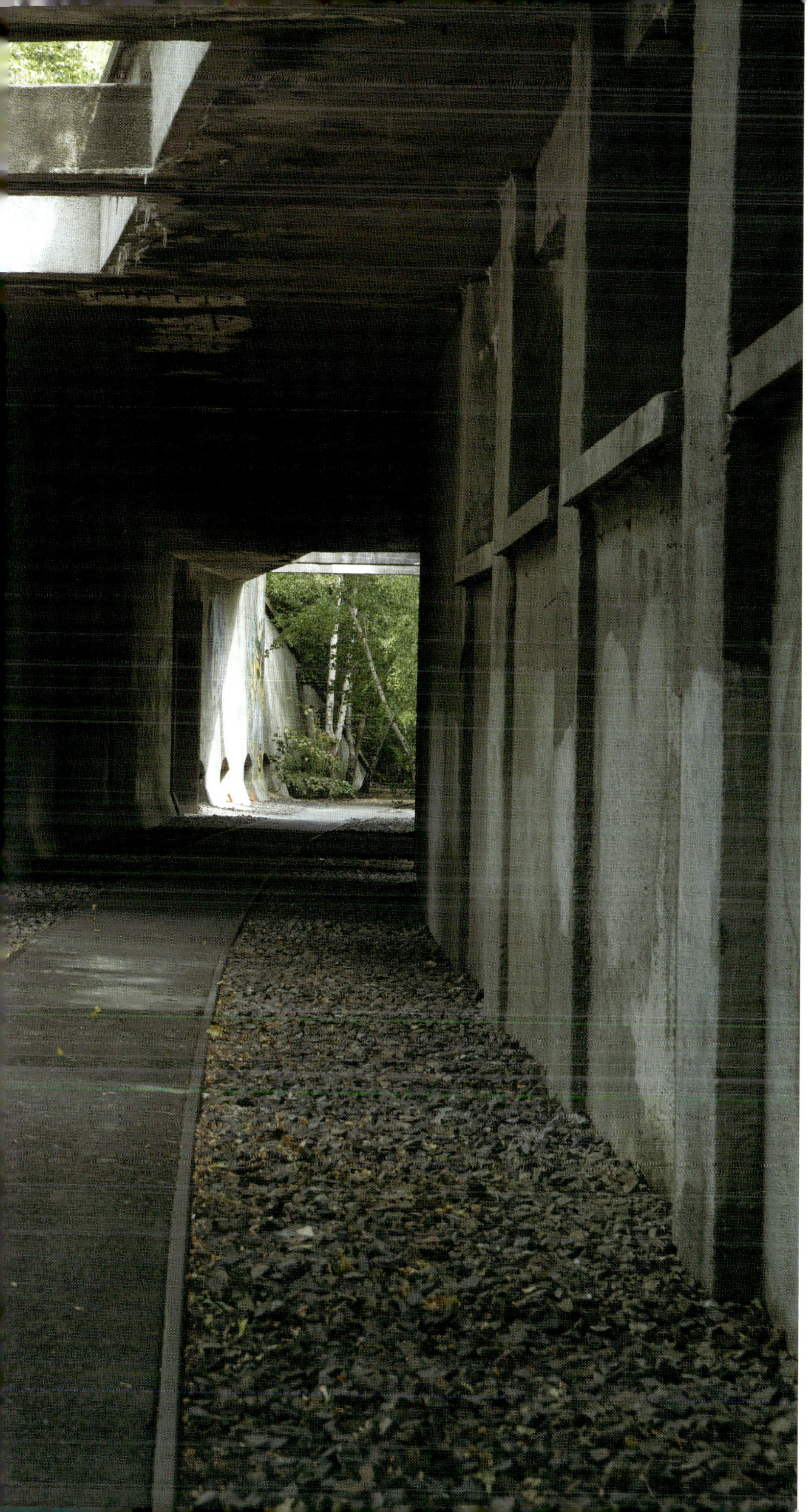

LEFT:

Natur-Park Sudgelände, Berlin, Germany

A vast extent of the former Tempelhof marshalling yards and a locomotive depot outside Berlin-Anhalt Station (closed in 1952) were transformed in the 1990s into a public amenity parkland embracing wild nature, contemporary art and relics of a former era of transport. Visitors encounter superannuated locomotives, concrete underpasses, modern sculptures and performance sites.

OVERLEAF:

Hangelsberg, Brandenburg, Germany

Trains of the DB Regio Nordost run past this station, but when it became unmanned in the late 20th century, its historic buildings were left to decay. Opened in 1842 on the line from Berlin to Cottbus and Frankfurt (Oder), it is one of Germany's oldest surviving stations. In recent years, a restoration group has acquired the buildings and is working to restore and use them as a cultural and educational centre.

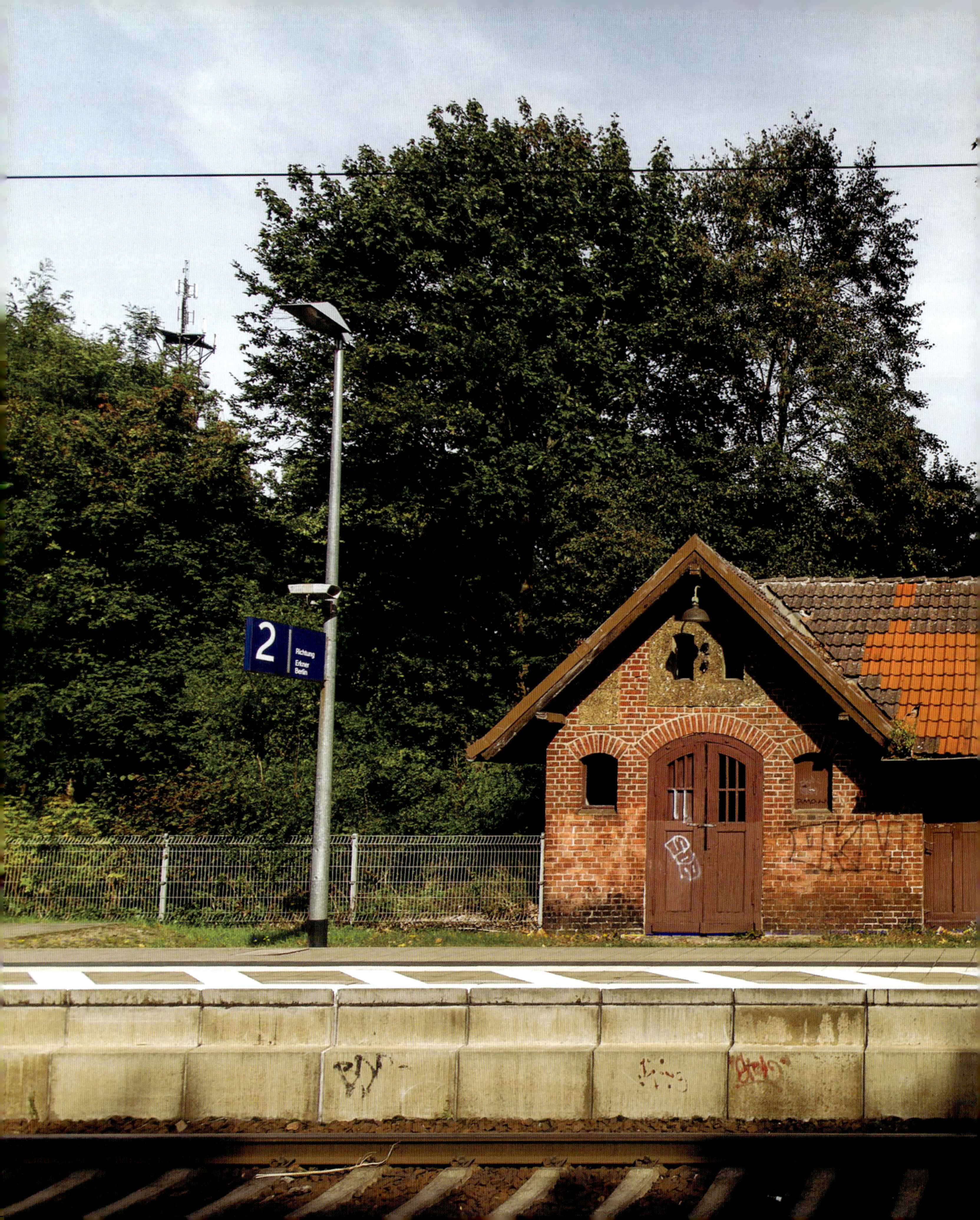
2
Richtung
Erkner
Berlin

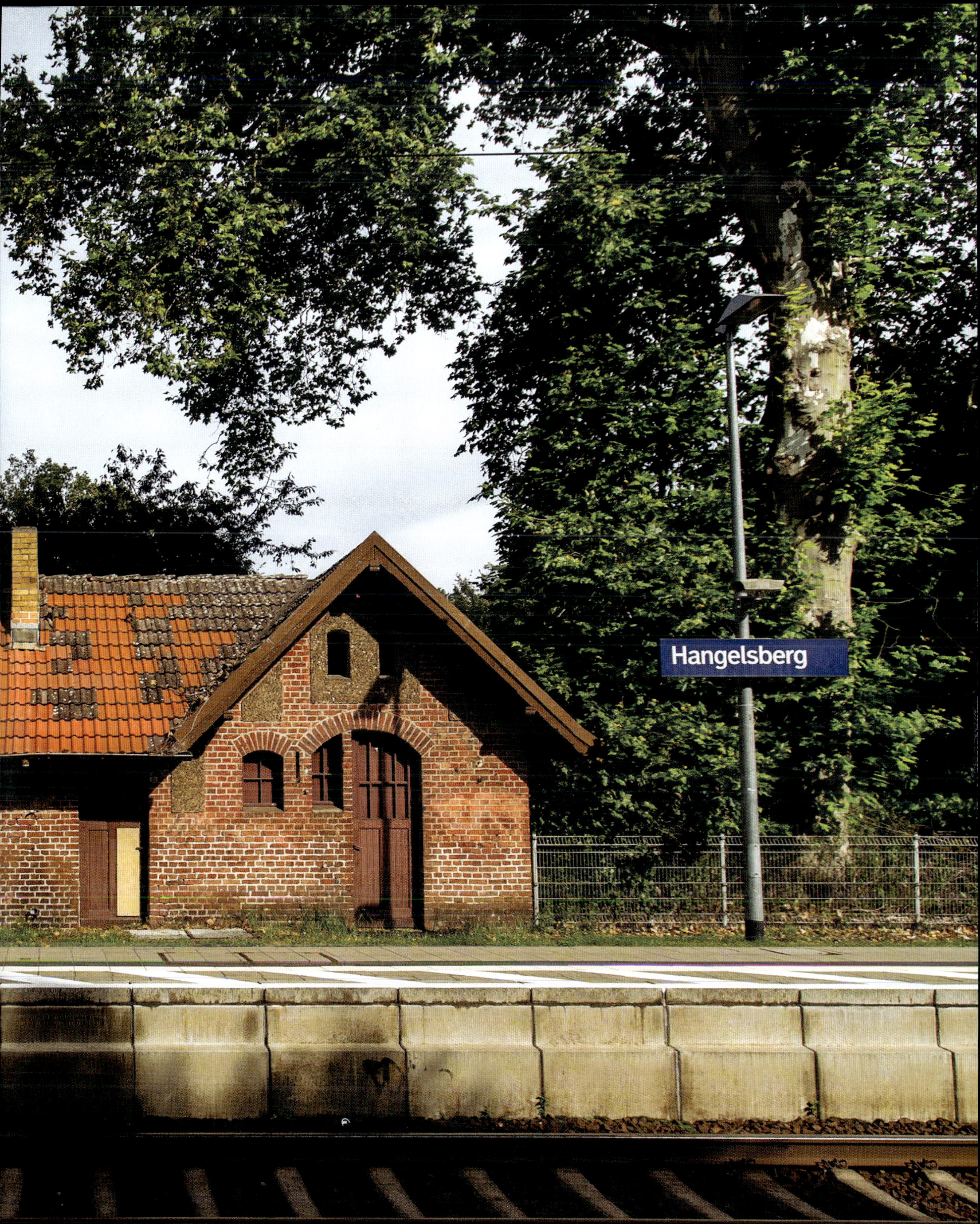
Hangelsberg

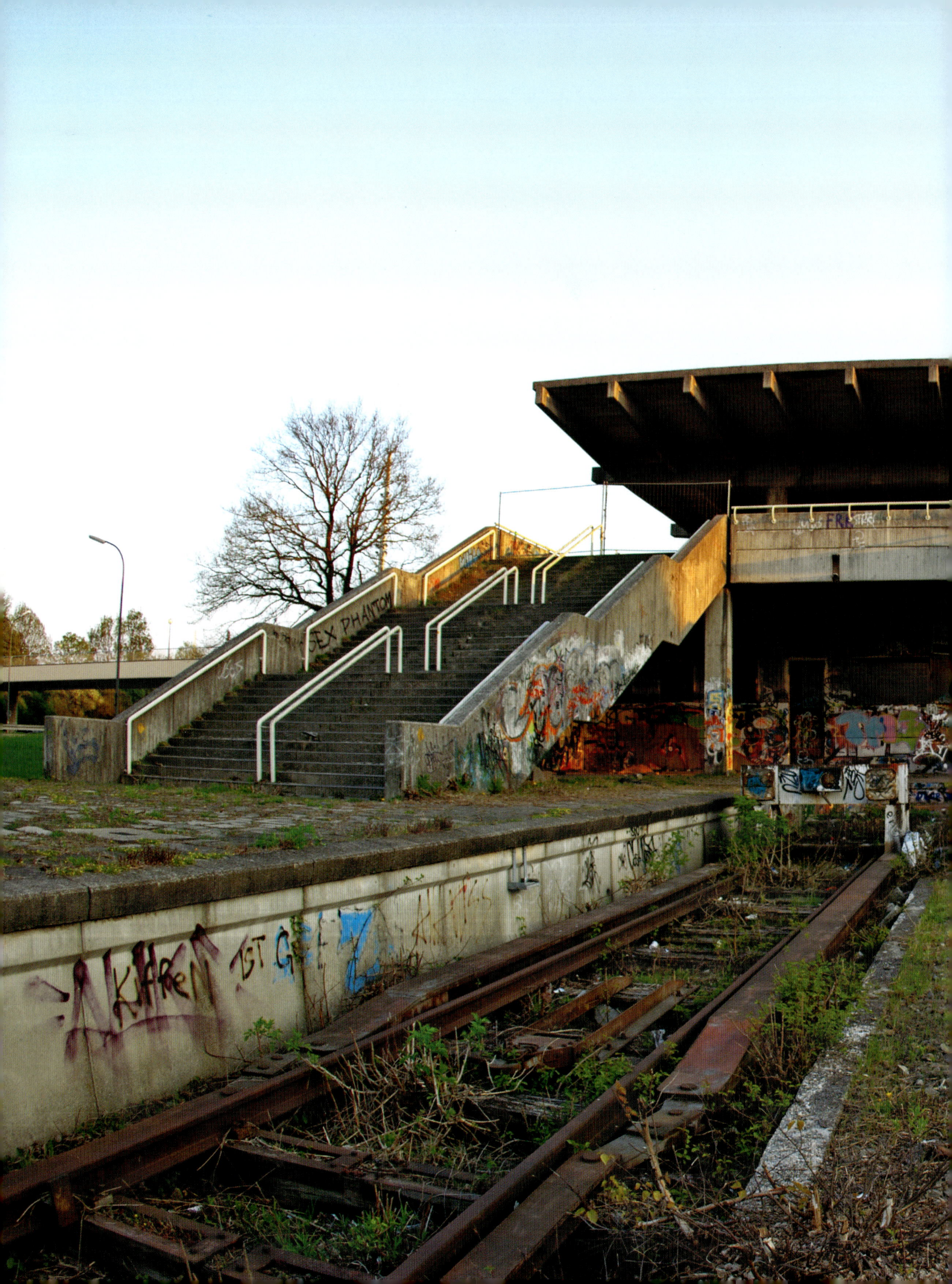
SEX PHANTOM

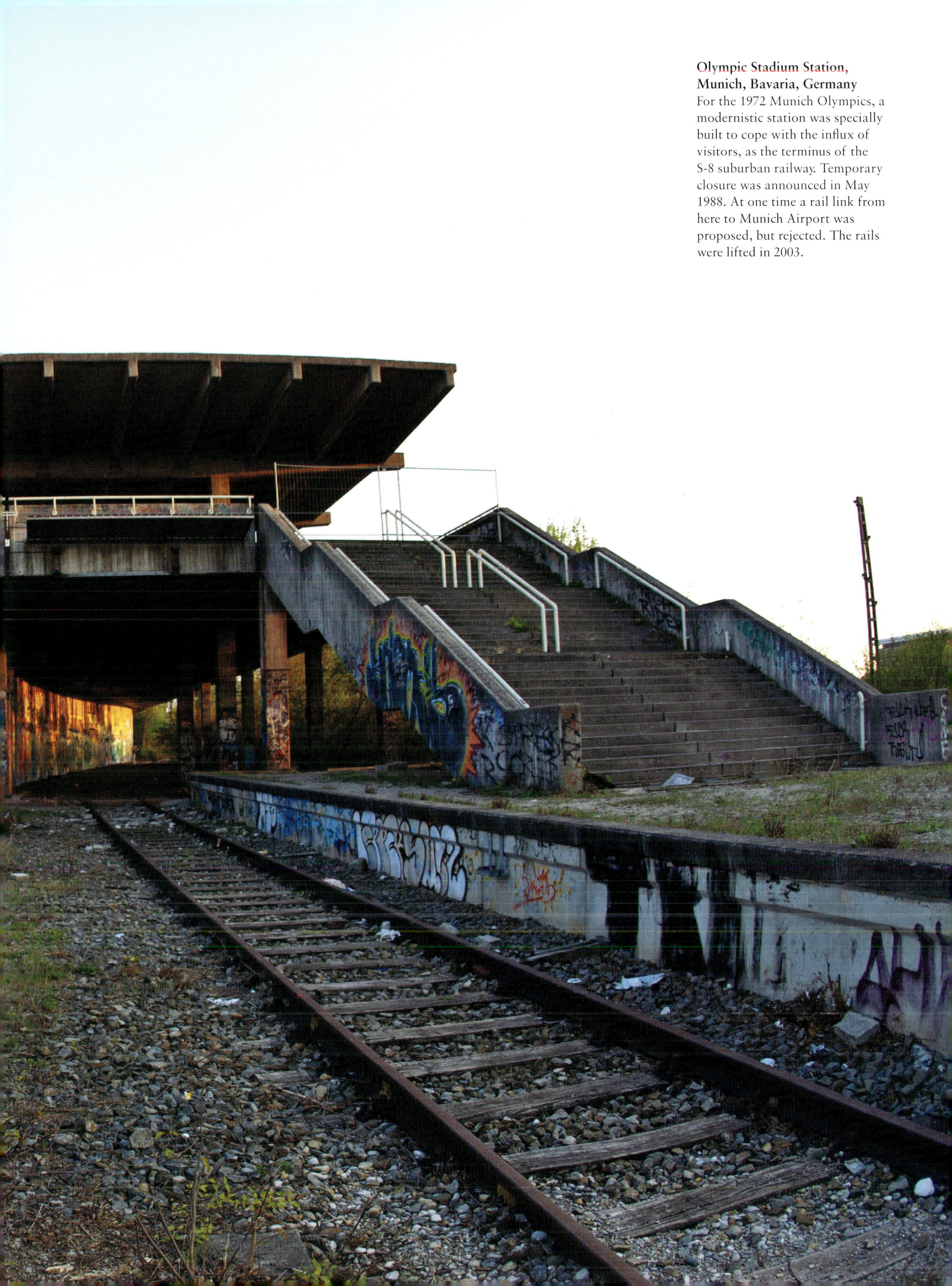

Olympic Stadium Station, Munich, Bavaria, Germany
For the 1972 Munich Olympics, a modernistic station was specially built to cope with the influx of visitors, as the terminus of the S-8 suburban railway. Temporary closure was announced in May 1988. At one time a rail link from here to Munich Airport was proposed, but rejected. The rails were lifted in 2003.

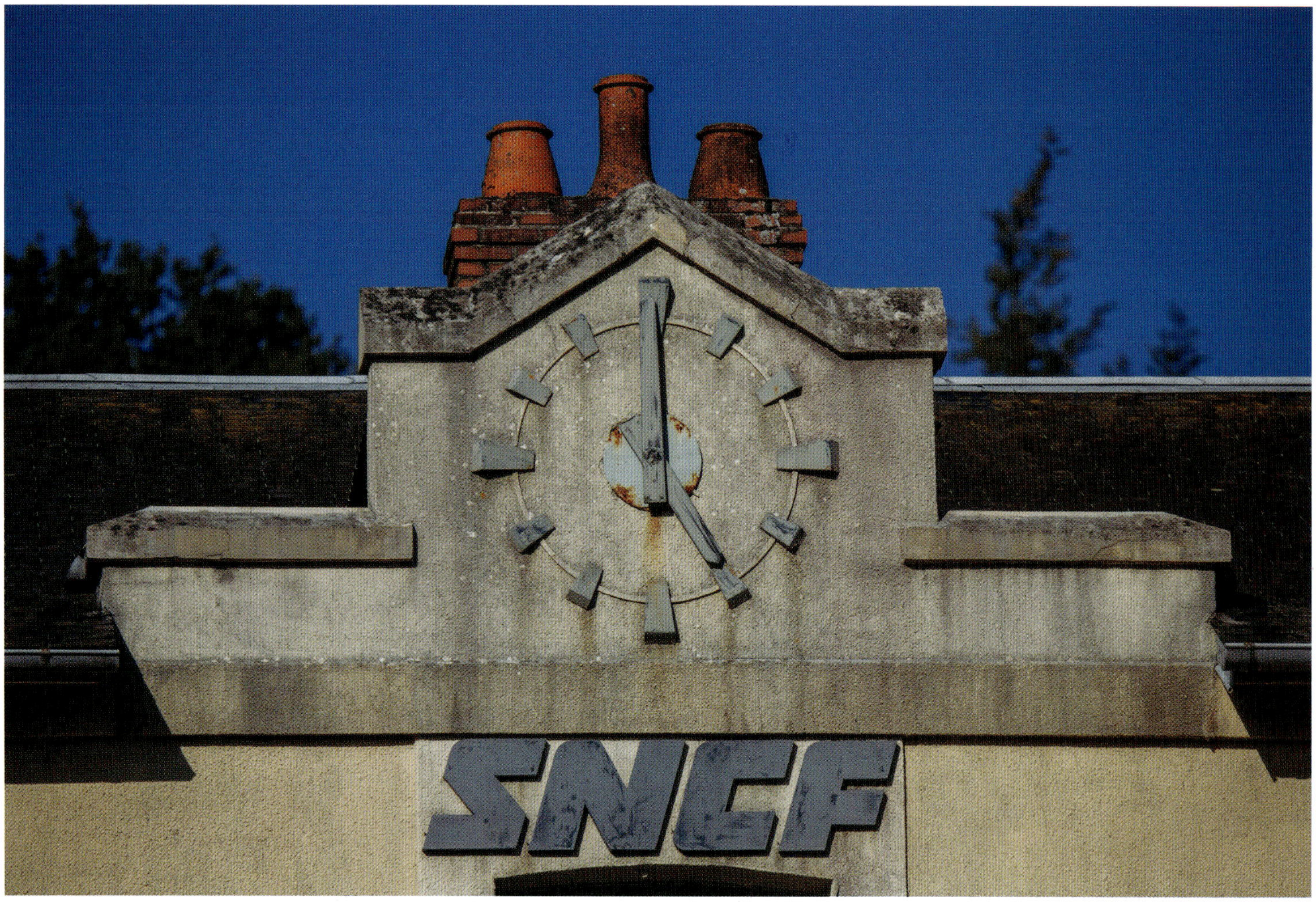
SNCF

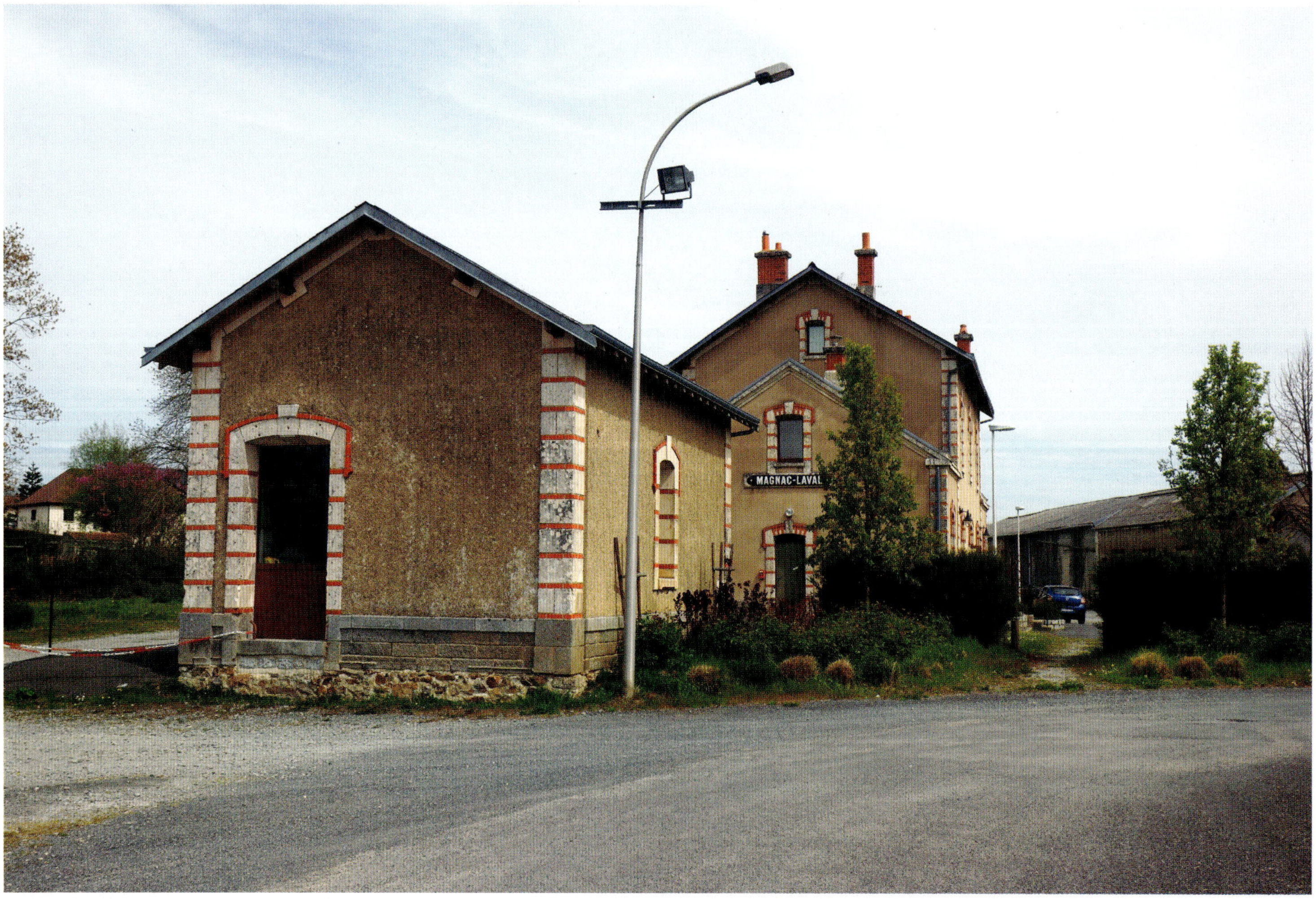
MAGNAC-LAVAL

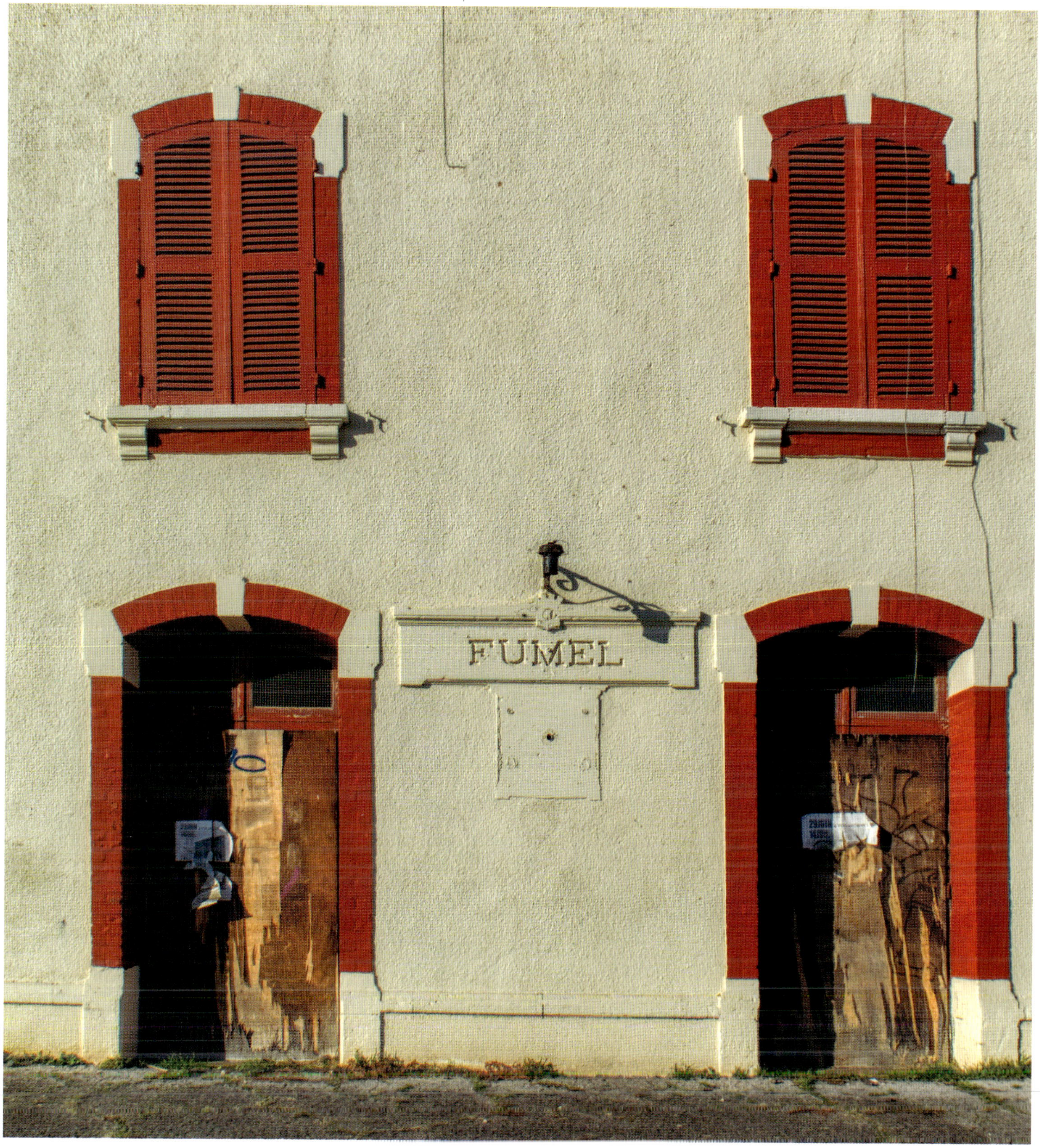

OPPOSITE ABOVE:
Bagnoles de l'Orne, Orne, France
Opened in 1881 on the coastal line between Briouze and Couterne, the station was closed in 2002. Acquired by the municipality in 2012, it has been resuscitated as a visitors' portal to this resort of thermal baths and green spaces. The clock however remains firmly at 5 o'clock.

OPPOSITE BELOW:
Magnac-Laval, Haute-Vienne, France
The Paris–Orleans Railway opened its branch from Dorat to here in 1908. Passenger traffic ceased in 1932 but a goods service was maintained until 1988. The line was dismantled after 1994, but there is still a Café de la Gare in the village.

ABOVE:
Fumel, Lot-et-Garonne, France
On the closed line between Monsempron–Libos and Cahors, the station was opened by the Paris-Orleans Railway in 1869, and closed to passengers in 1971. A goods service lasted until 2010. The trackbed here is now a greenway for ramblers.

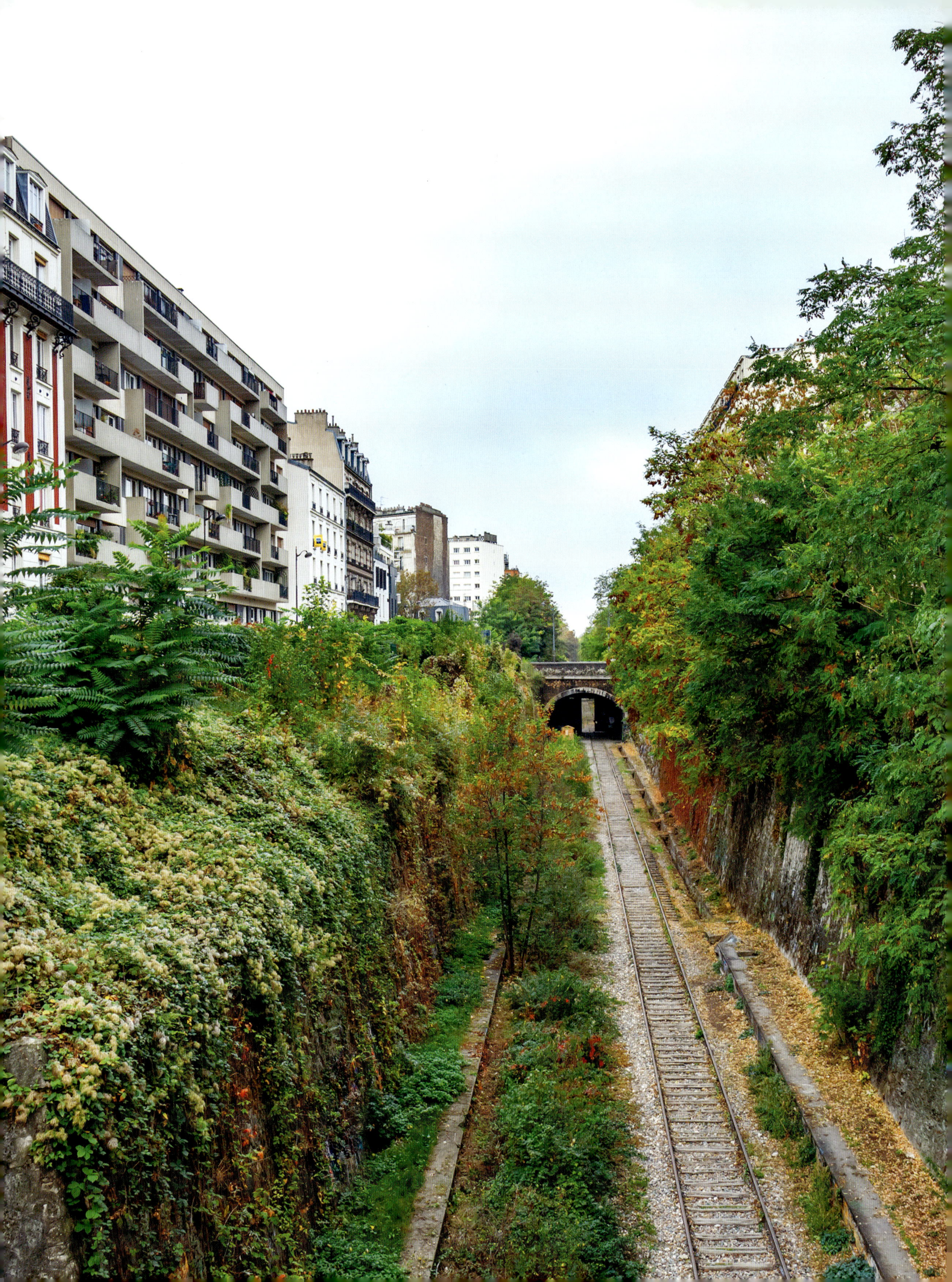

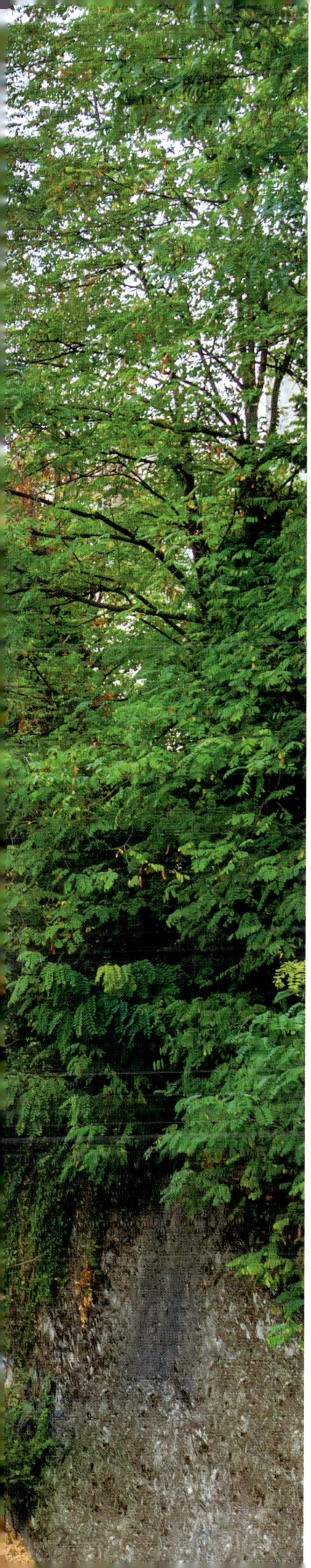

ALL PHOTOGRAPHS:

La Petite Ceinture, Paris, France

The 'Little Belt' is a railway that circles the inner city of Paris. Begun in 1852, its initial purpose was to transport materials for the fortifications then under construction. Passenger services began early, but the spread of métro lines made it redundant in 1934, though goods services continued to 1979. Today, only a short section on the west of the city is still in use by regional commuter trains.

LEFT:

Much of the line runs in cuttings, given over to the public as a green walking space, with pedestrian access.

BELOW:

Many stations on the line were built above the tracks. This typical Empire-style example may be Courcelles-Levallois.

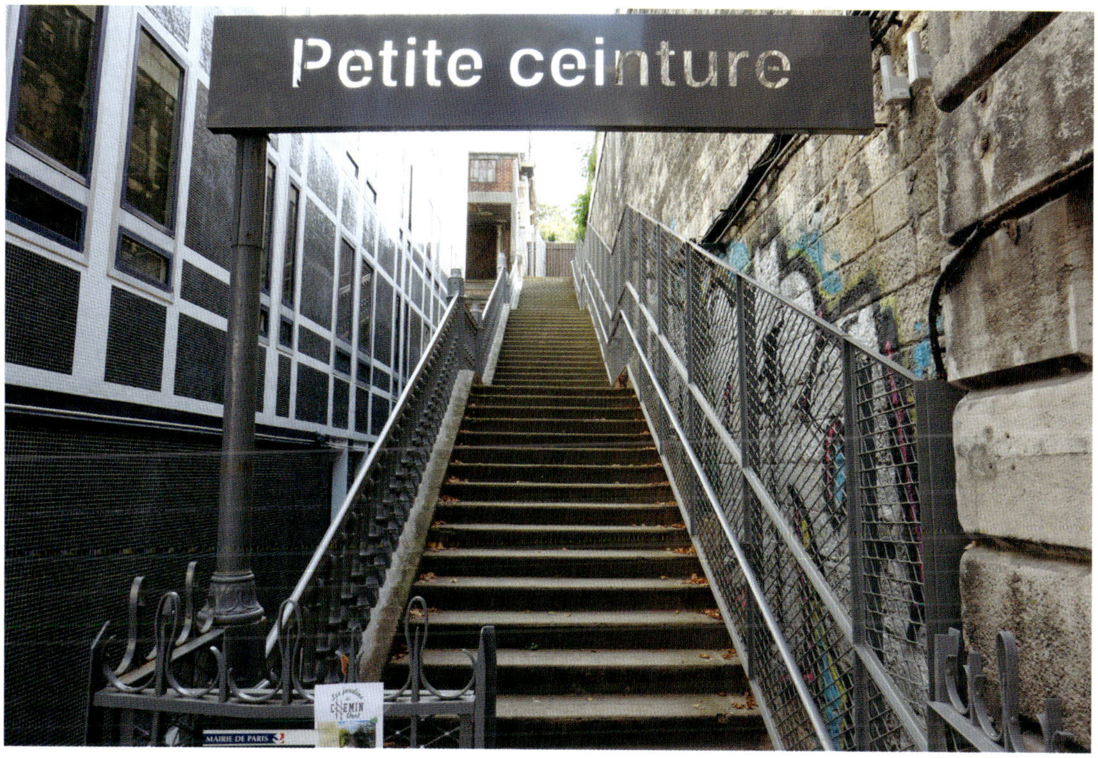

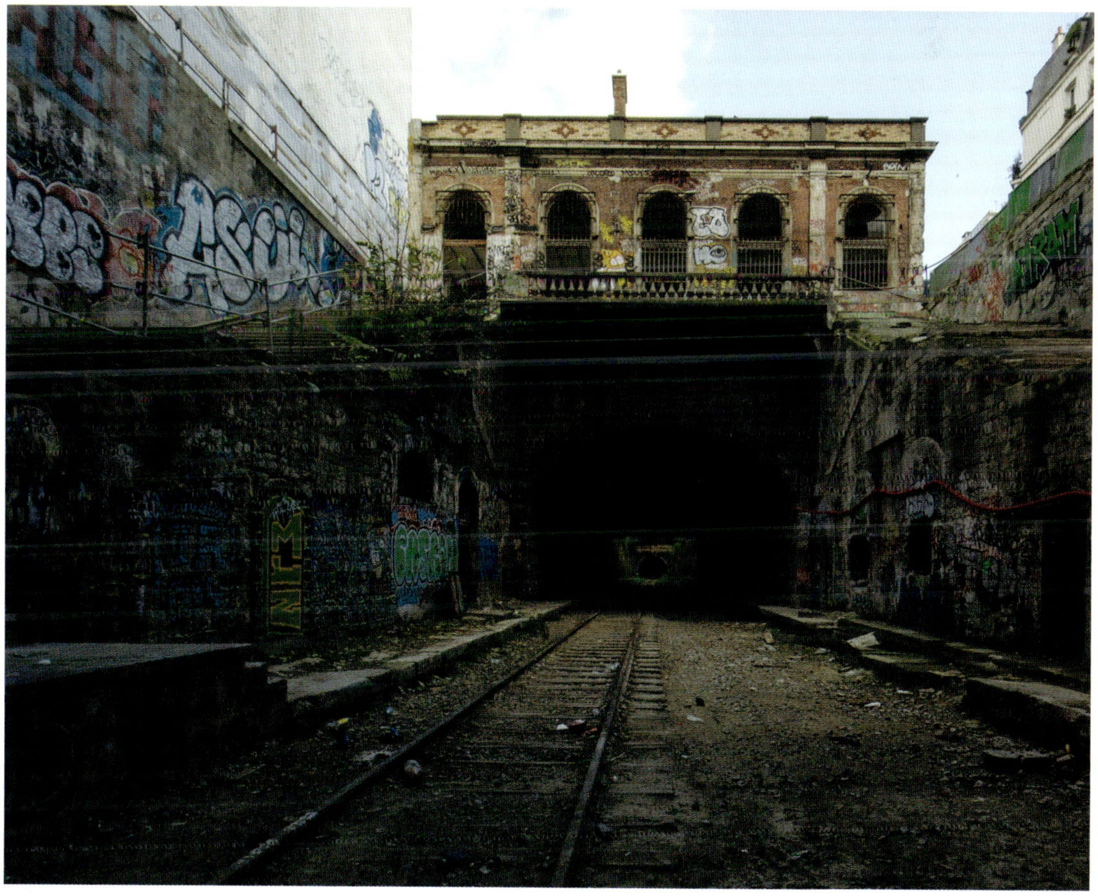

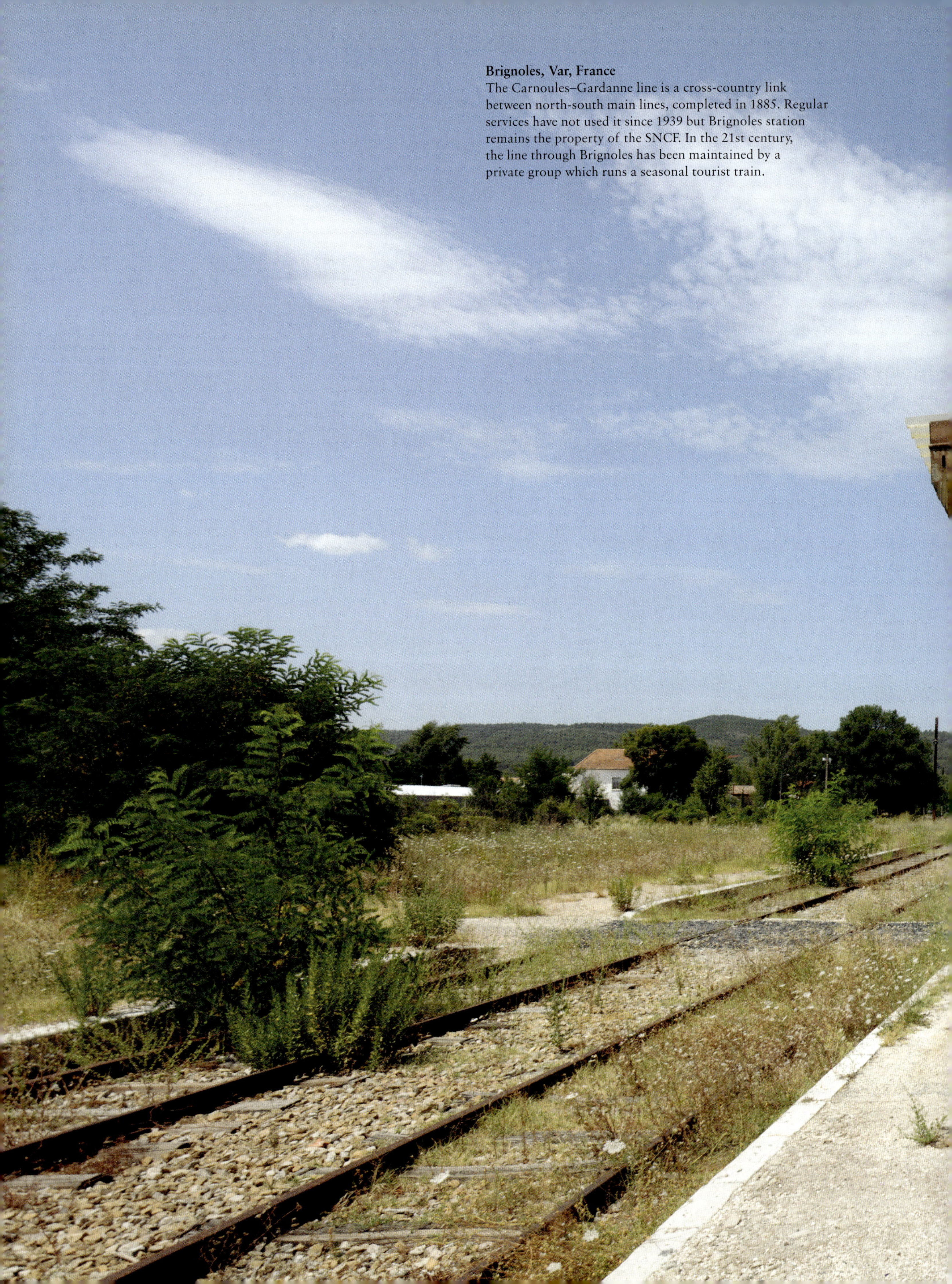

Brignoles, Var, France
The Carnoules–Gardanne line is a cross-country link between north-south main lines, completed in 1885. Regular services have not used it since 1939 but Brignoles station remains the property of the SNCF. In the 21st century, the line through Brignoles has been maintained by a private group which runs a seasonal tourist train.

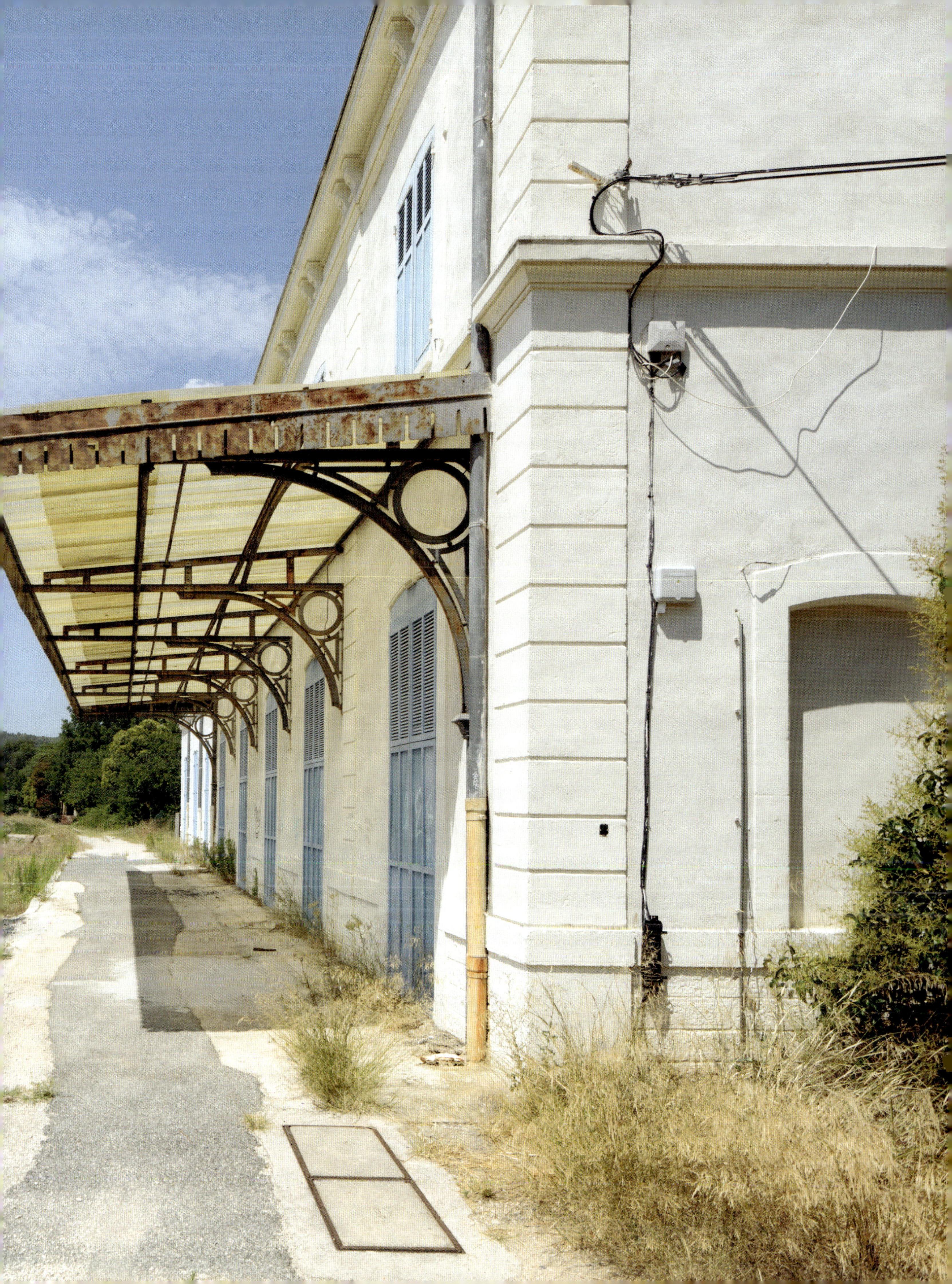

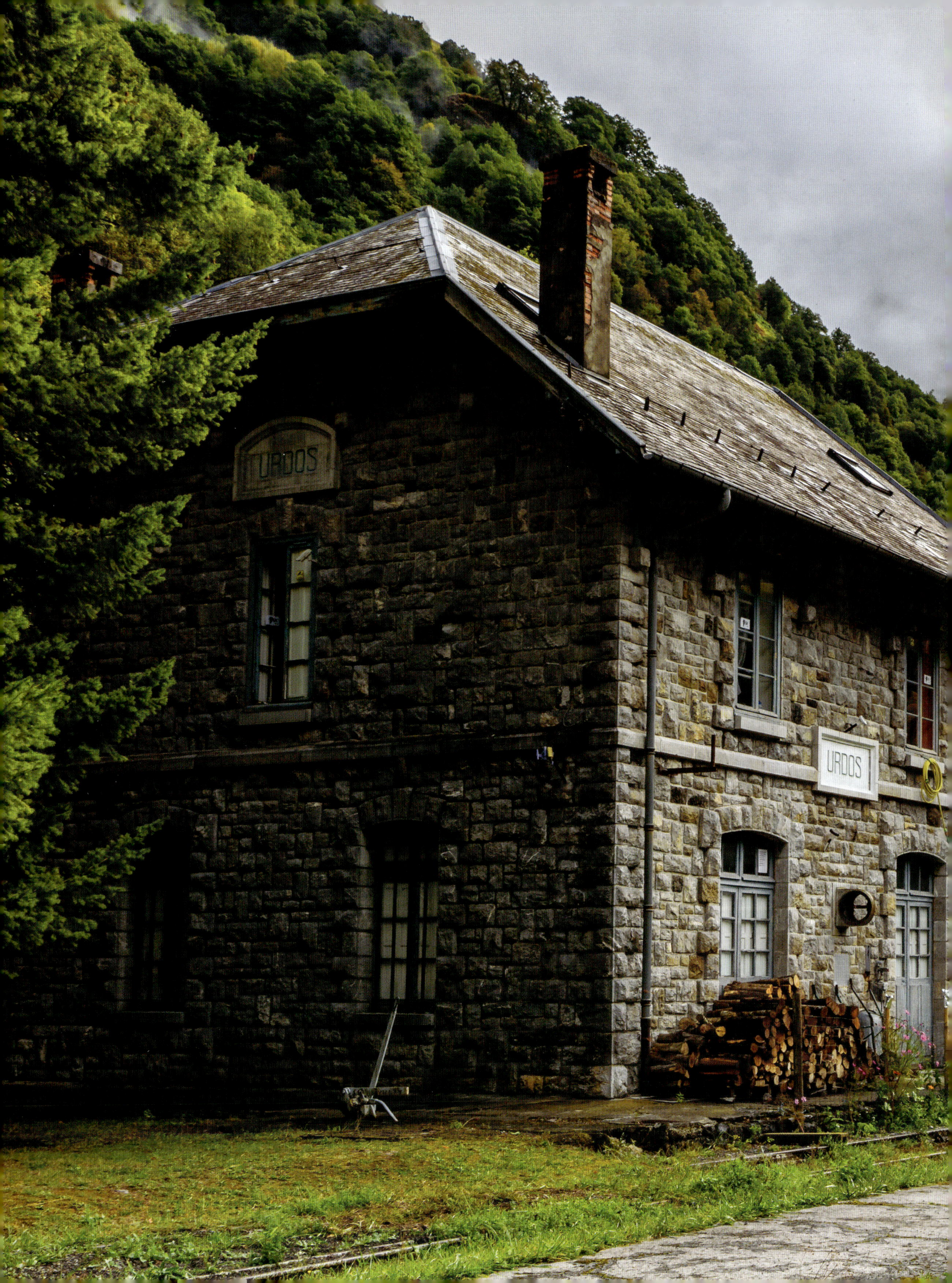
URDOS
URDOS

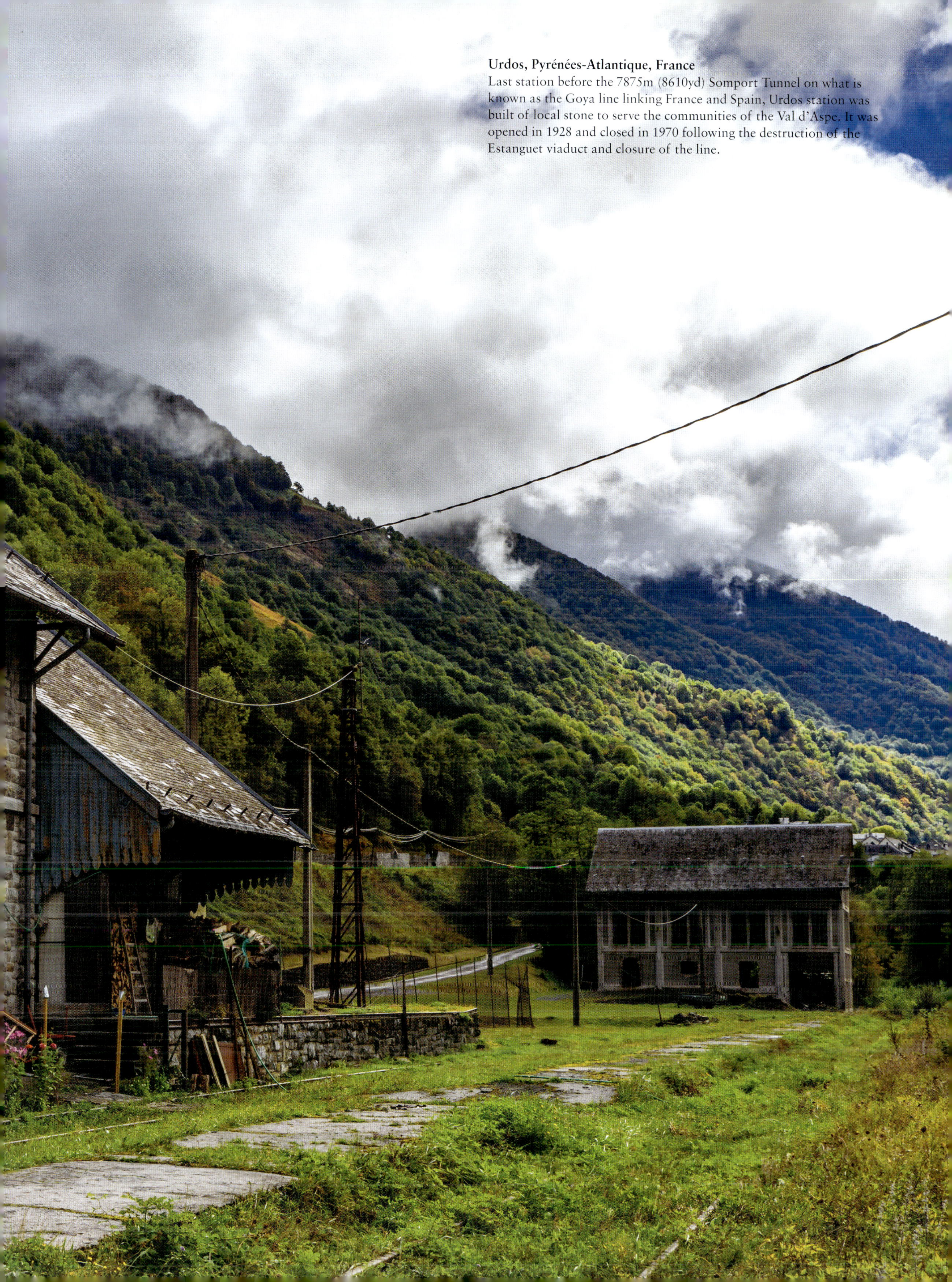

Urdos, Pyrénées-Atlantique, France
Last station before the 7875m (8610yd) Somport Tunnel on what is known as the Goya line linking France and Spain, Urdos station was built of local stone to serve the communities of the Val d'Aspe. It was opened in 1928 and closed in 1970 following the destruction of the Estanguet viaduct and closure of the line.

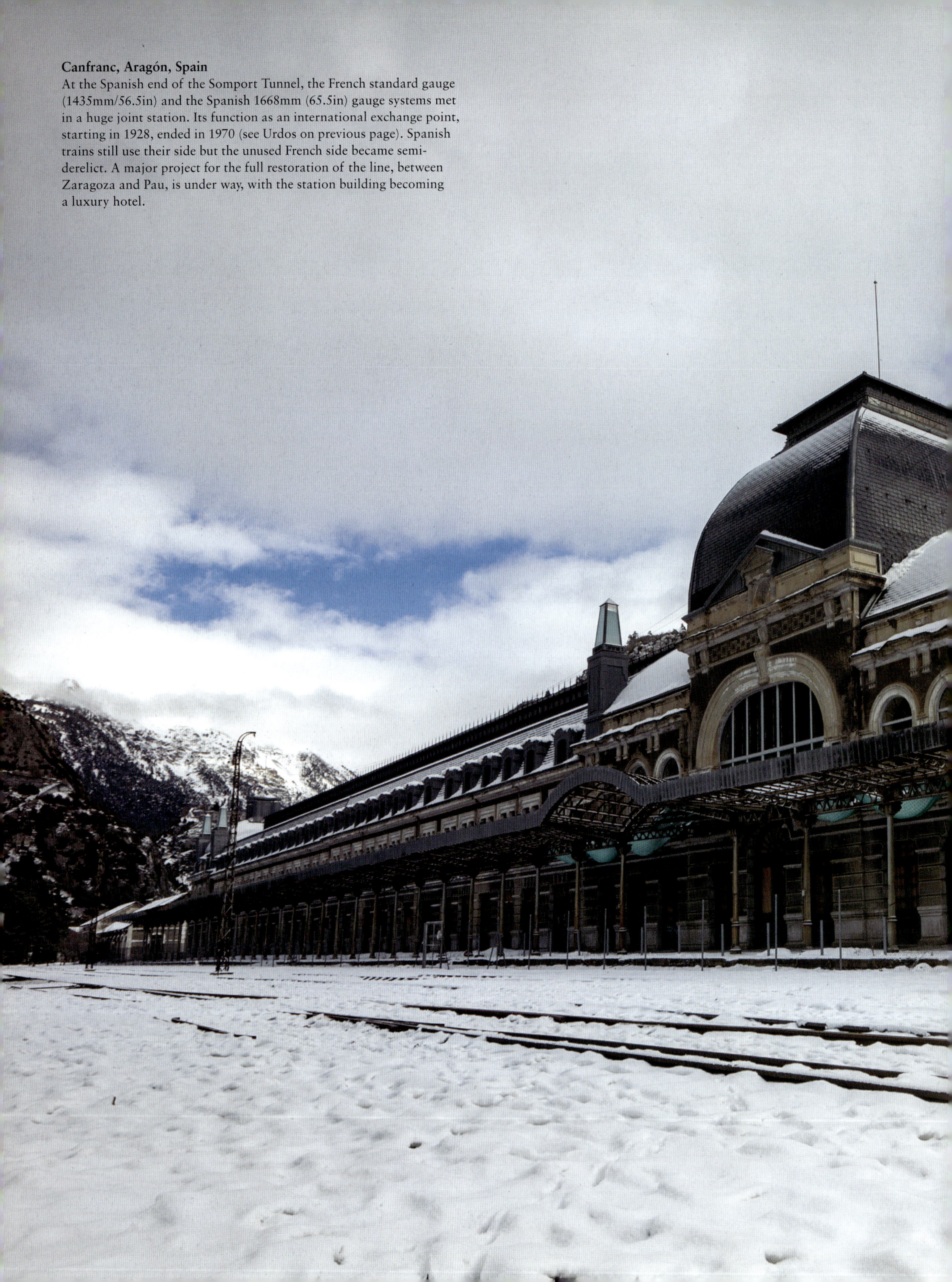

Canfranc, Aragón, Spain
At the Spanish end of the Somport Tunnel, the French standard gauge (1435mm/56.5in) and the Spanish 1668mm (65.5in) gauge systems met in a huge joint station. Its function as an international exchange point, starting in 1928, ended in 1970 (see Urdos on previous page). Spanish trains still use their side but the unused French side became semi-derelict. A major project for the full restoration of the line, between Zaragoza and Pau, is under way, with the station building becoming a luxury hotel.

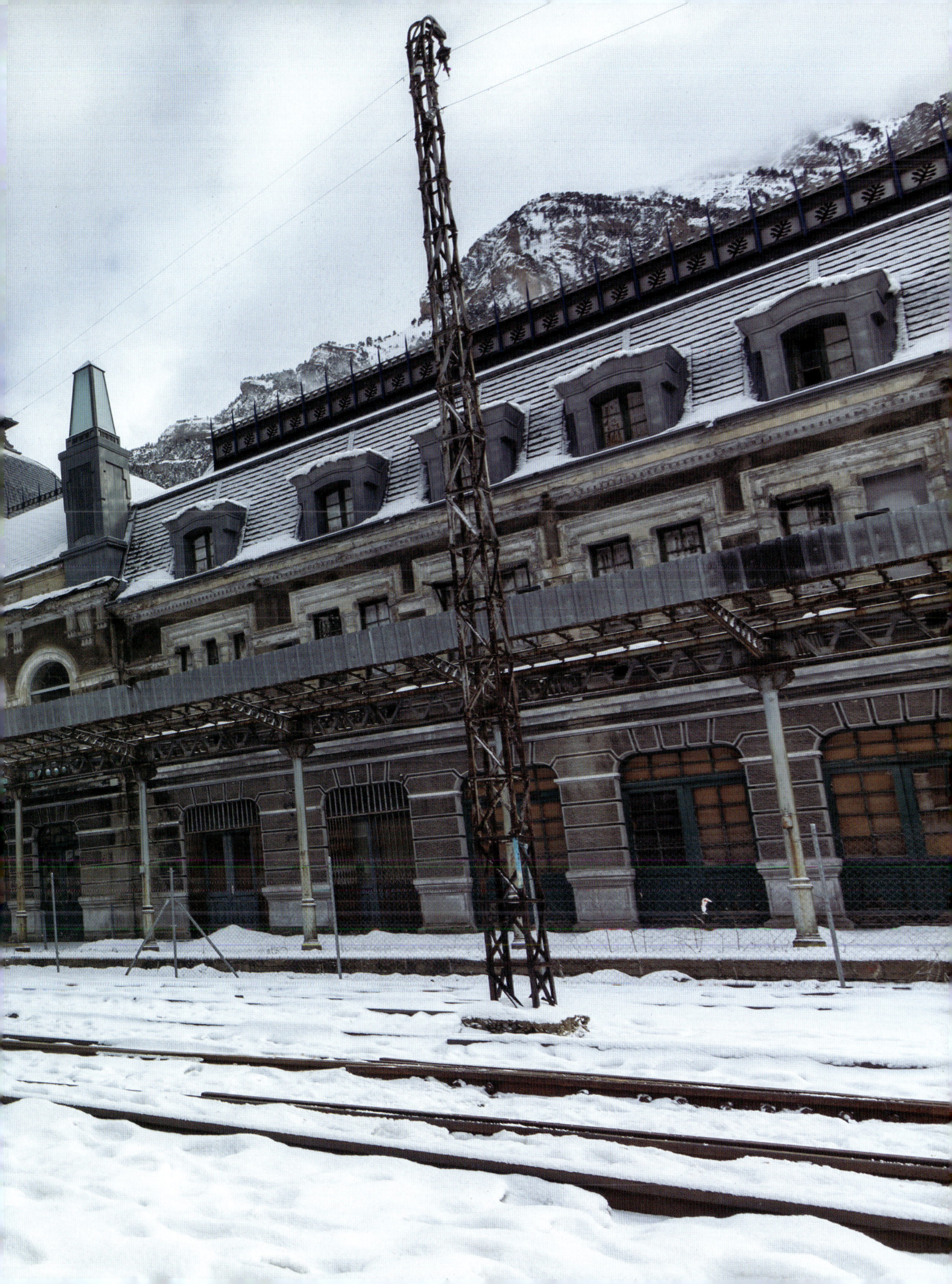

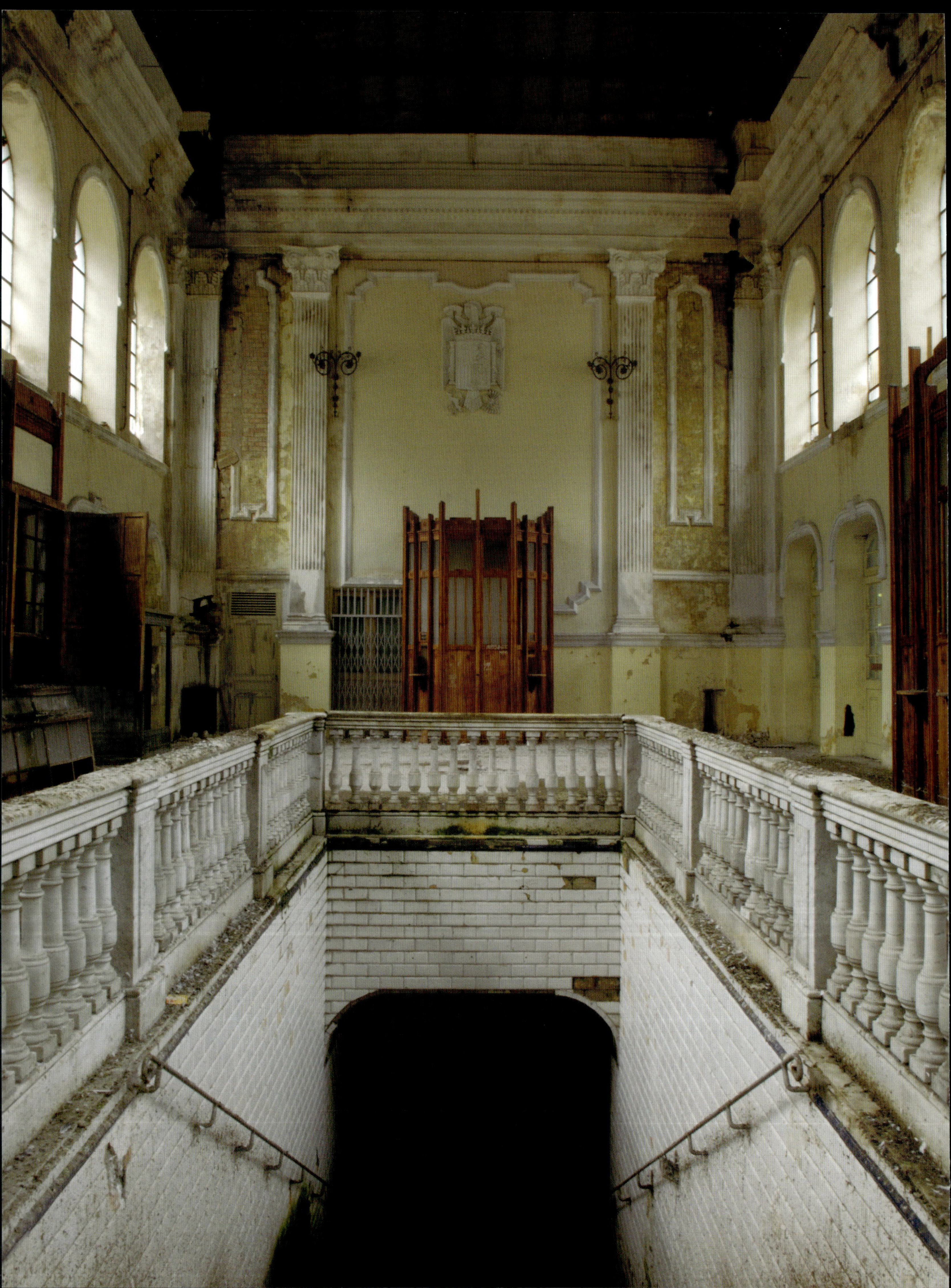

ALL PHOTOGRAPHS:

Canfranc, Aragón, Spain

The station was built as a prestige project, and its lavish decoration and grand scale were always out of proportion to the traffic. This 2005 image (opposite) shows the neglected state of what will be restored as a grand hotel. As a frontier point during World War II, it was the centre of many stories of espionage and of secret traffic between supposedly neutral Spain and Nazi Germany.

RIGHT:

A speed restriction plate lies among debris on the Spanish side.

BELOW:

The different rail gauges mean that in effect Canfranc is two stations in one, with most facilities duplicated.

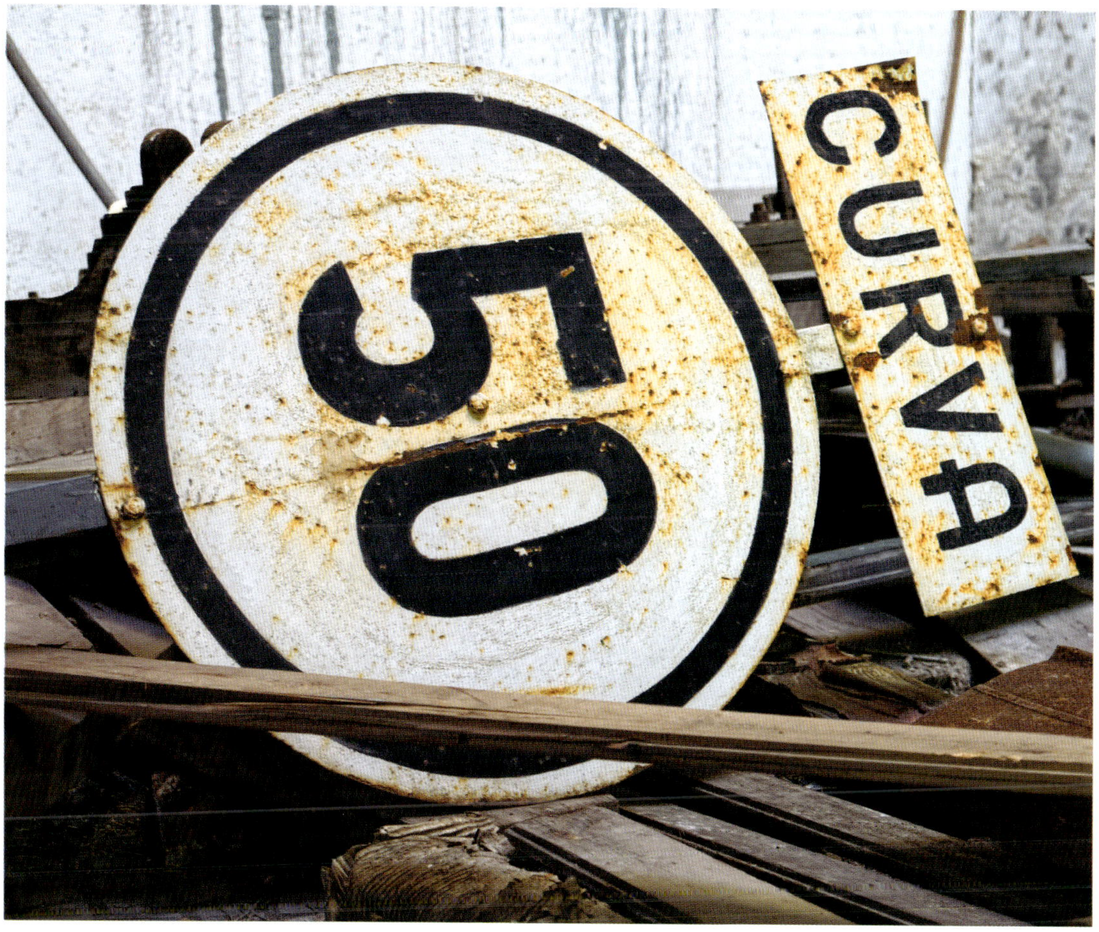

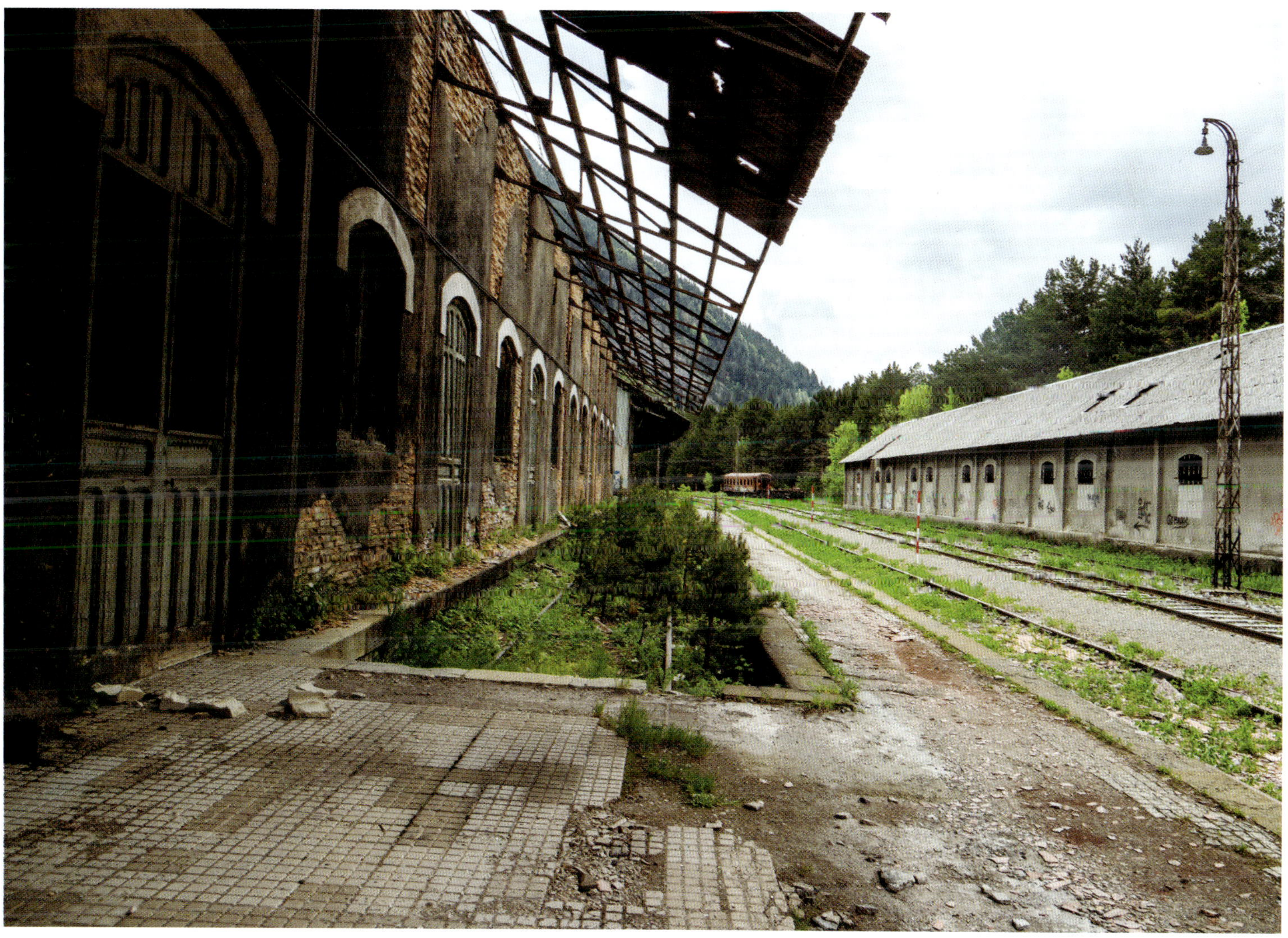

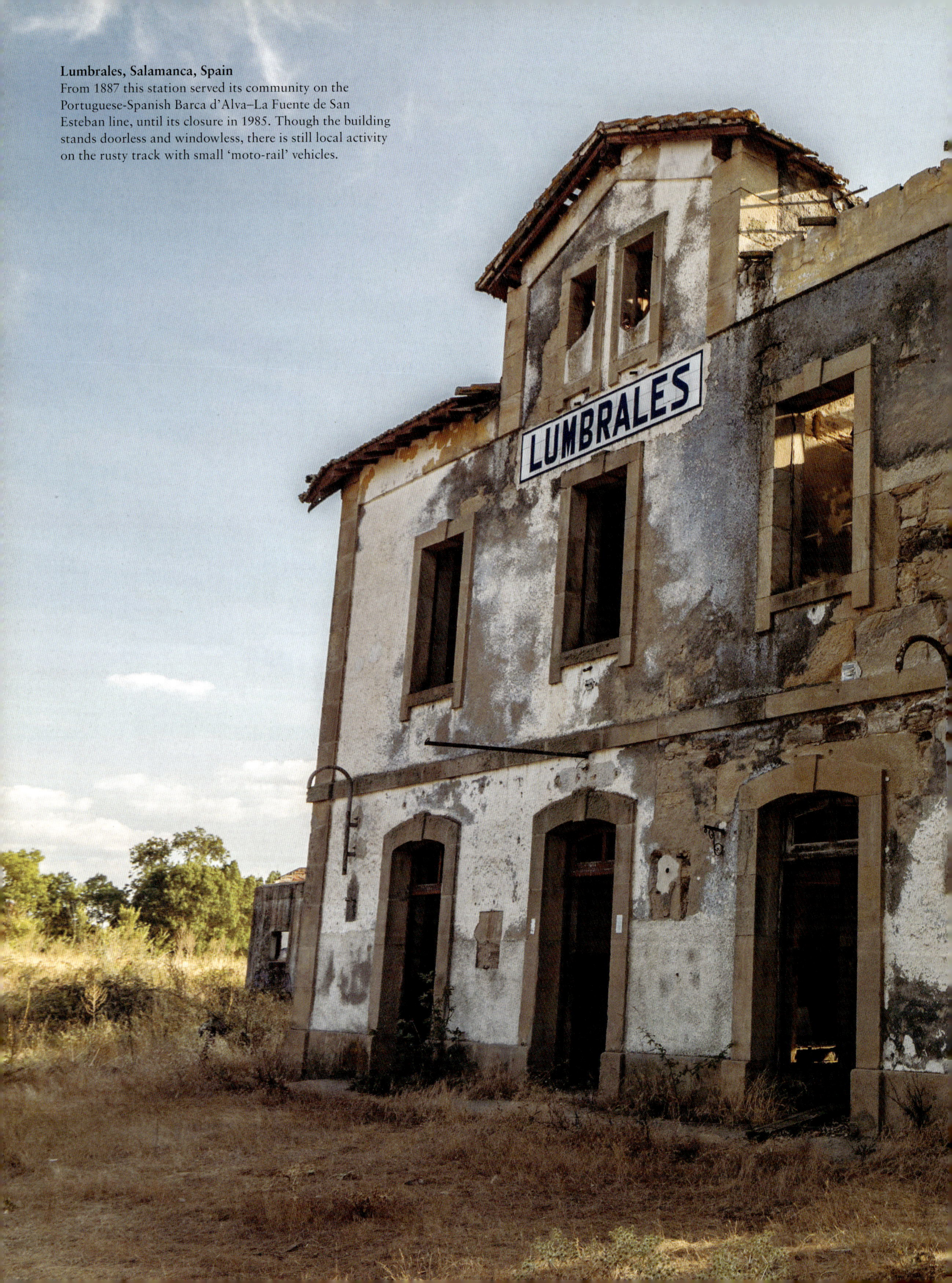

Lumbrales, Salamanca, Spain
From 1887 this station served its community on the Portuguese-Spanish Barca d'Alva–La Fuente de San Esteban line, until its closure in 1985. Though the building stands doorless and windowless, there is still local activity on the rusty track with small 'moto-rail' vehicles.

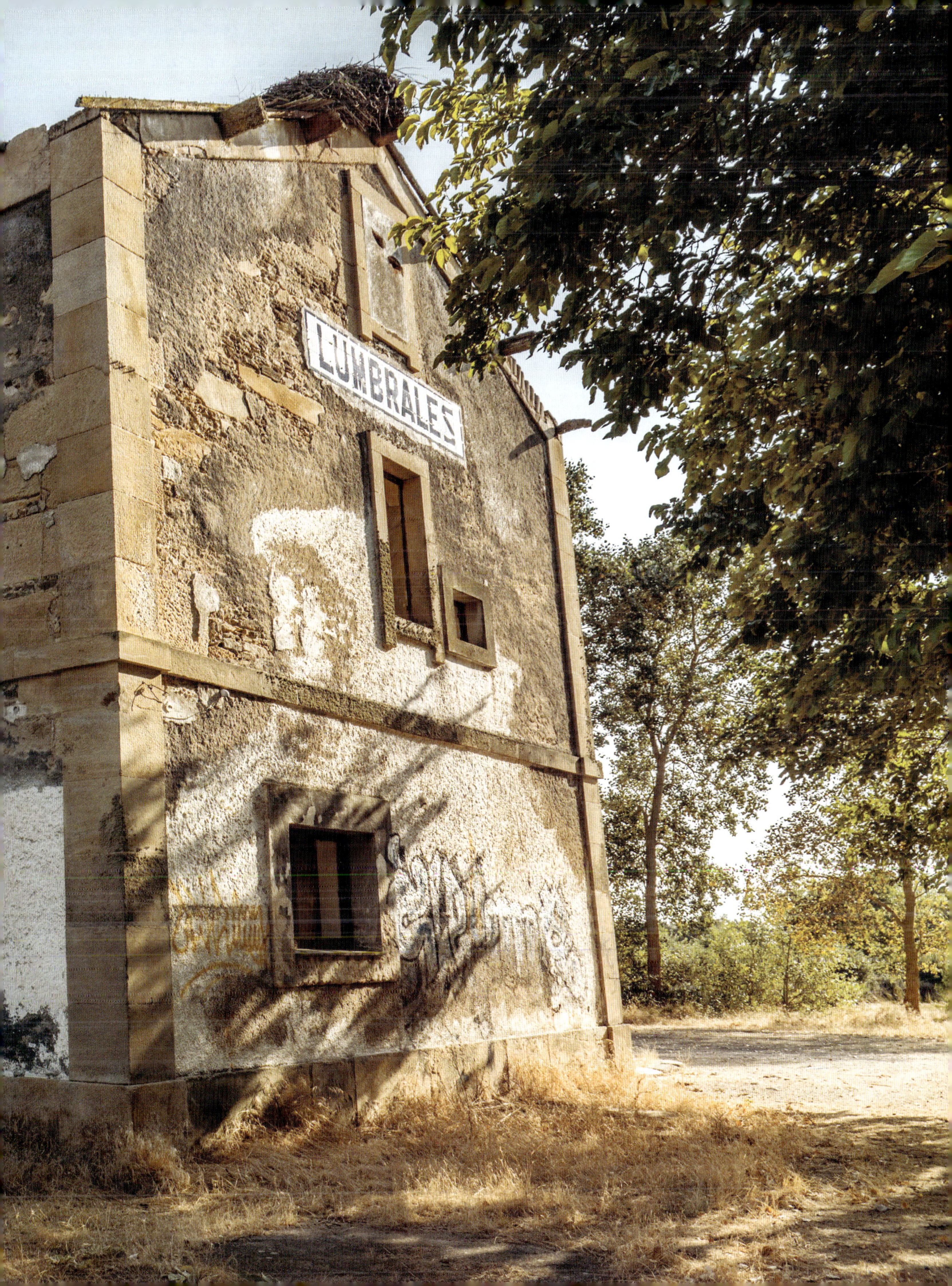
LUMBRALES

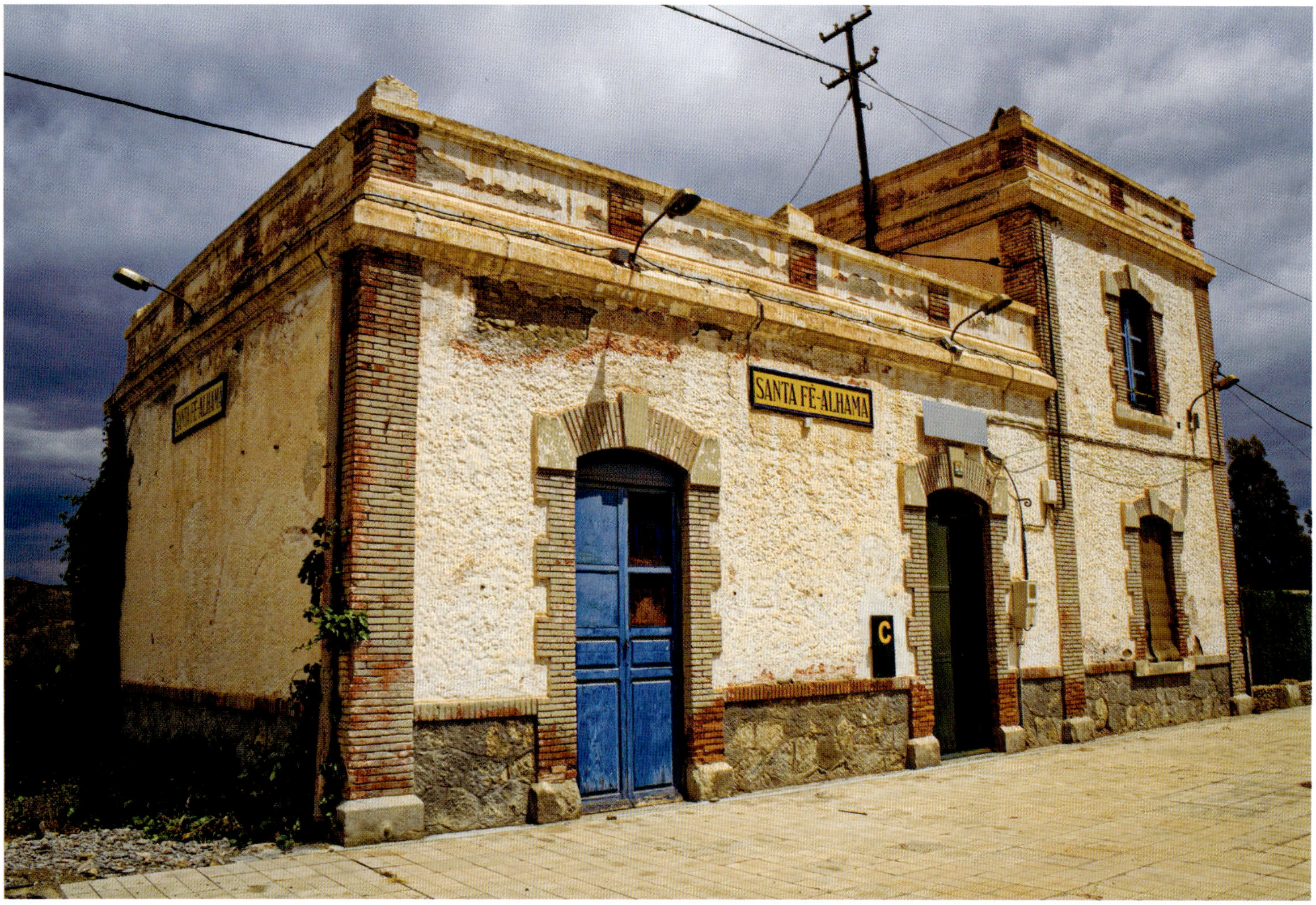
SANTA FE-ALHAMA
C

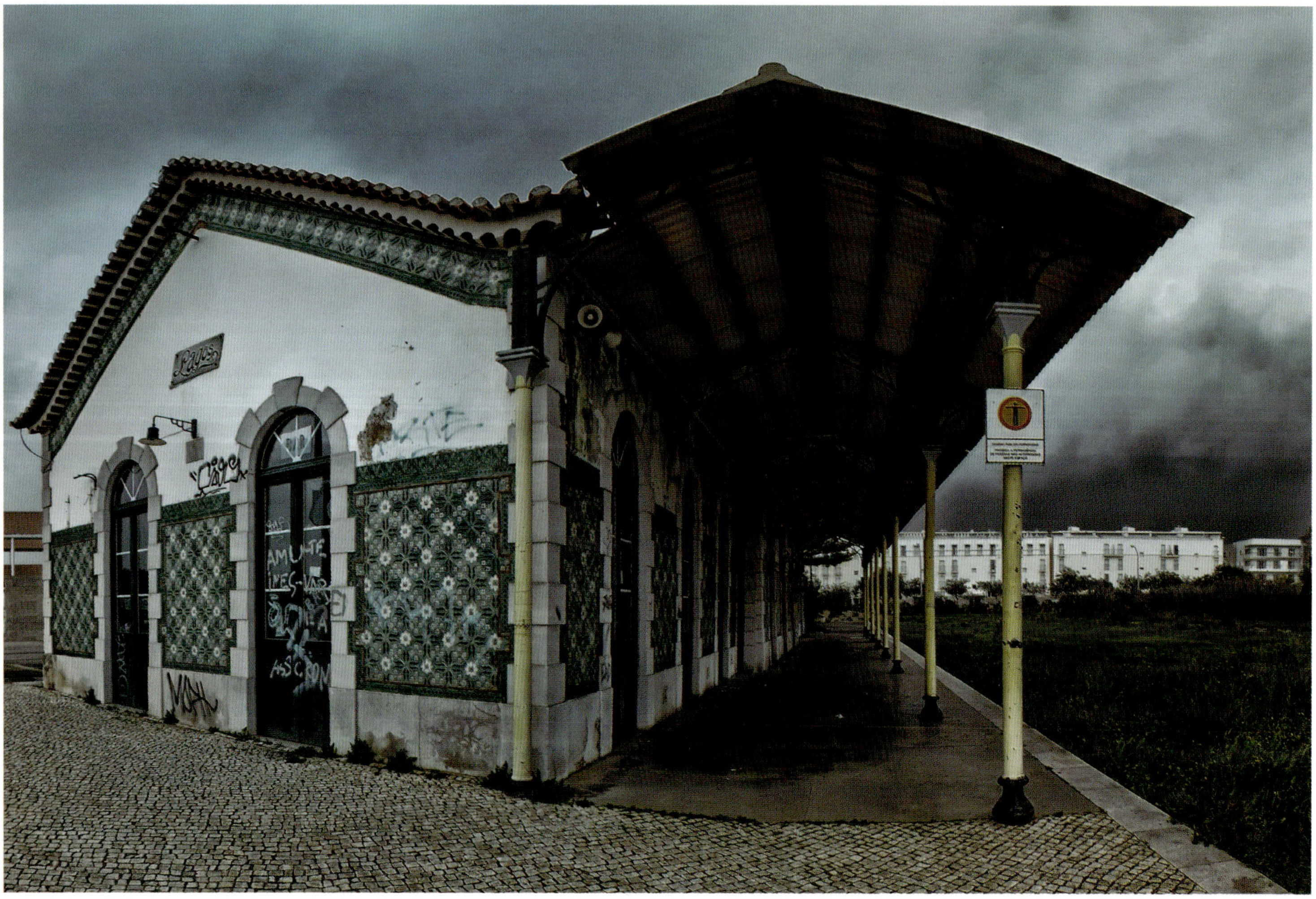

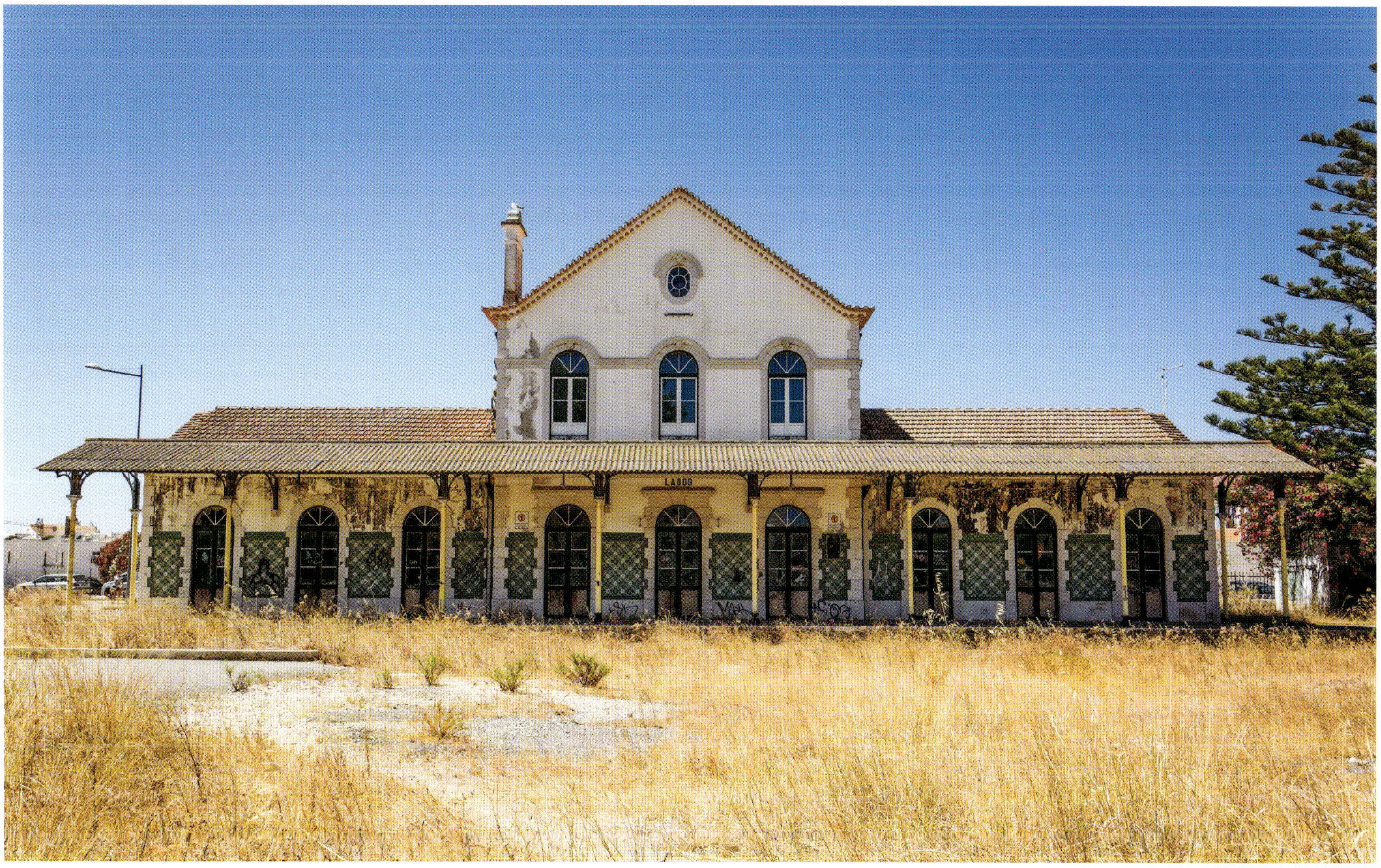

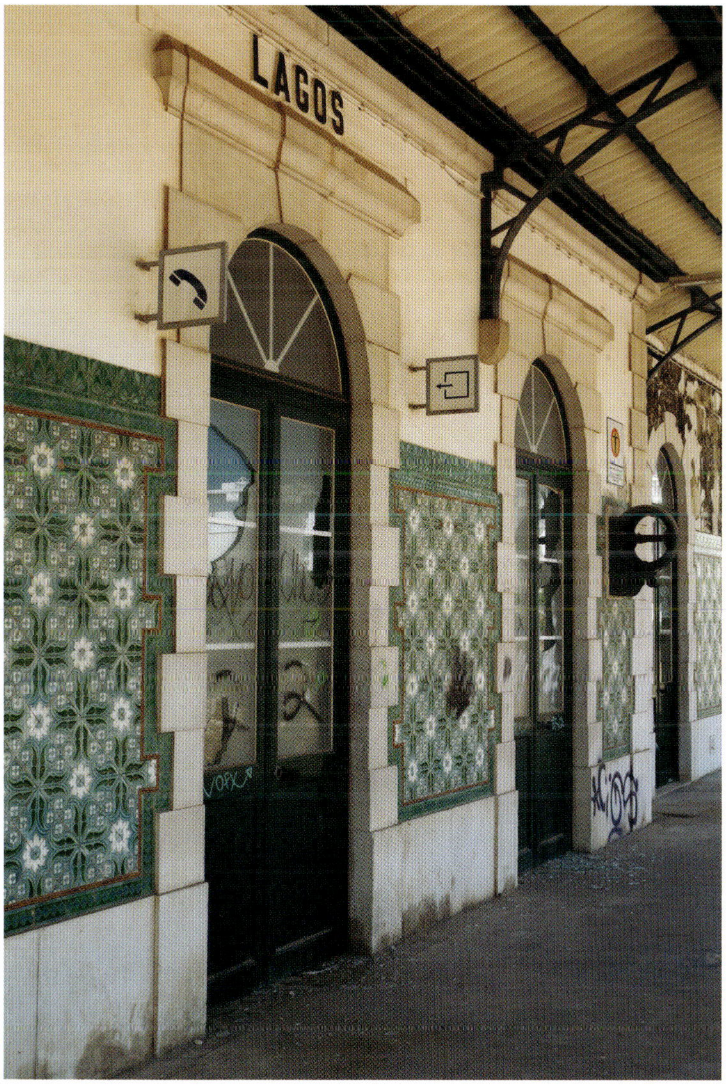

OPPOSITE ABOVE:
Santa Fe–Alhama, Almeria, Spain
On the Linares–Almeria line, this now-closed station opened in 1893. In 1912, the sector between here and Gergal became Spain's first electrified railway. Steam and electric locomotives were switched here, with the aim of using electric power on heavy mineral trains climbing the steep grades. In 1918 electrification was continued to Almeria and the locomotive depot was shut down.

OPPOSITE BELOW:
Crato, Portalegre, Portugal
An unusual feature of this station is the mosaic-tiled platform, with the name 'Crato' picked out in blue. It was opened in 1863 by the Linha do Leste.

Freight services still run but the station has had no passenger trains since 2012.

ABOVE & LEFT:
Old Station, Lagos, Faro, Portugal
This was the terminus of the Algarve line and when opened in 1922 it was the most westerly station in mainland Europe. With decorative tiling and ironwork, it is an attractive building, though assailed by vandals since its closure in 2006. Replaced by a new station, it now stands in grass-grown solitude and is currently for sale

Corchiano, Viterbo Province, Italy
A typical stucco-walled single-platform wayside station, Corchiano stands on the Civitavecchia–Capranica-Orte line in the Lazio region, which opened in 1928 and was closed down in 1994. The building remains intact and a 'Save the Station' campaign is under way.

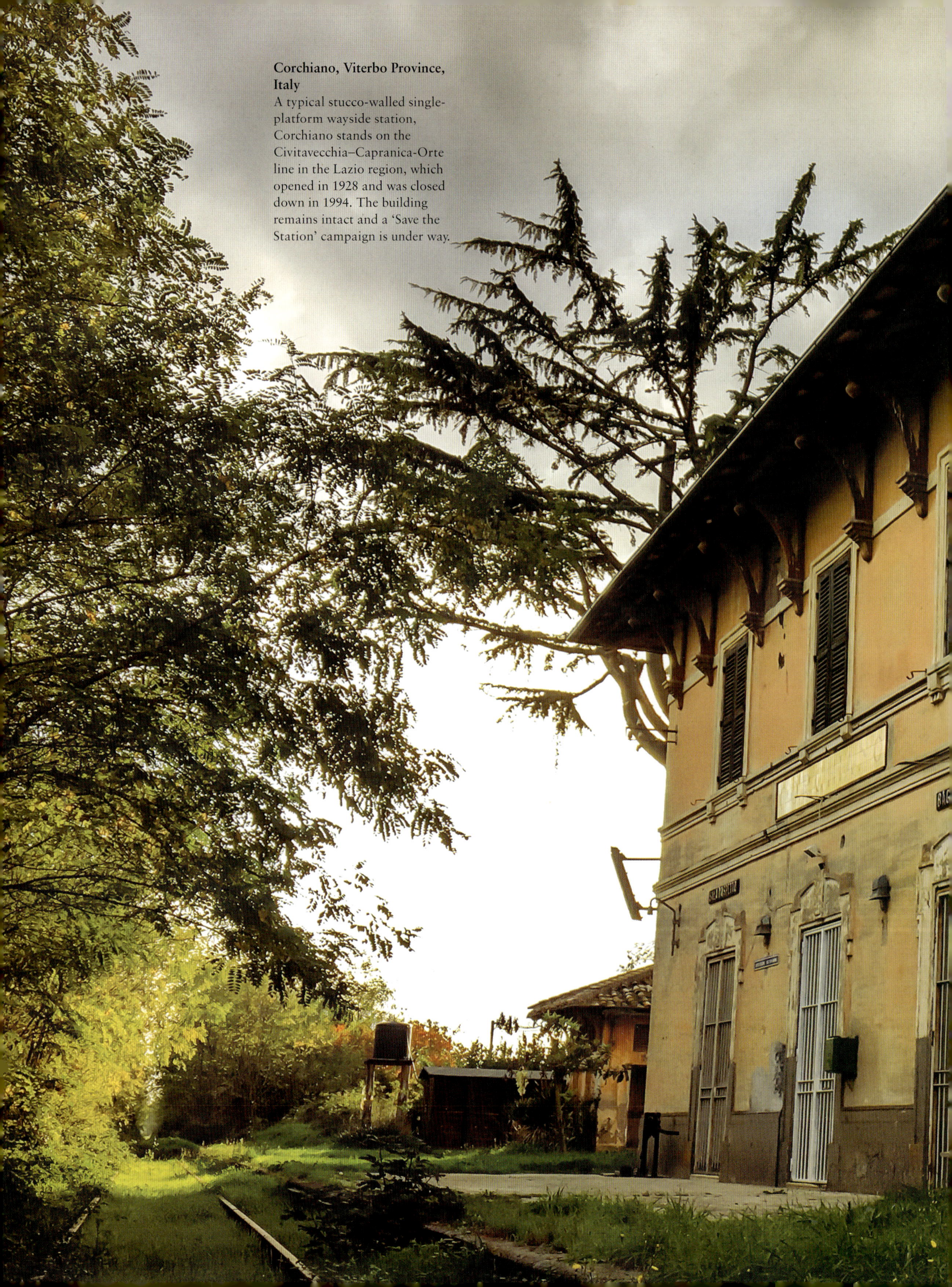

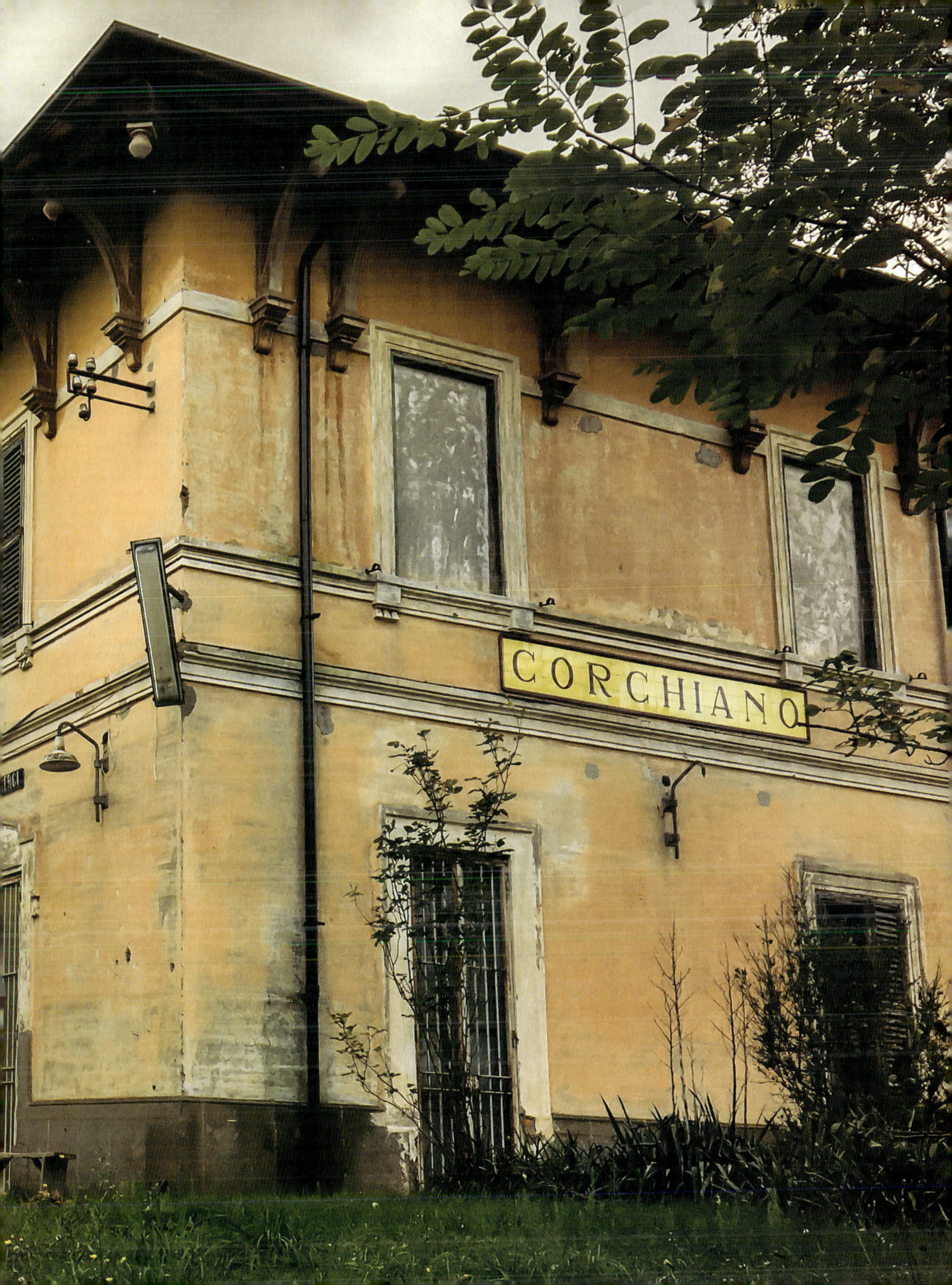
CORCHIANO

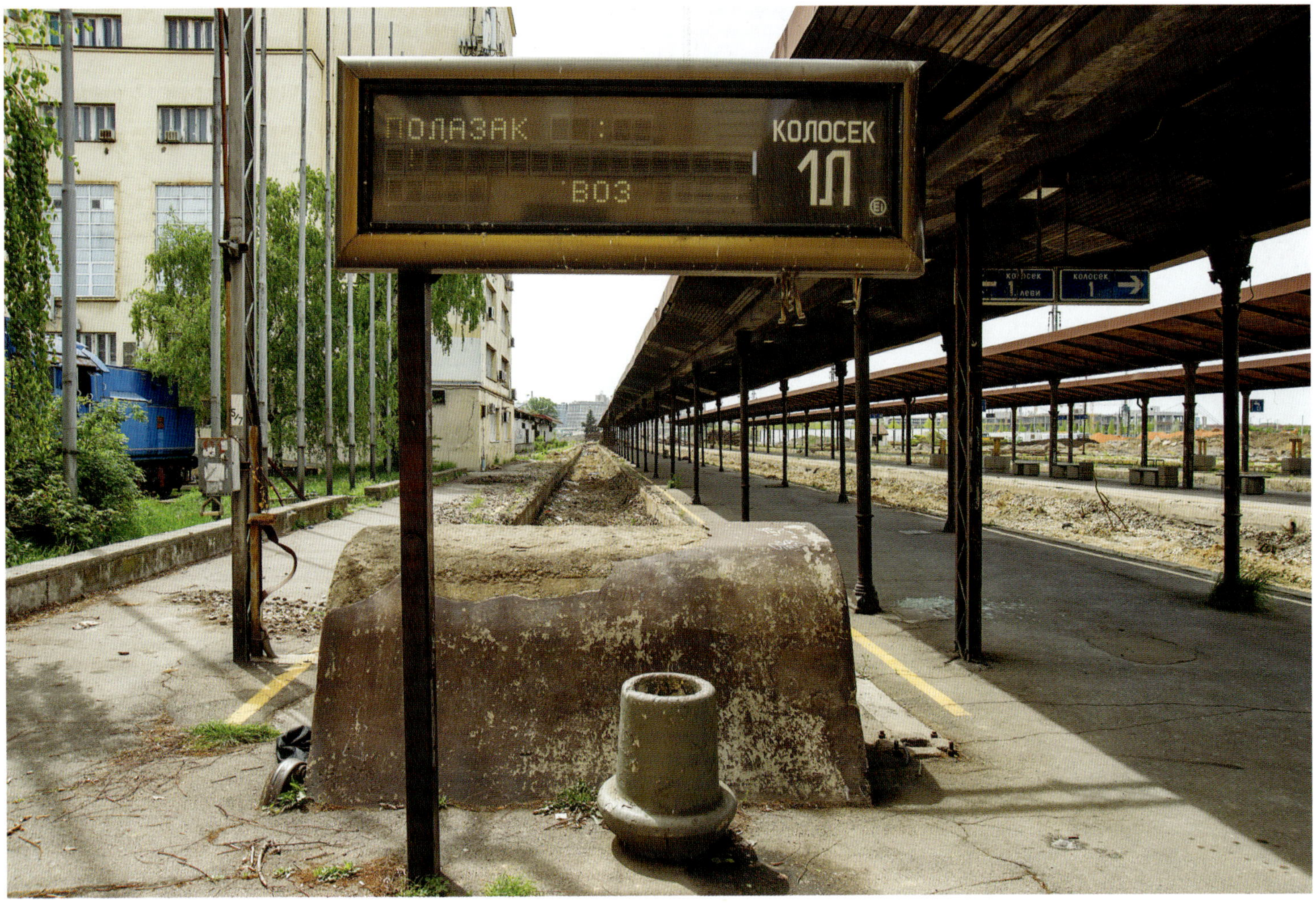
ПОЛАЗАК
КОЛОСЕК
1Л
ВОЗ

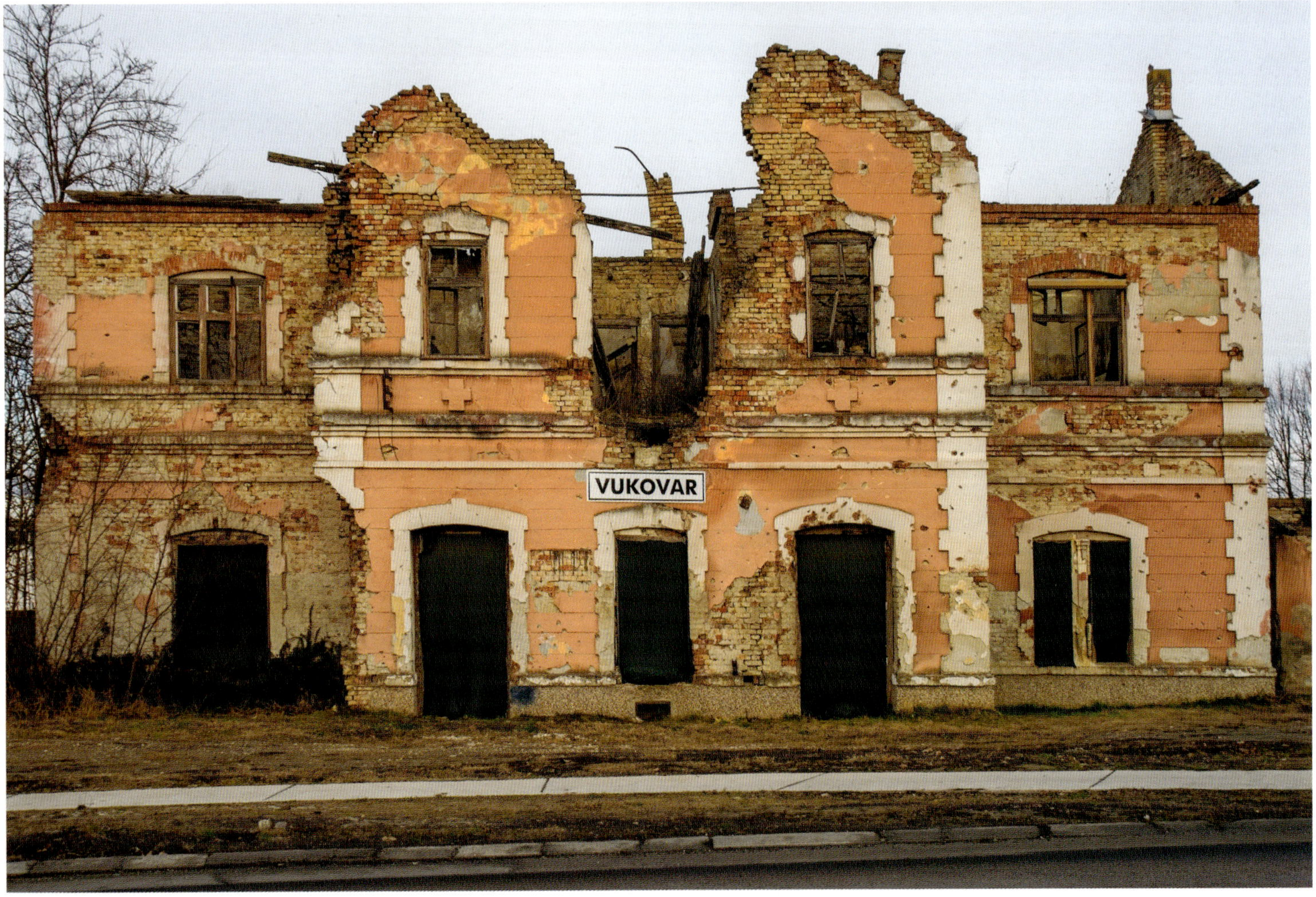
VUKOVAR

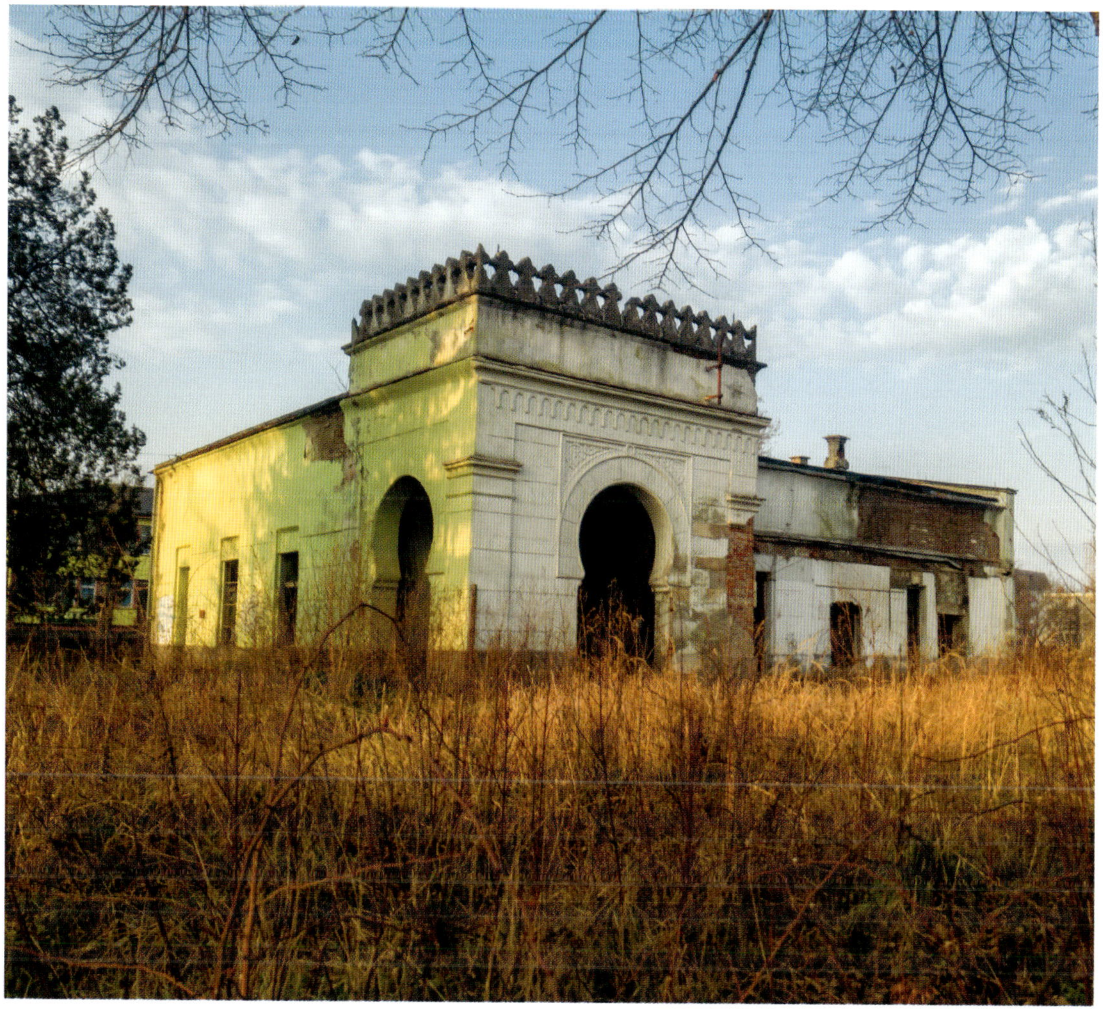

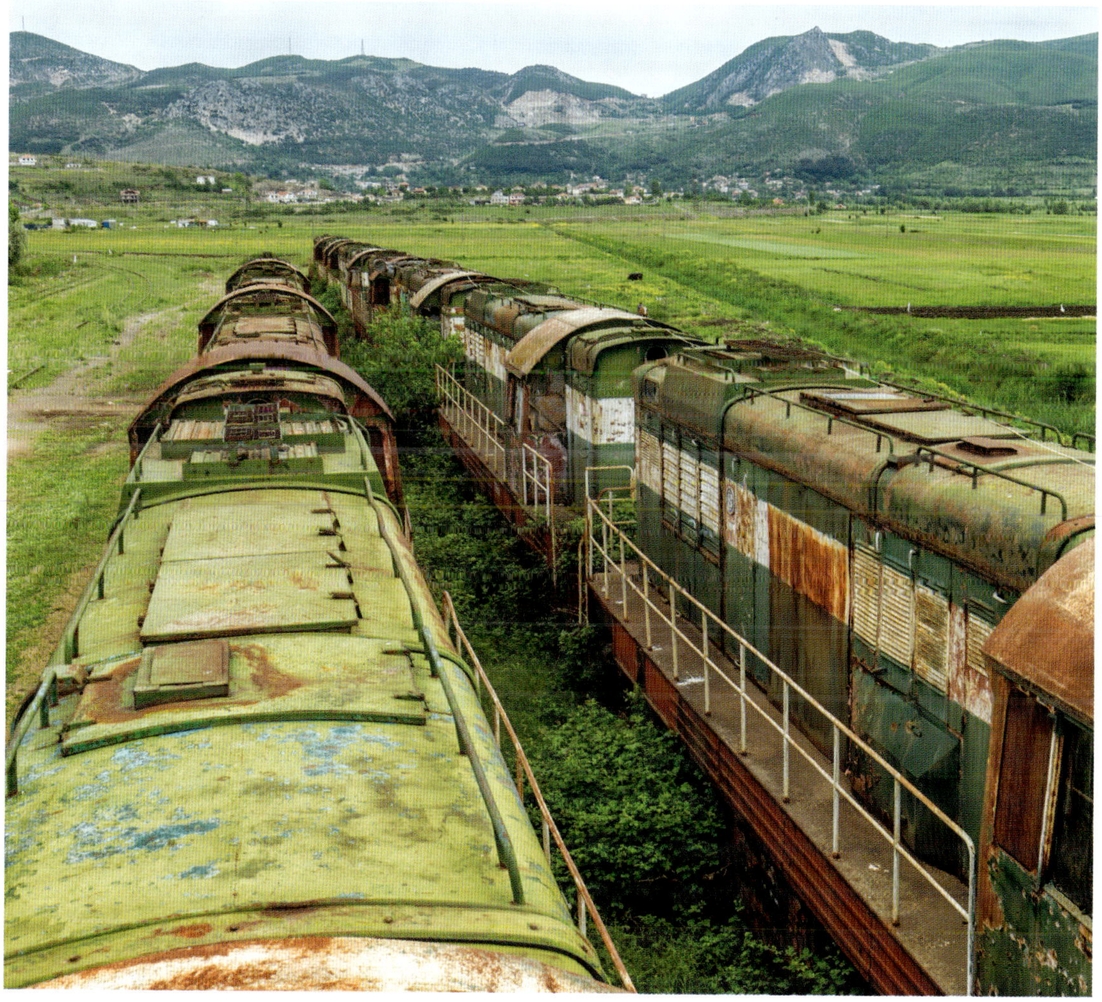

OPPOSITE ABOVE:
Old Main Station, Belgrade, Serbia
From 1884 to 2016, this was Belgrade's main station, handling both internal and international services. Heavily damaged in World War II, it was rebuilt by 1953. In the 1970s, it was despatching up to 150 trains a day. Since the break-up of Yugoslavia in the early 1990s, services were much reduced and by 2016 trains were being diverted to the new Prokop station. The last train from here went in 2018.

OPPOSITE BELOW:
Vukovar, Vukovar-Syrmia County, Croatia
The railway came to this Danube river-port in 1878. The system in the Balkan countries suffered badly from the warfare and political disintegration of the 1990s. Vukovar was the scene of a siege, battle and bloody aftermath in 1991. The shattered station building was abandoned, and is now a 'dark tourism' destination for some.

ABOVE LEFT:
Bosanski-Brod, Bosnia
Allied bombs in 1945 destroyed all but the south wing of this splendid Moorish-style station on the 760mm (30in) gauge line between Bosanski-Brod and Sarajevo. The line was completed under the auspices of the Austro-Hungarian Empire in 1882. Erected in 1897, the station was lavishly appointed and decorated both inside and out. In reduced form it continued in operation until 1969.

LEFT:
Prrenjas, Elbasan Province, Albania
Lines of disused locomotives, mostly Czech-built T-669s and carriages once owned by German, Italian and Austrian railways stand at this closed station, relics of the Albanian railways' decline since 1990. The line from Elbasan to Prrenjas was built as recently as 1974 and extended into Montenegro in 1986, but is no longer in use.

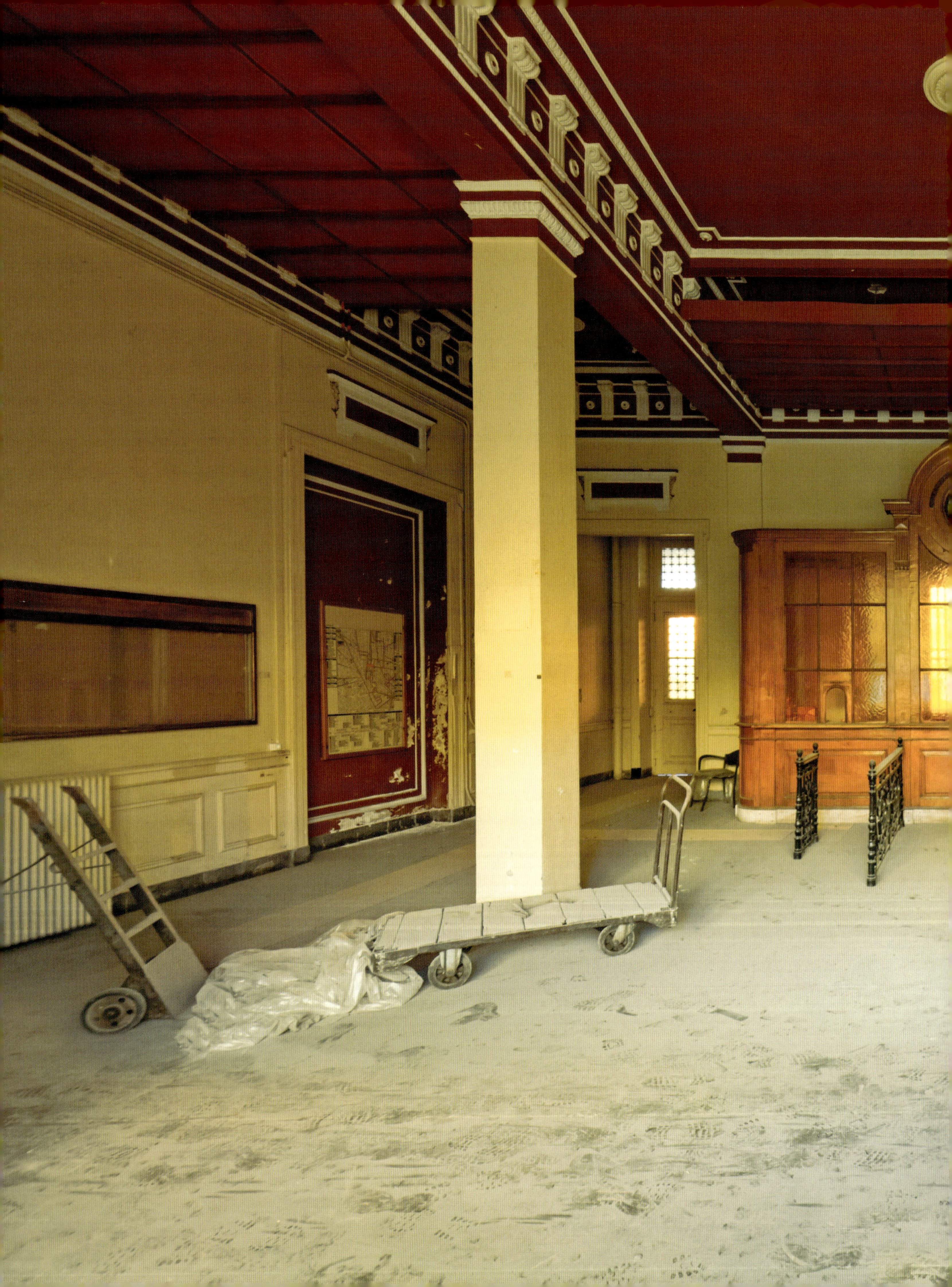

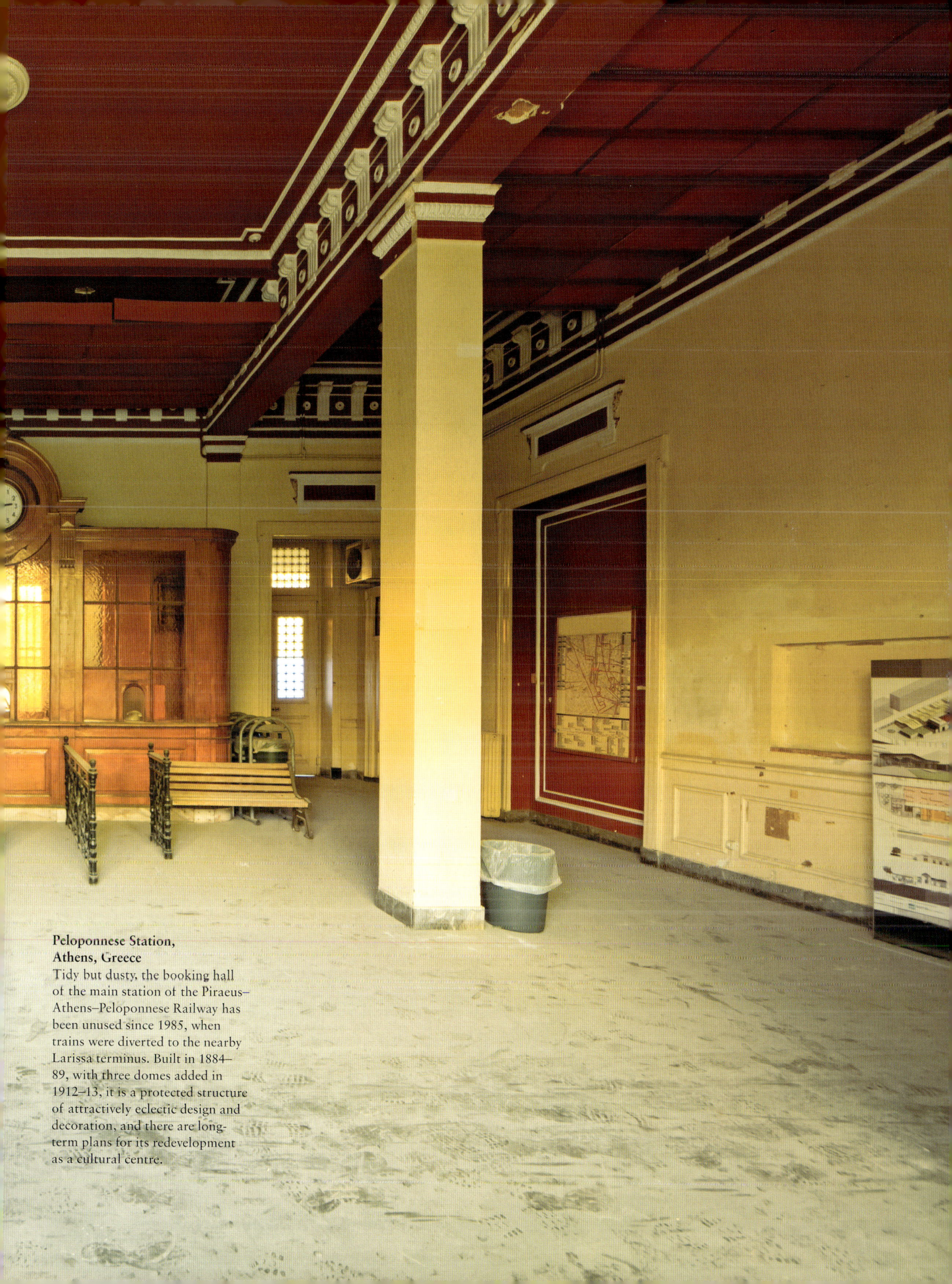

Peloponnese Station, Athens, Greece
Tidy but dusty, the booking hall of the main station of the Piraeus–Athens–Peloponnese Railway has been unused since 1985, when trains were diverted to the nearby Larissa terminus. Built in 1884–89, with three domes added in 1912–13, it is a protected structure of attractively eclectic design and decoration, and there are long-term plans for its redevelopment as a cultural centre.

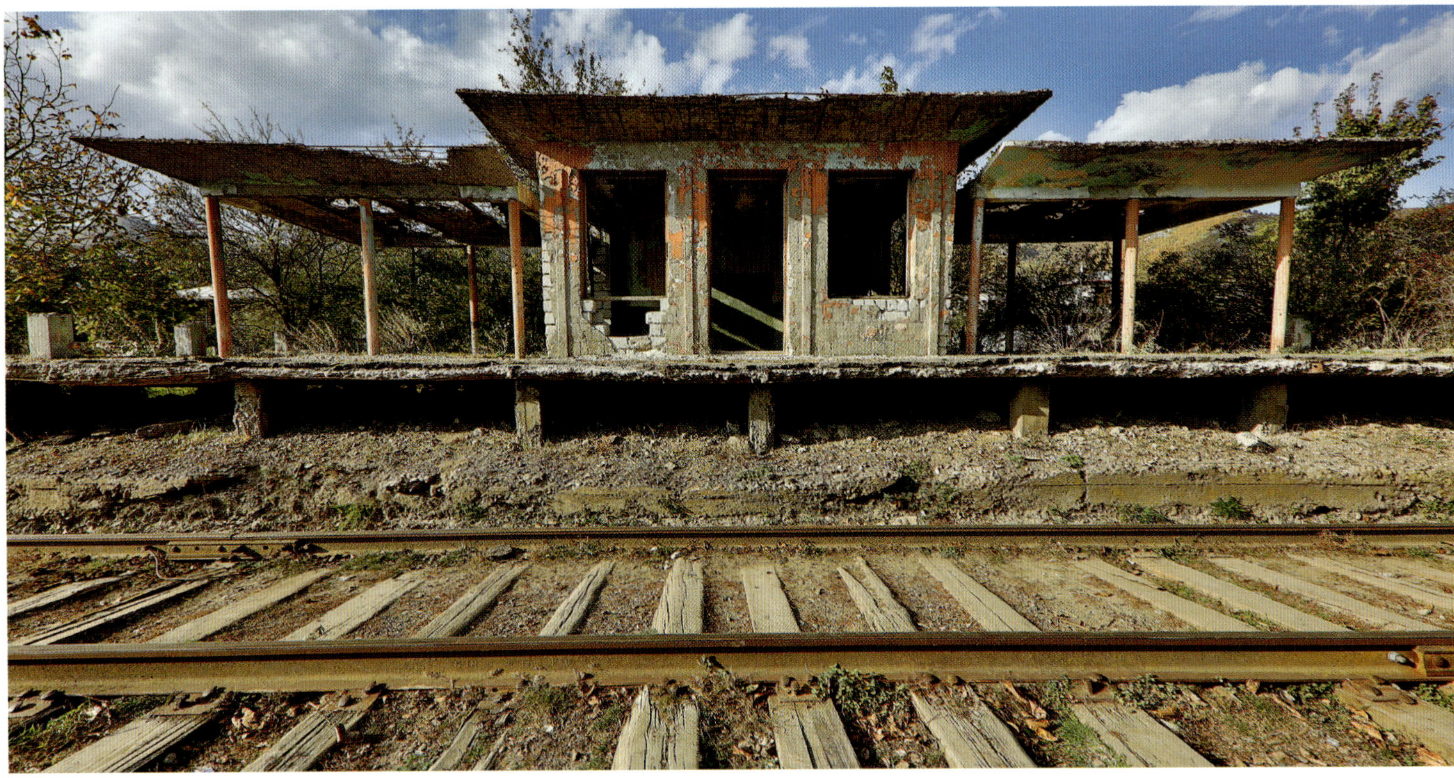

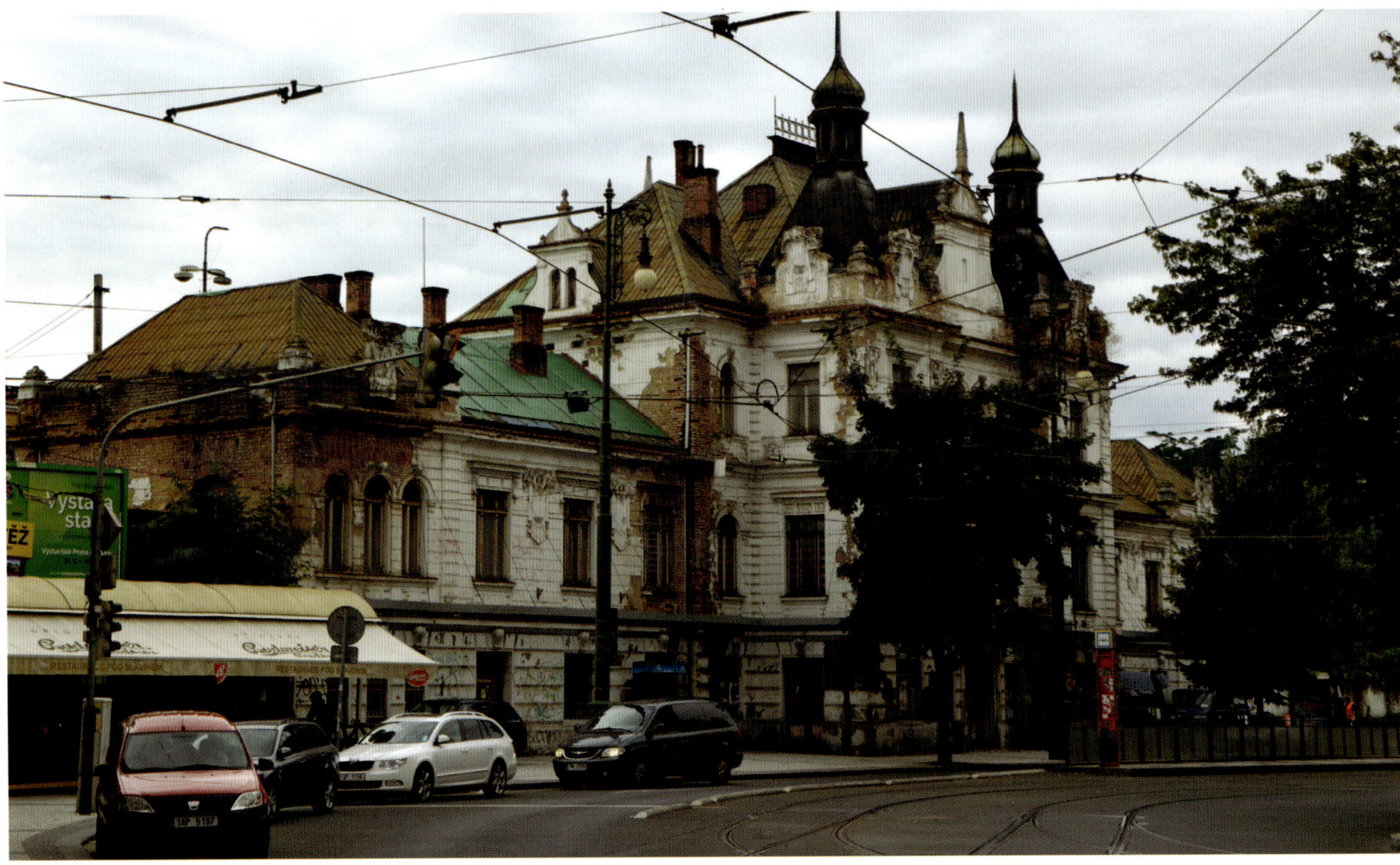

ABOVE TOP:

Abandoned station, Georgia

Track laid to the Russian railways' gauge of 1520mm (just under 5ft) fronts an abandoned station in Georgia. Until the collapse of the Soviet system in 1990, Georgia was part of the USSR.

ABOVE BELOW:

Vysehrad Station, Prague, Czech Republic

In 1901, this art-nouveau building replaced the original station of 1874 on the link line between Prague's Western and Main stations. Closed in 1960, the station has been listed as a heritage site since 2001. Privately purchased in 2007, it has become increasingly decrepit while negotiations between successive owners and the city have failed to reach agreement.

OPPOSITE & OVERLEAF:

Budapest, Hungary

These abandoned carriage sheds alongside the Hungarian Railways' Istvántelek Maintenance Main Depot, just outside Budapest, are known locally as the 'locomotive graveyard'. Numerous locomotives as well as derelict carriages are kept here. They are not on public view, which gives the place a strong appeal to urban explorers and ruin-photographers. The buildings, dating from the early 20th century, are as dilapidated as the vehicles. Some parts are in a dangerous condition and the site is closed to the public.

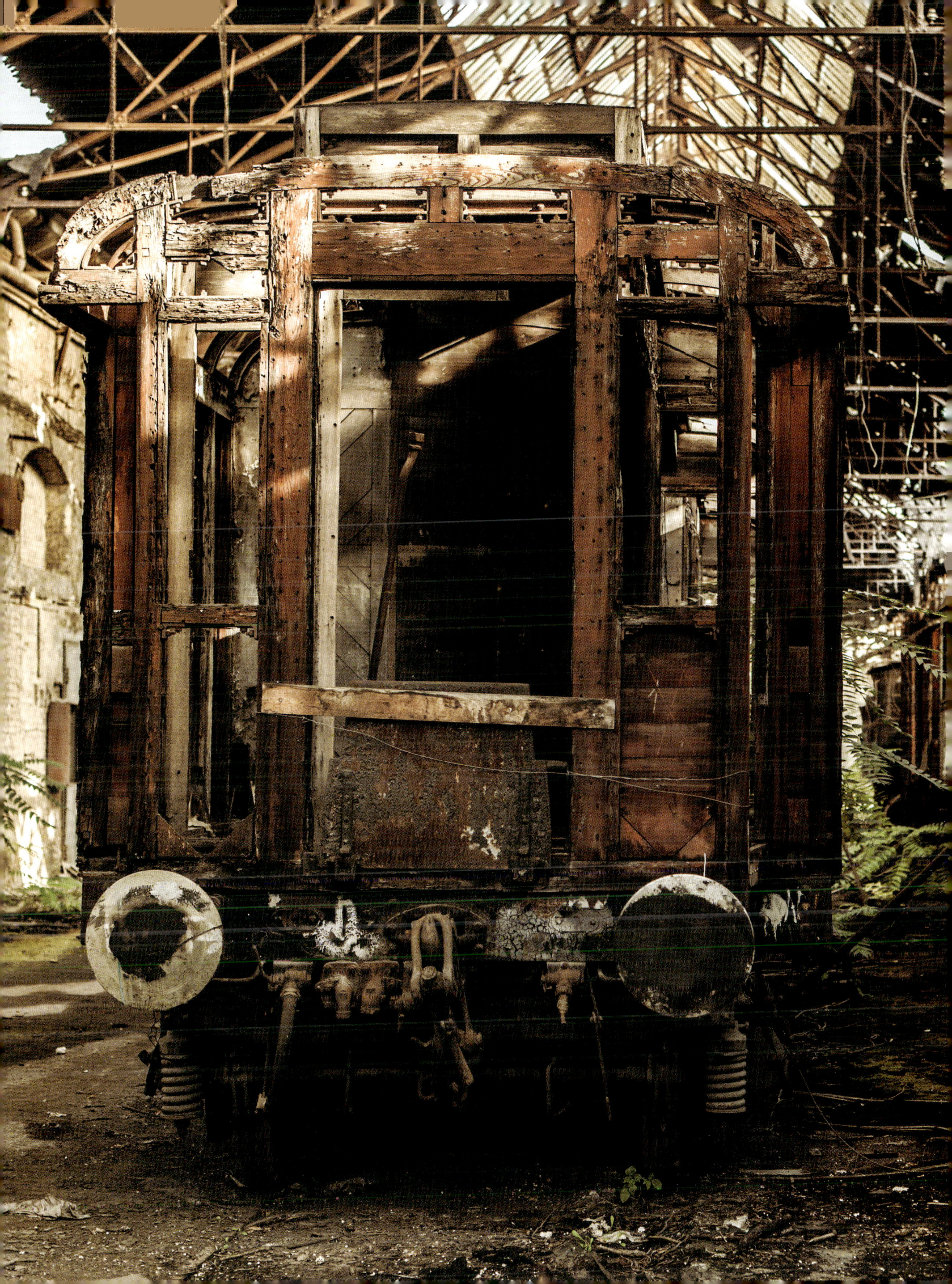

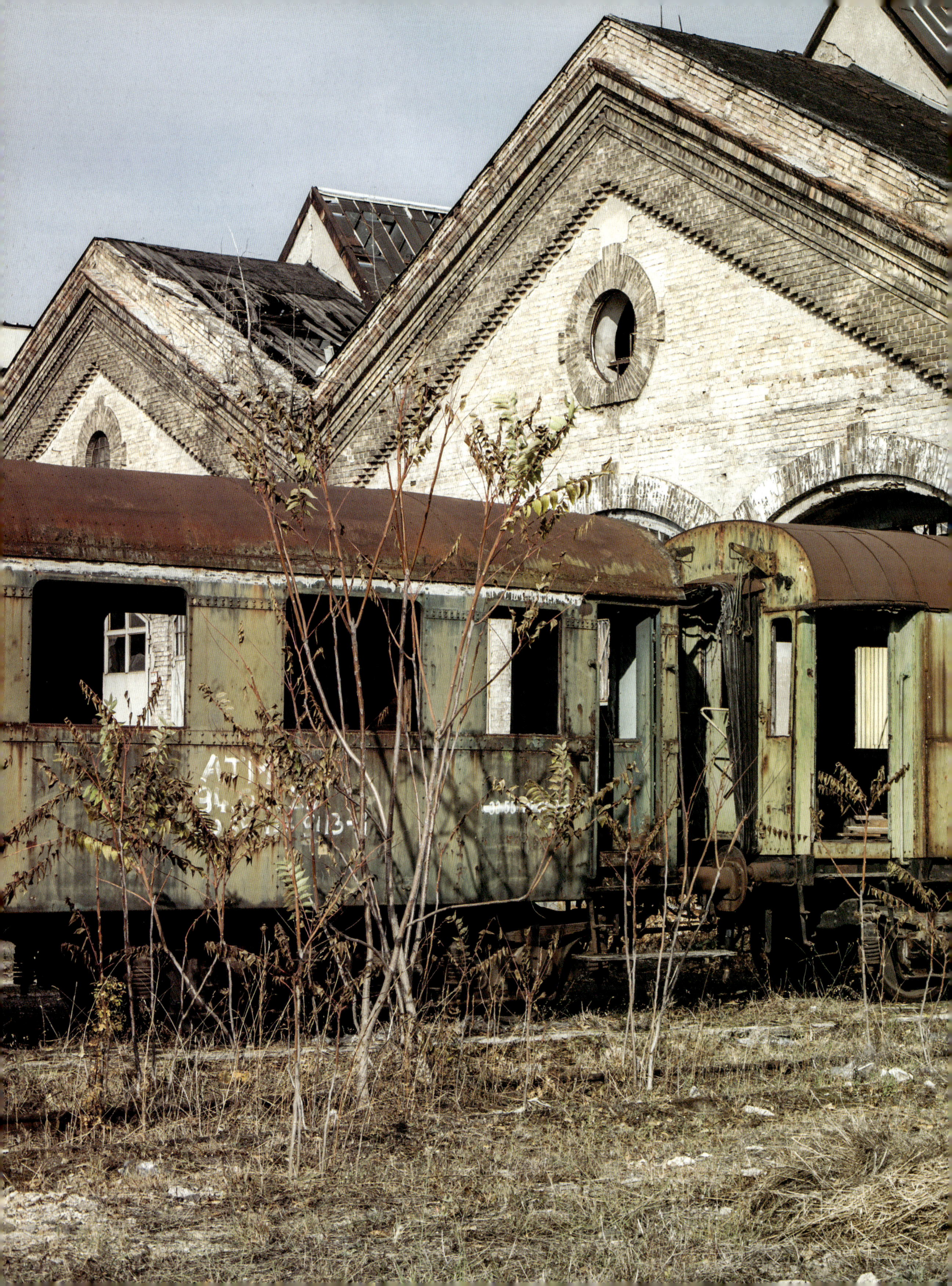

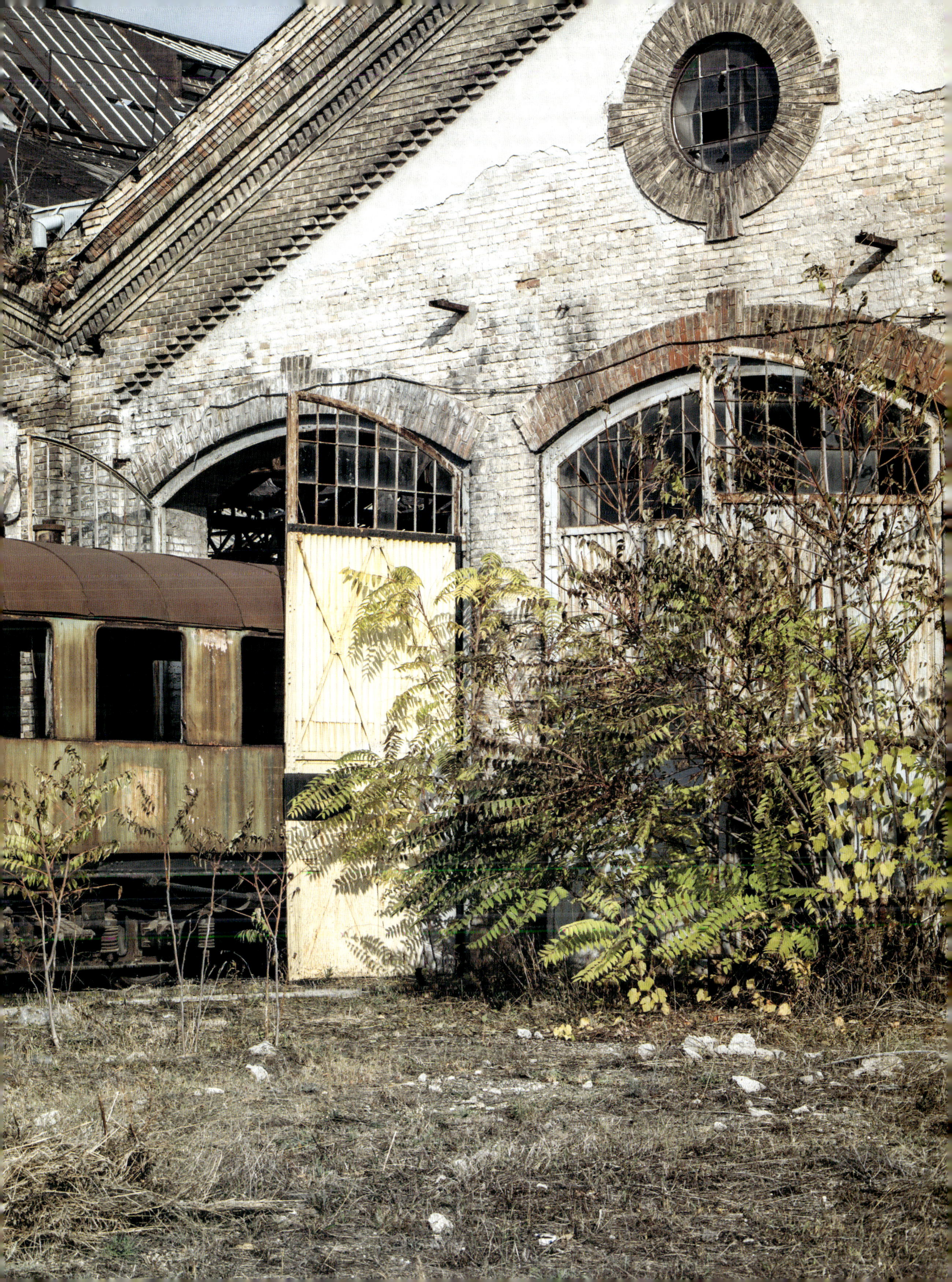

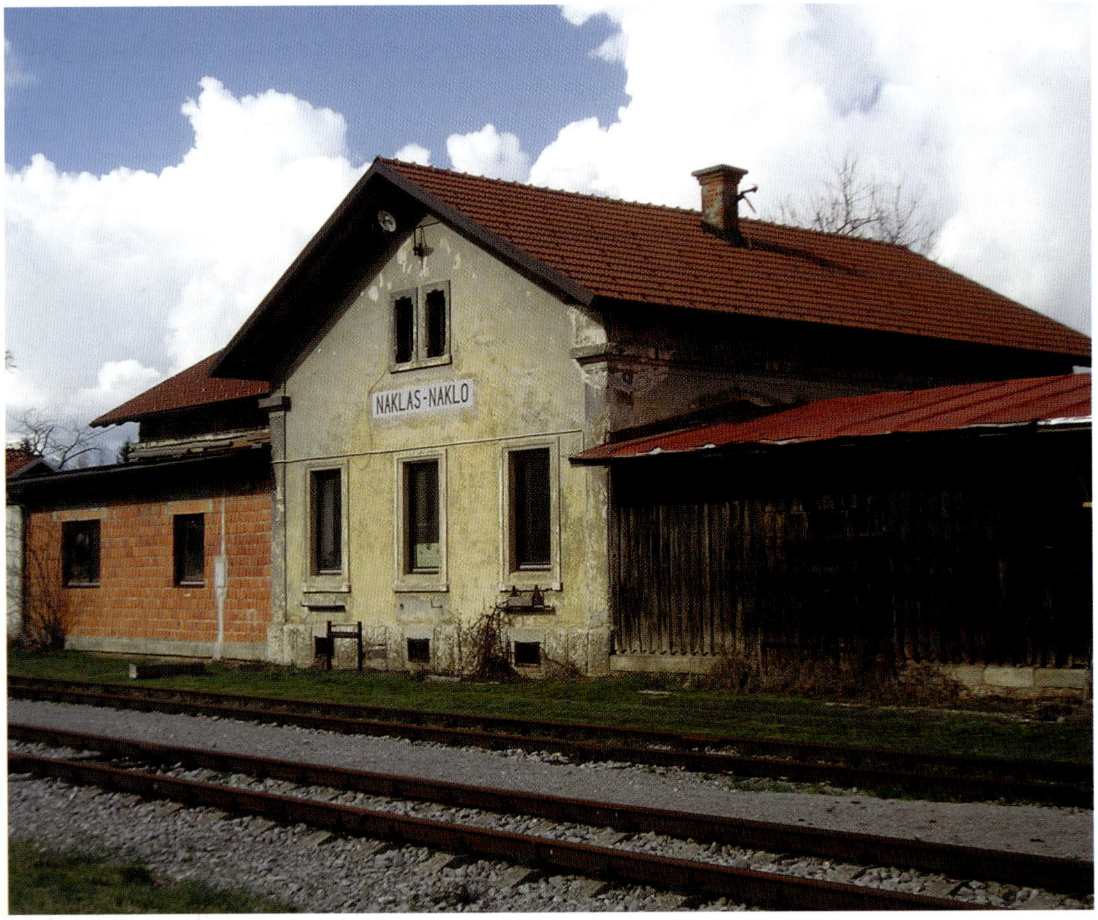

LEFT:

Naklo, Gorenjska Region, Slovenia

This village station on the line between Tržic and Kranj was closed in 1966. The Naklo–Kranj section was opened in 1905, when Slovenia was part of the Austro-Hungarian Empire. An unmaintained track remains between Kranj and Naklo; the rest of the line has been lifted.

BELOW:

Tolkmicka, Elblag County, Poland

The coastal railway from Elblag to Braniewo was opened as an independent concern in 1899. The section from Tolkmicka to Braniewo was shut down in 2012 and the rest of the route has seen no trains since 2015. This harbour town on the Vistula estuary was clearly one of the line's more important stations.

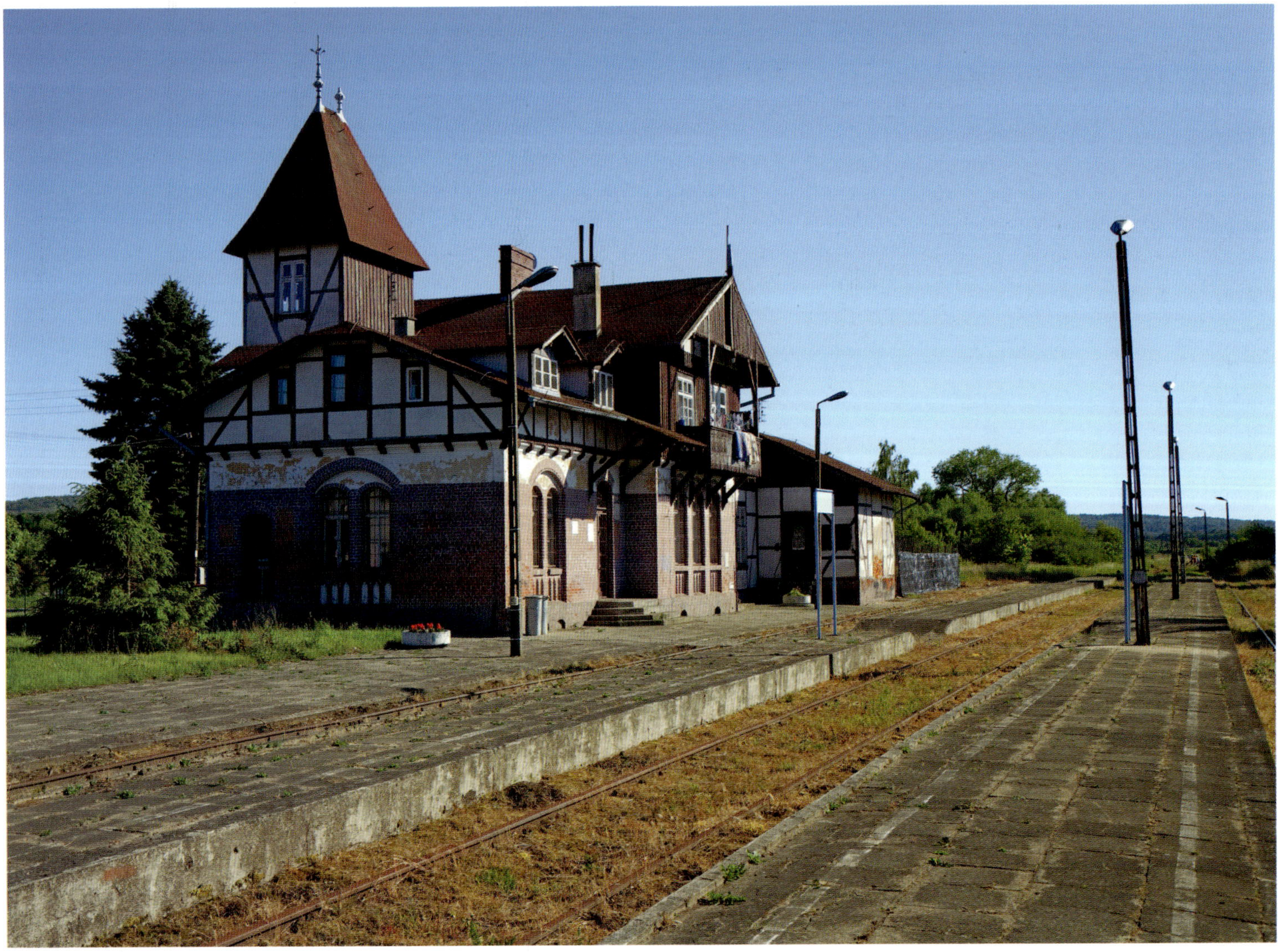

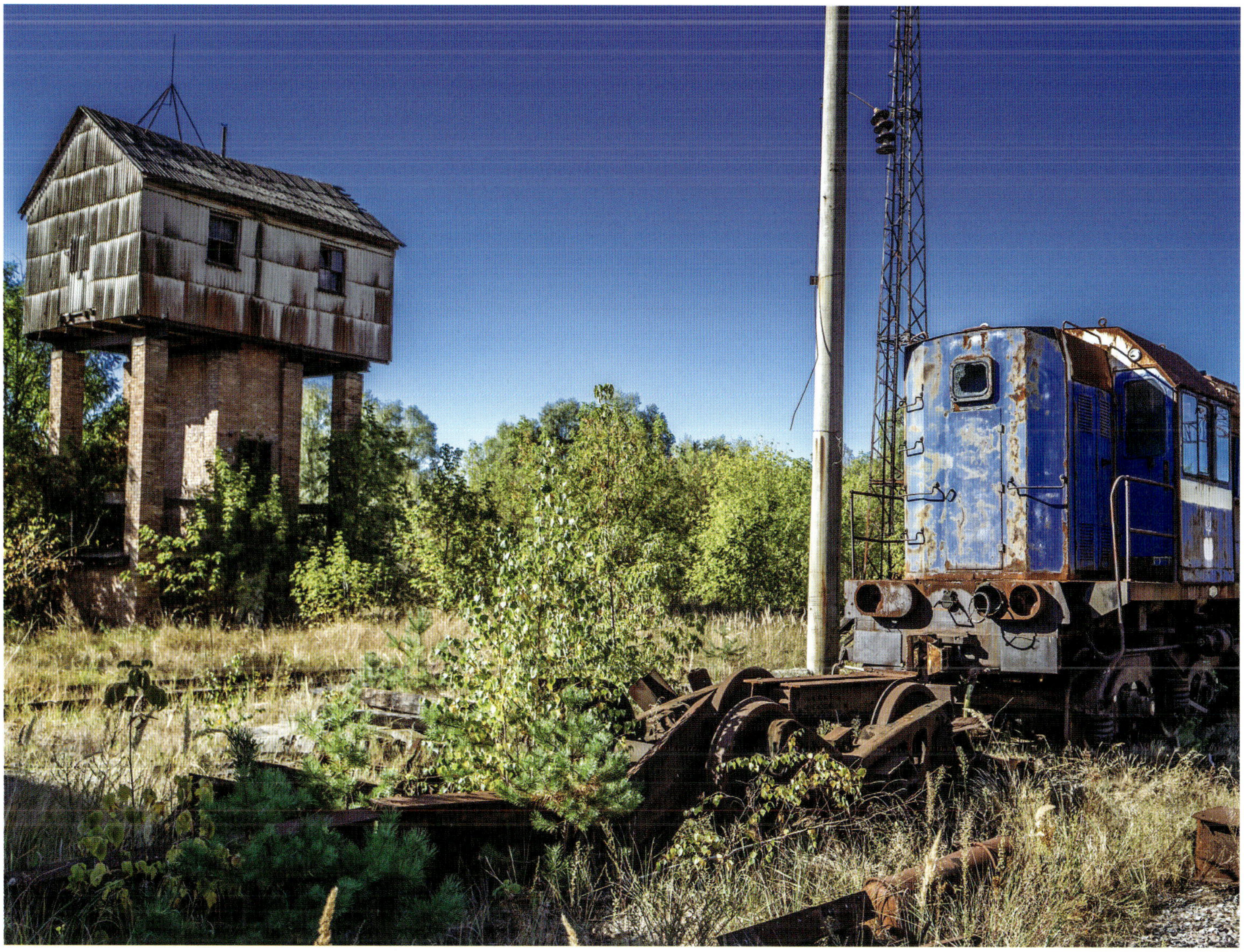

ABOVE:
Yaniv Station, Chernobyl, Ukraine
Following the Chernobyl nuclear disaster in 1986, Yaniv was included in the Exclusion Zone and its staff was evacuated, leaving the buildings and equipment behind. One track remains functional but the station is abandoned to all except workers engaged in metal recycling.

RIGHT:
Varshavsky Station, St Petersburg, Russia
The 'Warsaw' station was opened in 1851 for a line to the Tsar's palace at Gatchina. By 1862 it was extended to Warsaw in Poland. Located south of the city centre, the station closed in 2001, and the main building and train shed are now a shopping centre.

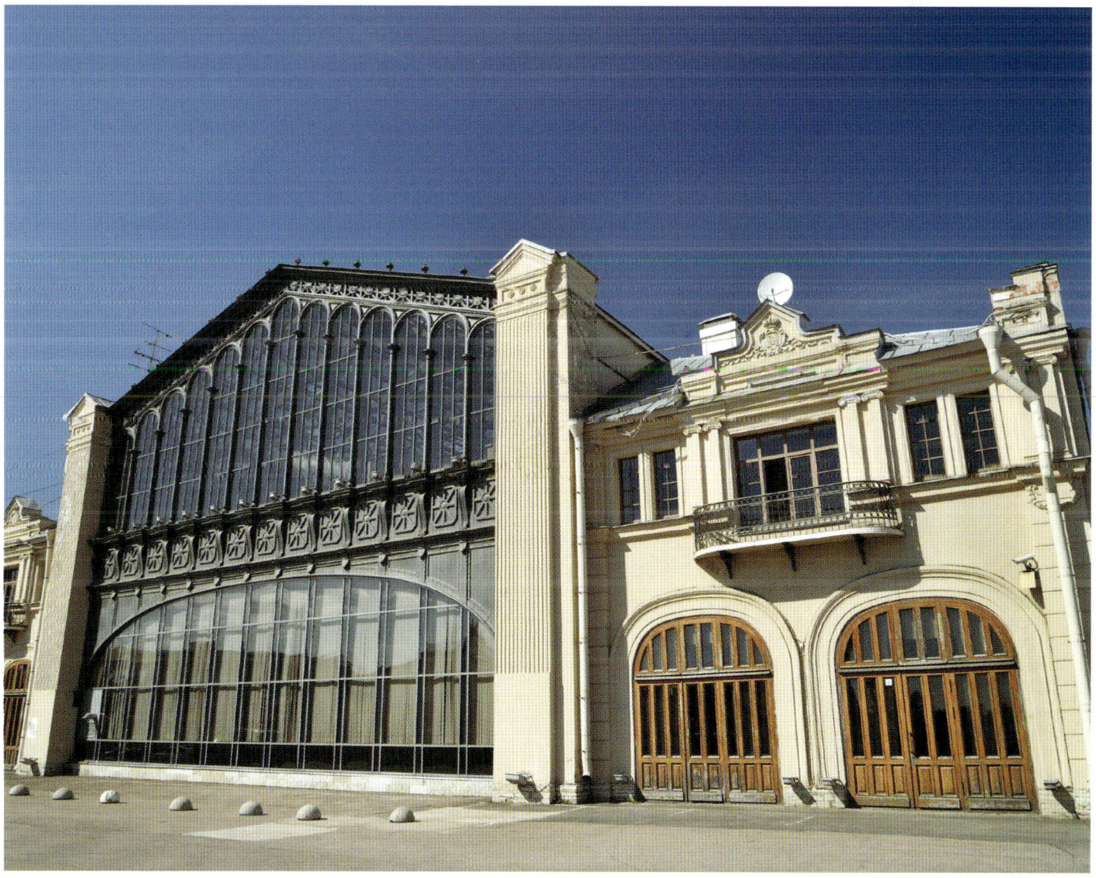

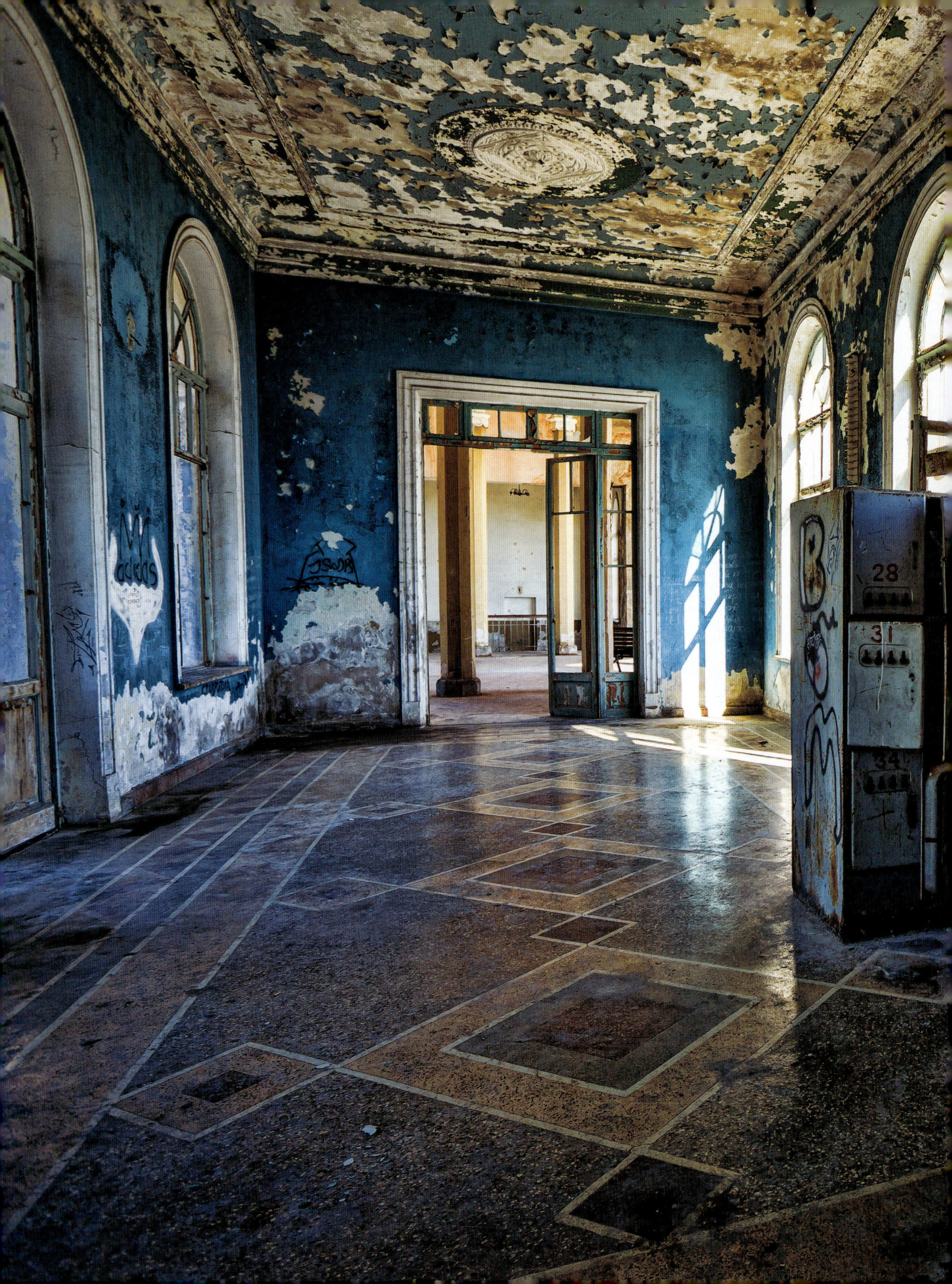
adidas
28
31
34

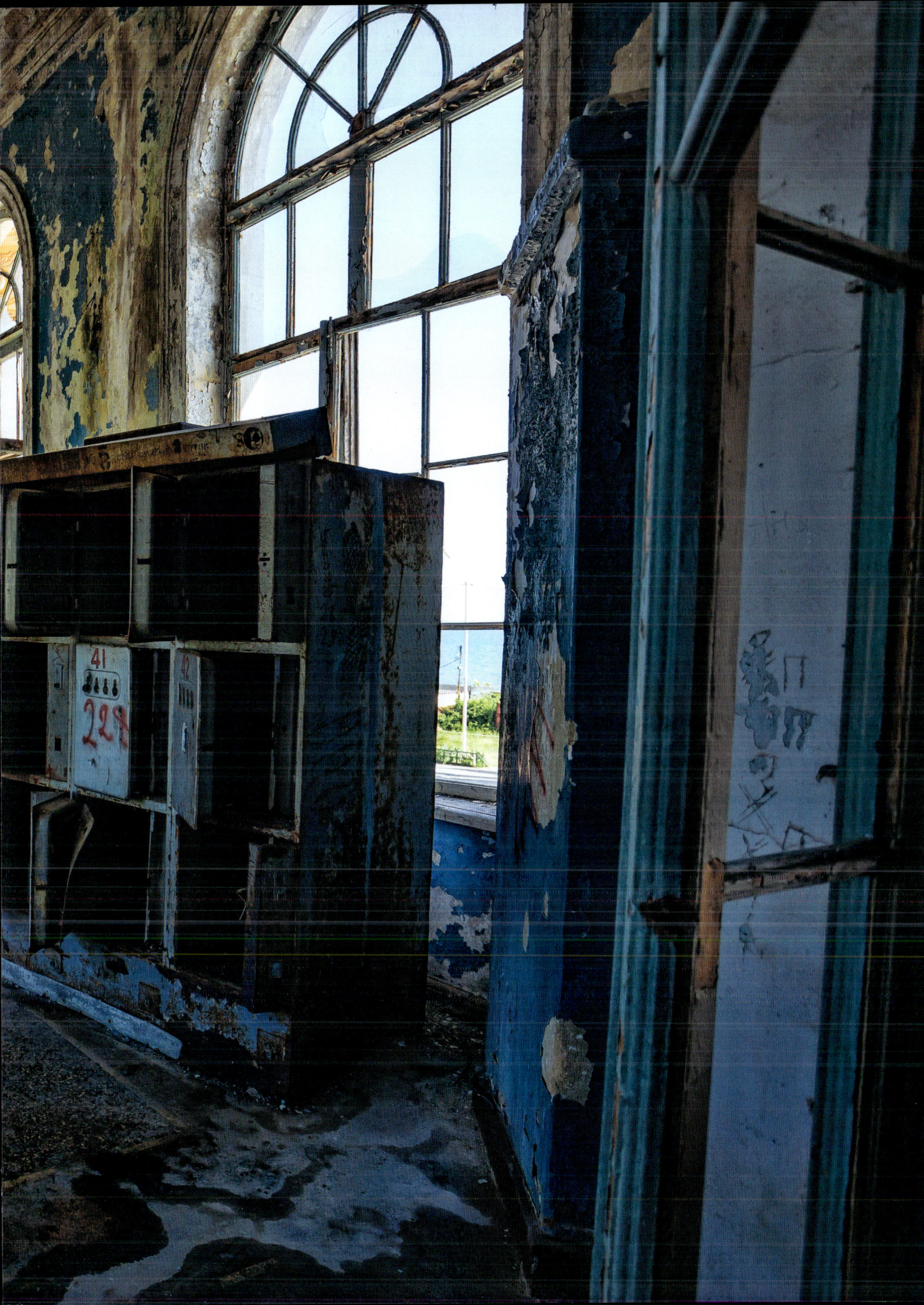
41
42

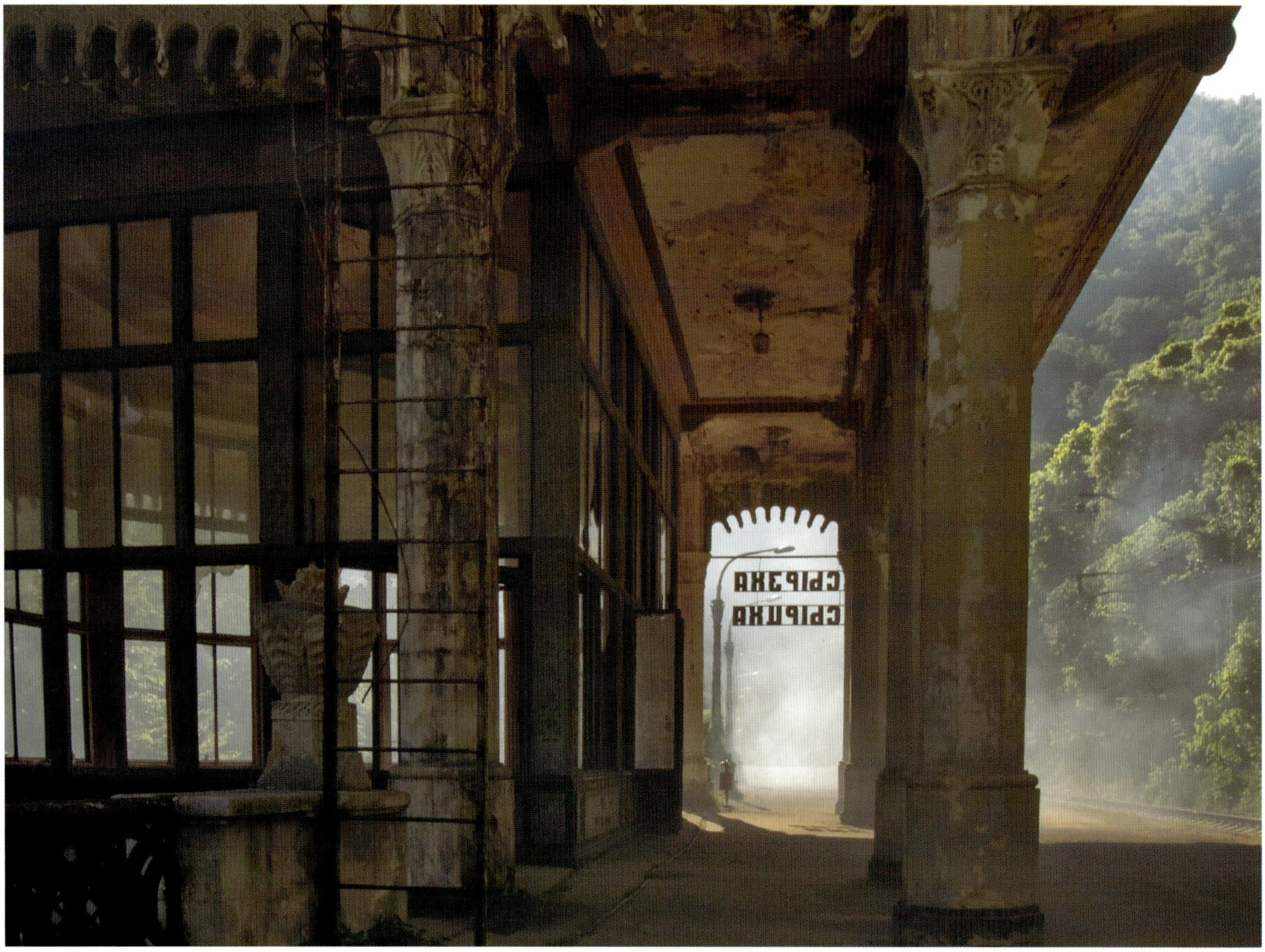

PREVIOUS PAGES:

Baratashvili Station, Sukhumi, Abkhazia

The North Caucasus Railway was built by the Soviet Railways in 1930-49, reaching Sukhumi in 1938. This station, east of the city's main station and derelict since 1993, displays the kind of Beaux-Arts grand style favoured by designers of public buildings in the Stalinist period.

ALL PHOTOGRAPHS ON THESE PAGES:

Psyrtsha Station, New Athos, Abkhazia, Georgia

Perched above a lagoon, in picturesque scenery, Psyrtsha Station, with its pillars and palm trees, gently decays. Warfare in 1992-93, followed by Abkhazia's secession from Georgia, ended its traffic. Walkways link it to the nearby monastery of St Simon the Caananite. The electrified line itself remains in use from Sukhumi into Russia.

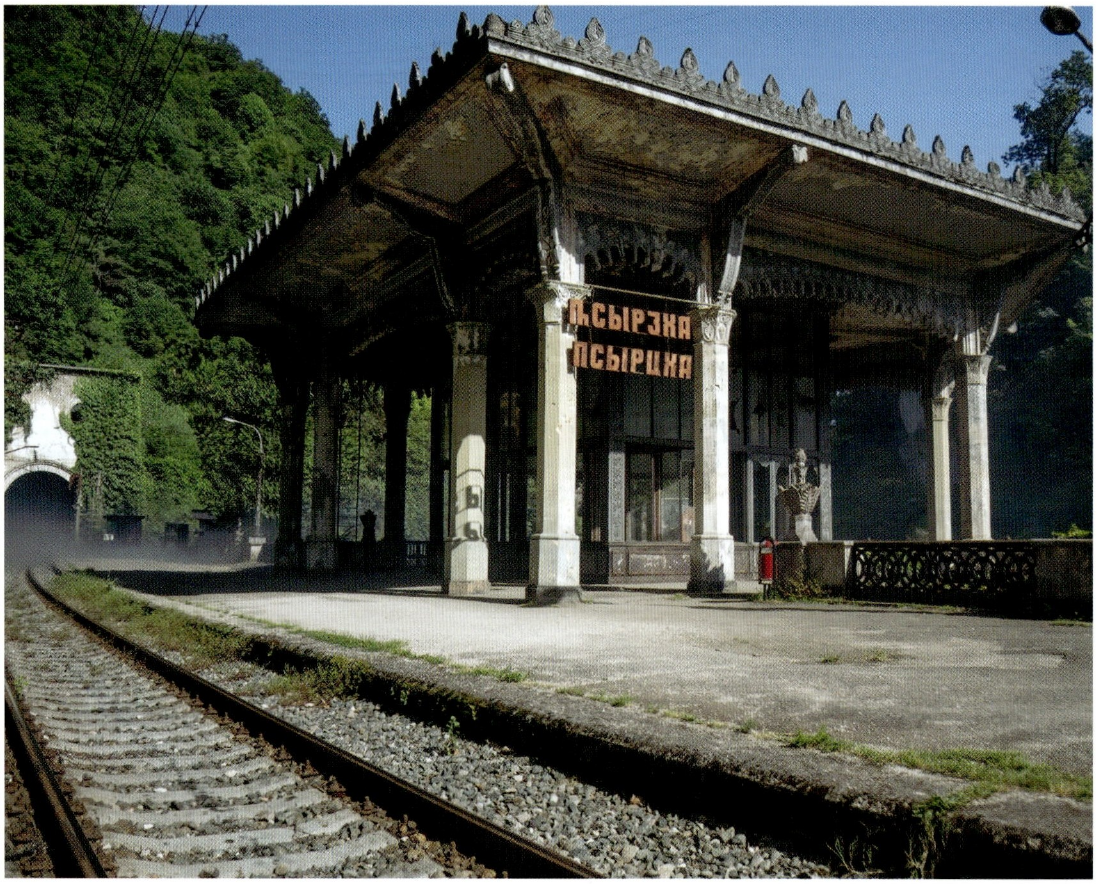

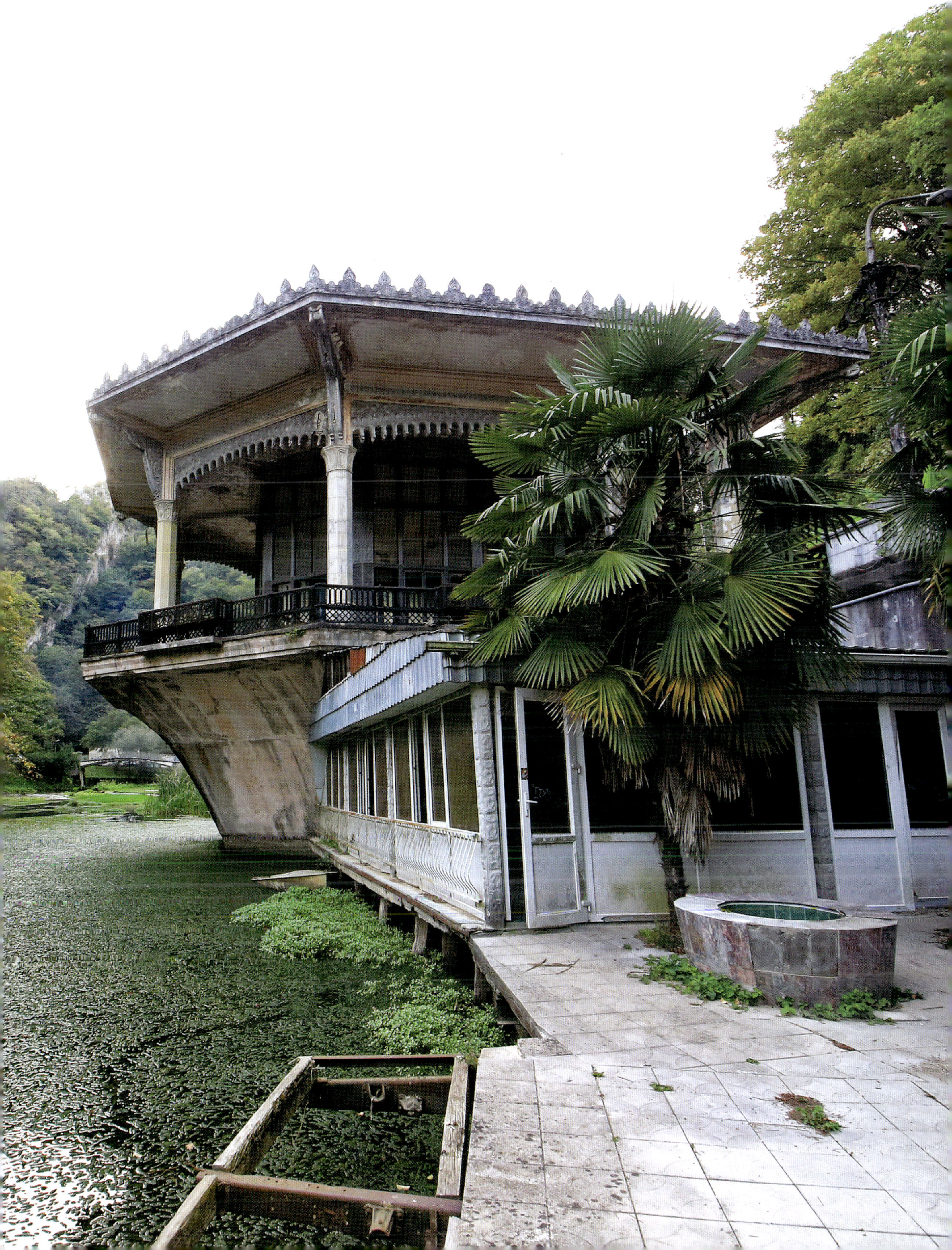

Haapsalu, Lääne County, Estonia
A Baltic seaside resort town, Haapsalu was served by a branch railway from 1904. Passenger traffic ceased in 1995, and goods in 2004, and the station is now a museum. A notable feature is the 216m (236yd) canopy extending the entire length of the platform. The near locomotive is a 2-10-0 built by the Russian Kolomna Works in 1951, and a German Krauss-Maffei of 1943 can also be seen.

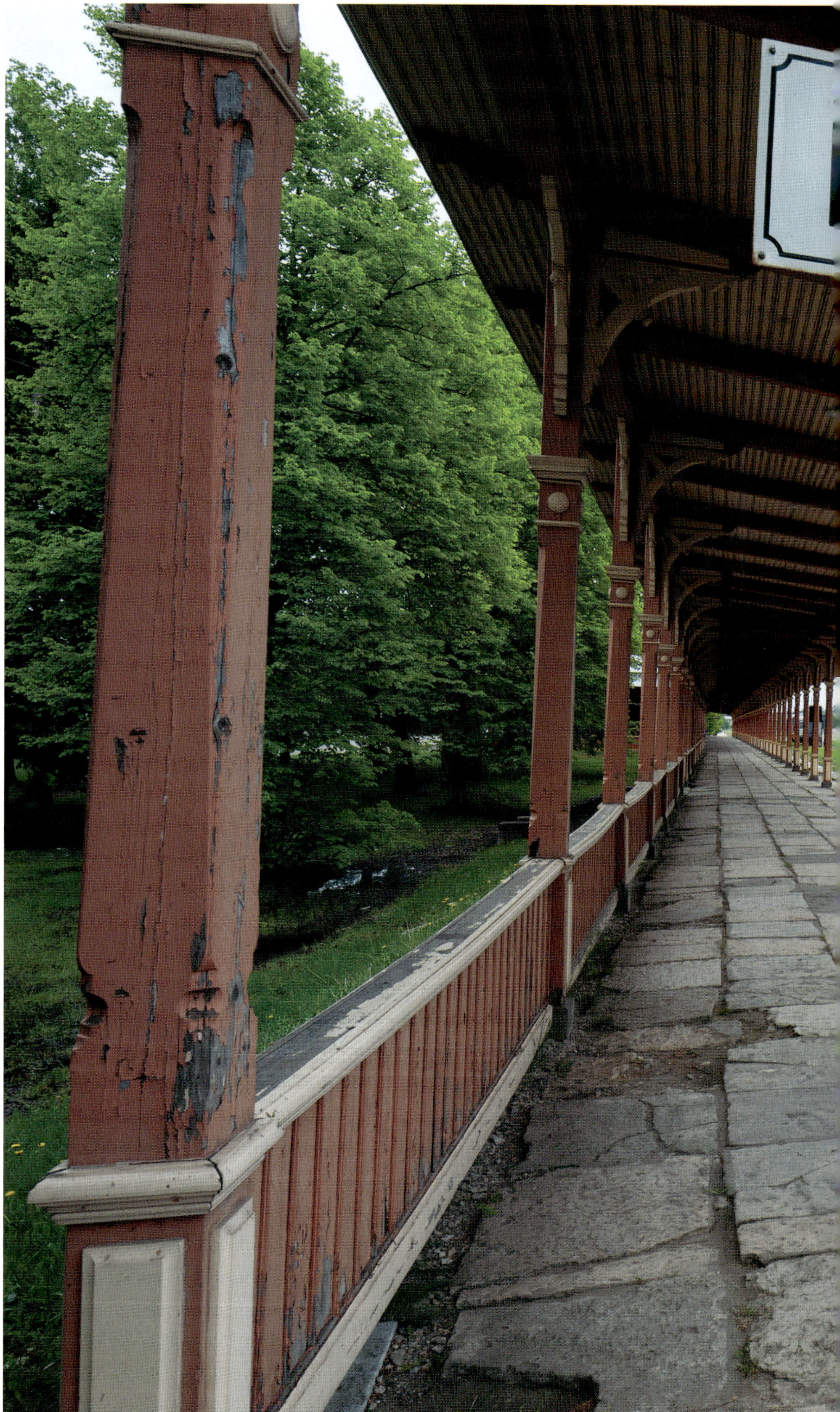

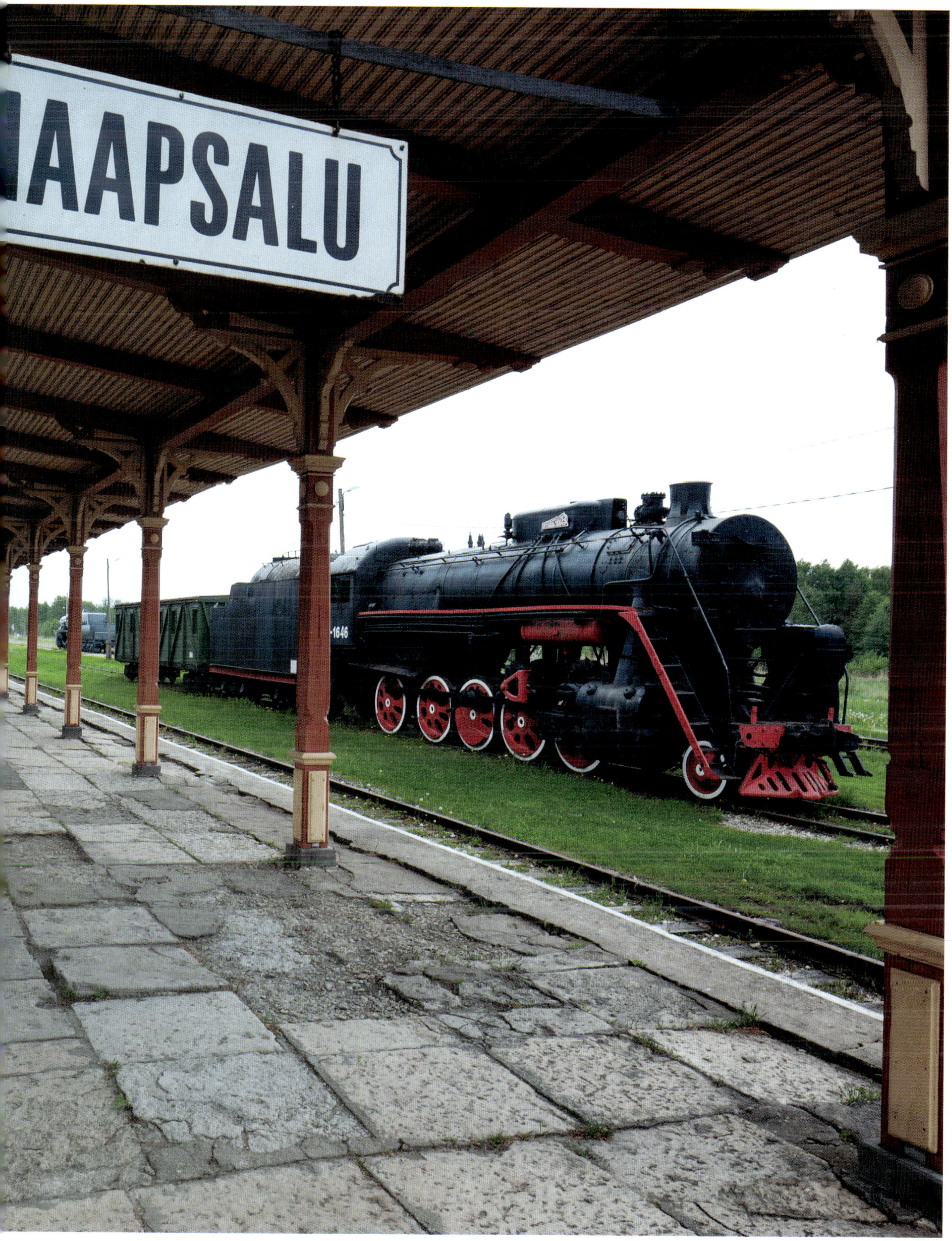
HAAPSALU
-1646

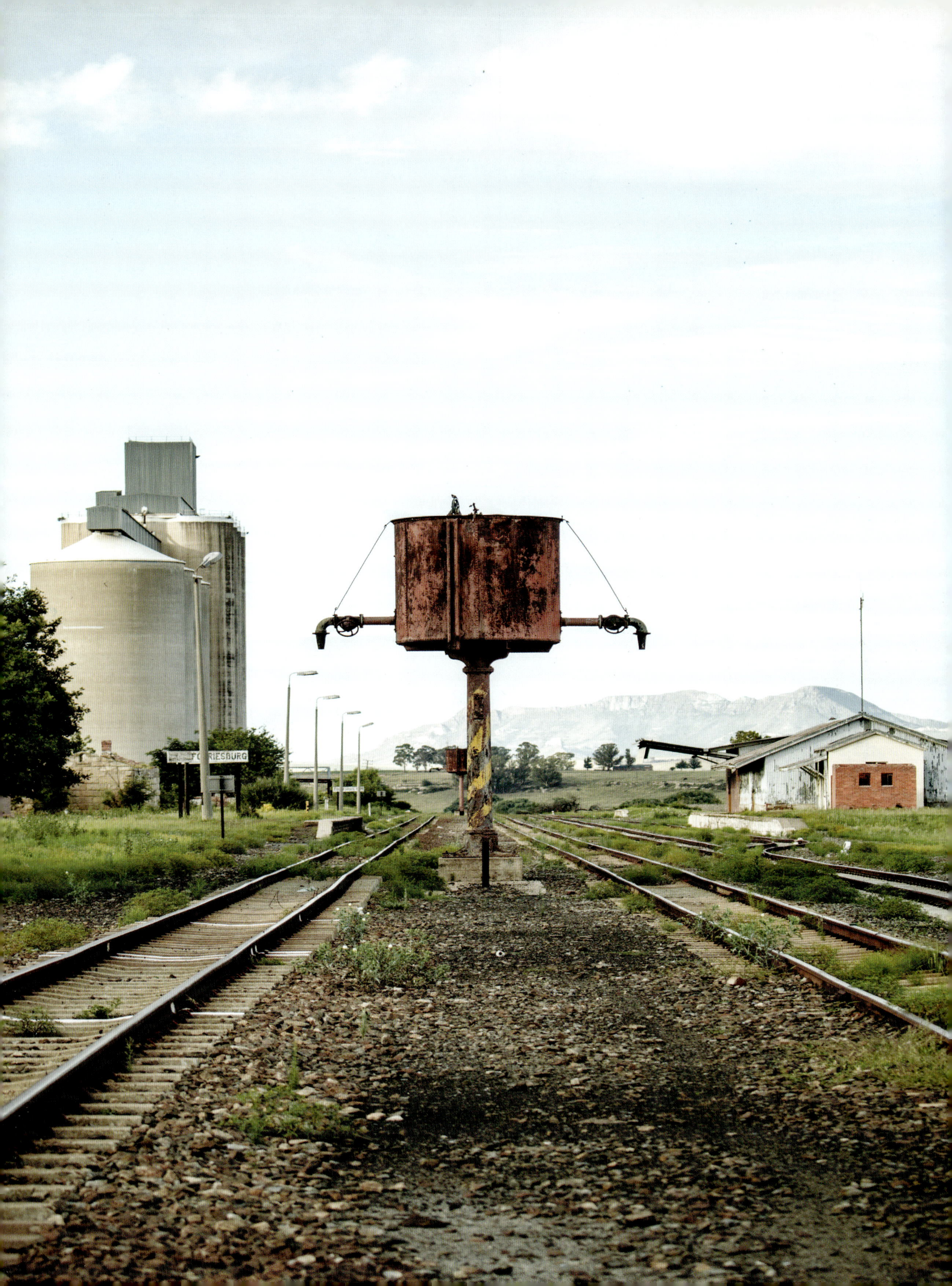
FO RIESBURG

Africa and the Middle East

The first railway on the African continent was opened in Egypt in 1852, and by the 1960s there were many systems, mostly constructed by colonial powers for transport of coal and minerals, for bulk haulage of cash crops, and also to allow for rapid movement of troops in the event of 'troubles'. This discrete pattern of development, with a variety of rail gauges, worked against the construction of a pan-African network of railways, which never came into being. In the 20th century, many lines fell into disuse as road networks improved and air transport spread. As a result, large numbers of stations, especially in thinly populated country districts, were closed. Some stations, especially in South Africa, are preserved for heritage services, while the majority stand empty or have been pulled down. In the Middle East, 20th-century warfare has had a disastrous effect on railway systems, cutting off international routes like the Hejaz Railway from Syria to Saudi Arabia and those through Lebanon from Turkey to Iraq, leaving wrecked or abandoned stations as monuments.

OPPOSITE:
Fouriesburg, Free State, South Africa
On the scenic line from Bloemfontein to Bethlehem, steam trains ran regularly until 1987. Now the line is little used and there are no regular services. Fouriesburg had a reservoir and so was used as a watering point as well as a place where trains in opposite directions could pass each other. Now road trucks transport grain from the elevators.

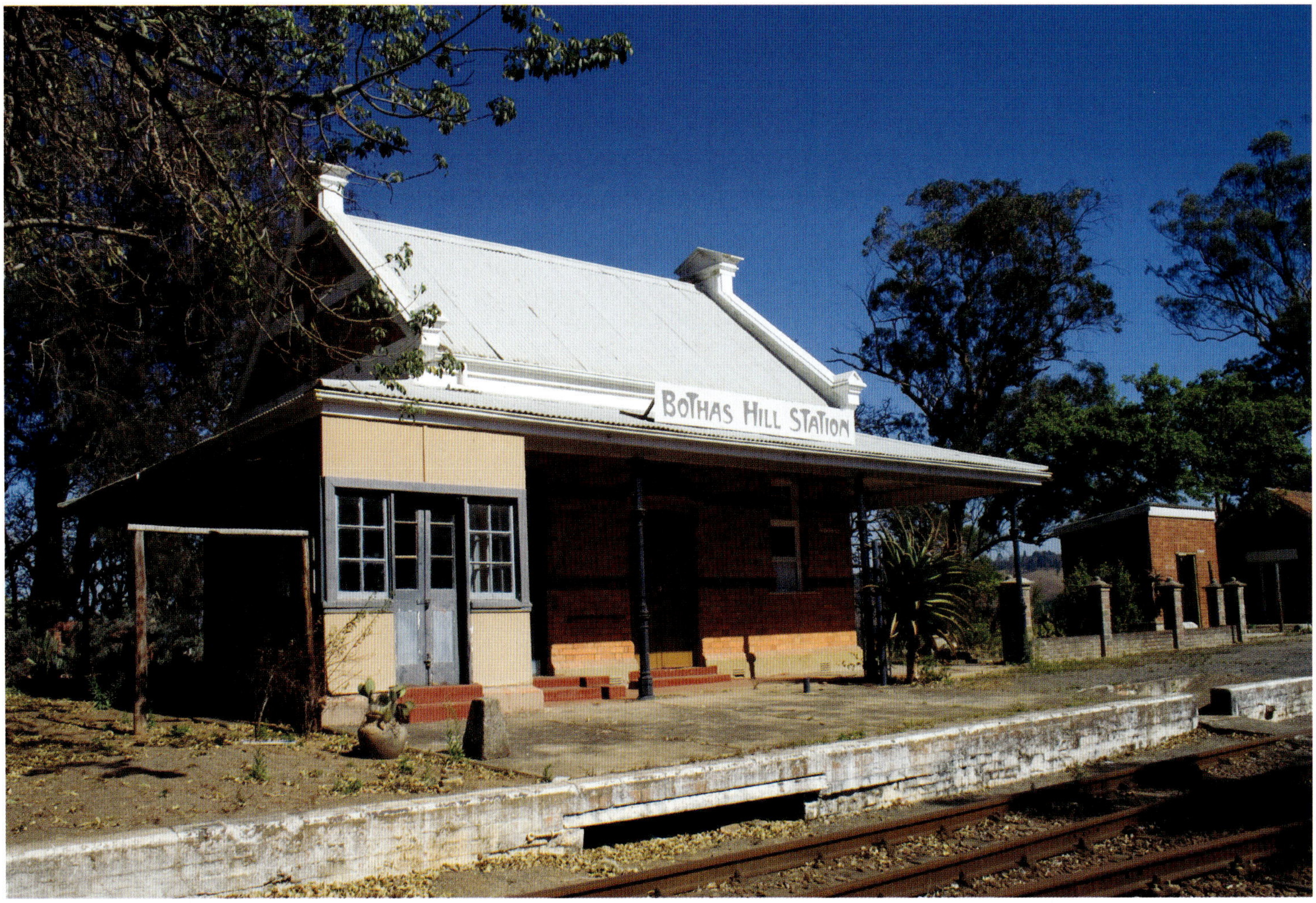

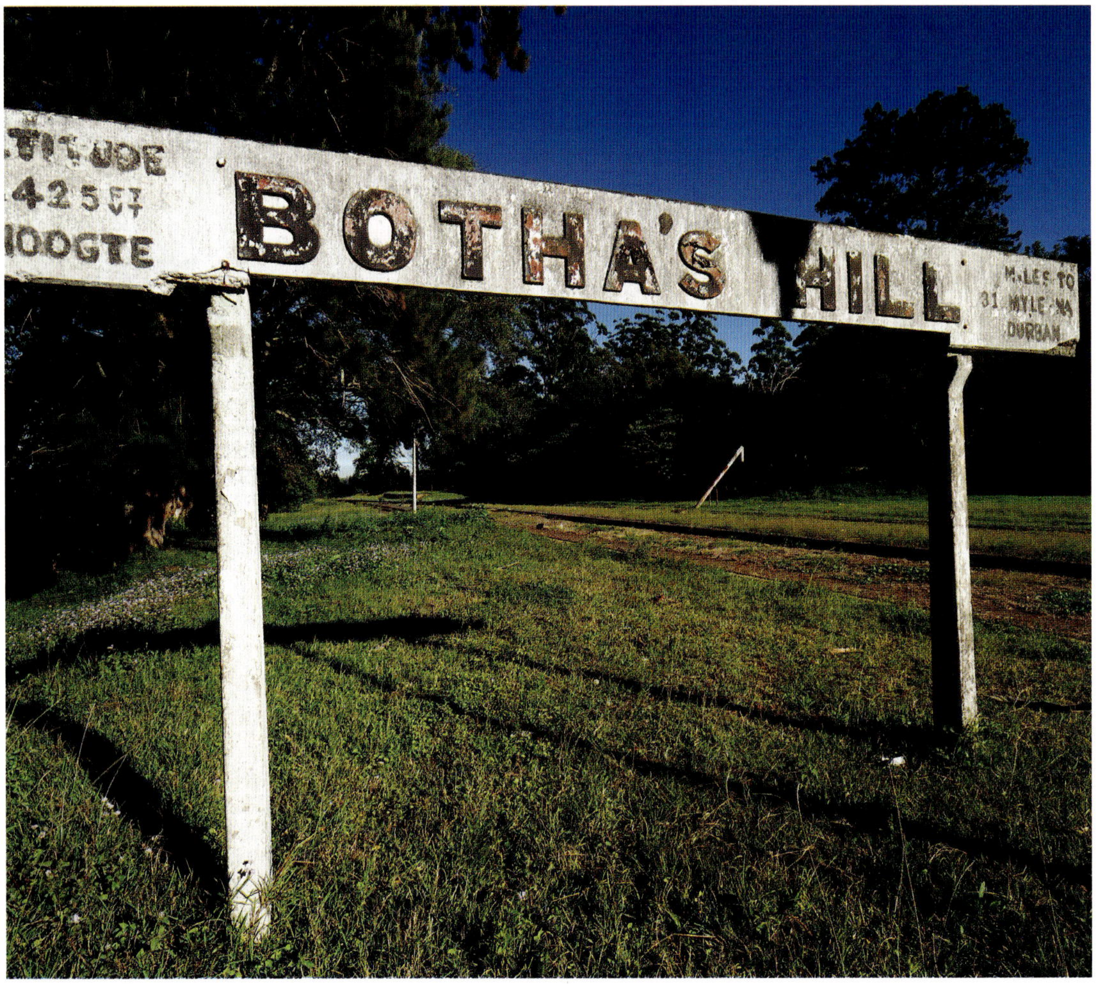

ALL PHOTOGRAPHS:

Botha's Hill, Kwa-Zulu Natal, South Africa

The Natal Government Railway reached Botha's Hill in 1879, and the station's original wooden and iron buildings were replaced by an attractive brick pavilion around 1900. Original equipment remains in the signal compartment and the track plan shows that the station was a crossing point on the single line. Since the 1990s, the condition of the building has deteriorated.

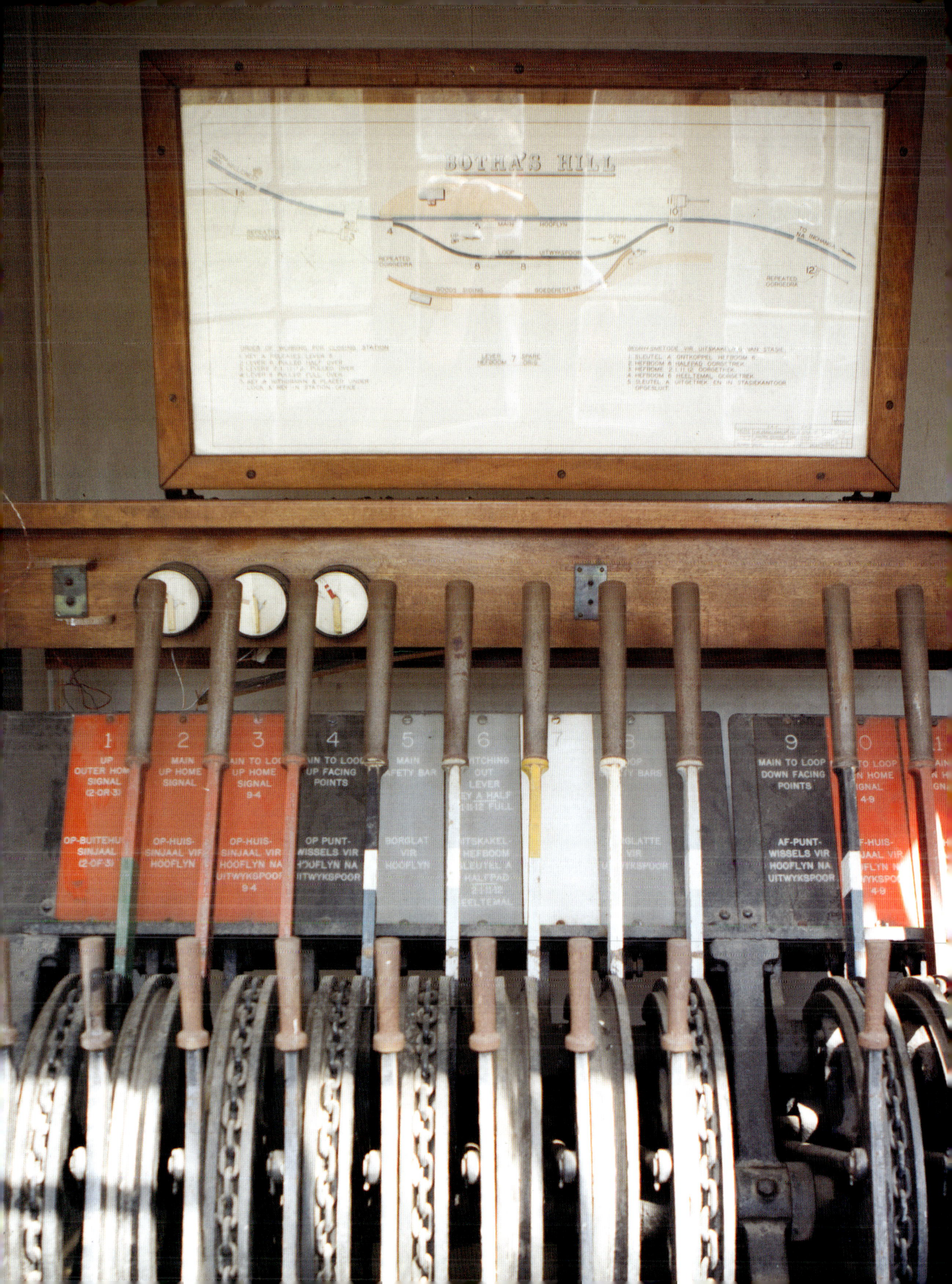
BOTHA'S HILL
2
MAIN
UP HOME
SIGNAL
OP-HUIS-
SINJAAL VIR
HOOFLYN
9
MAIN TO LOOP
DOWN FACING
POINTS
AF-PUNT-
WISSELS VIR
HOOFLYN NA
UITWYKSPOOR

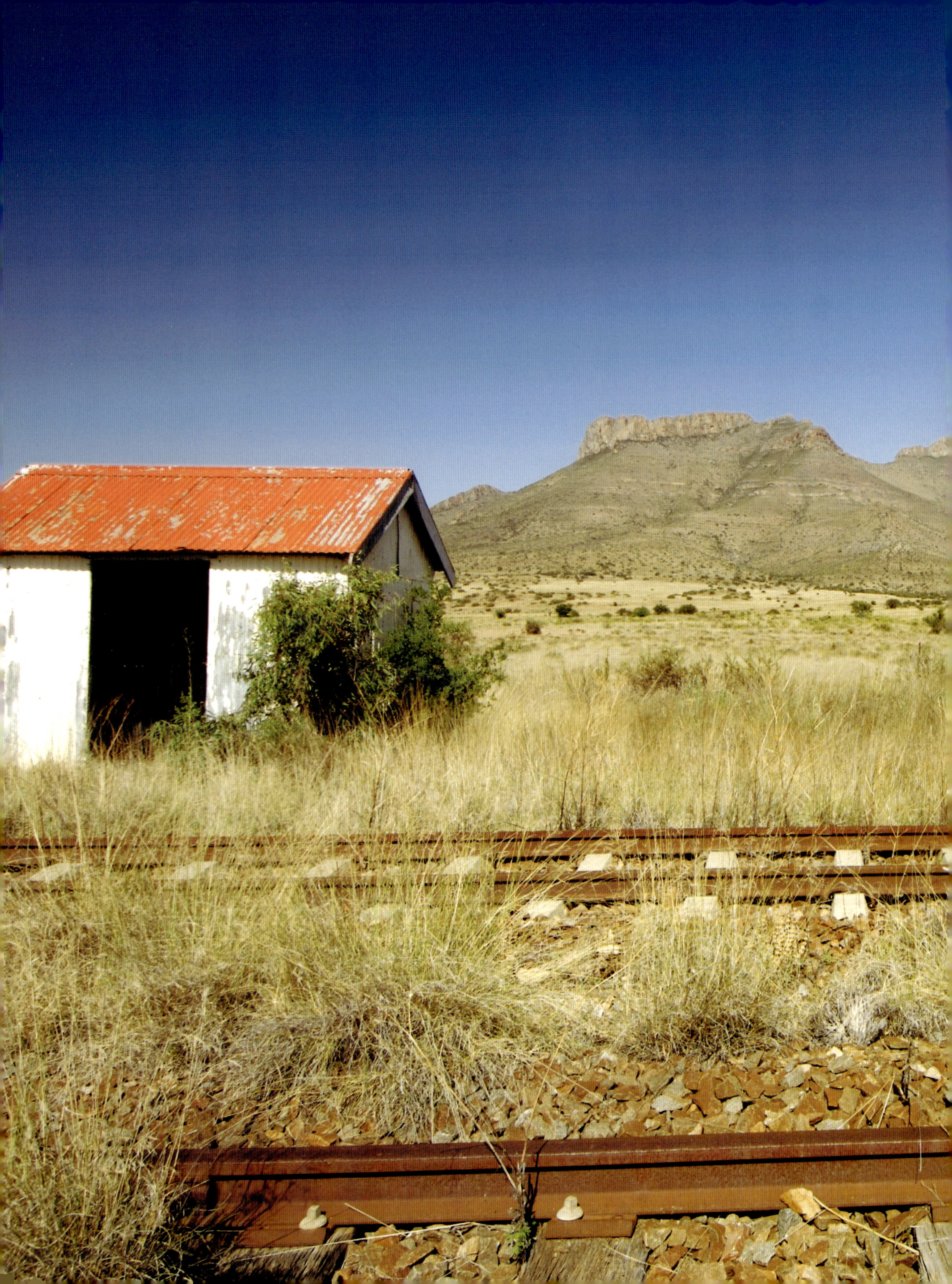

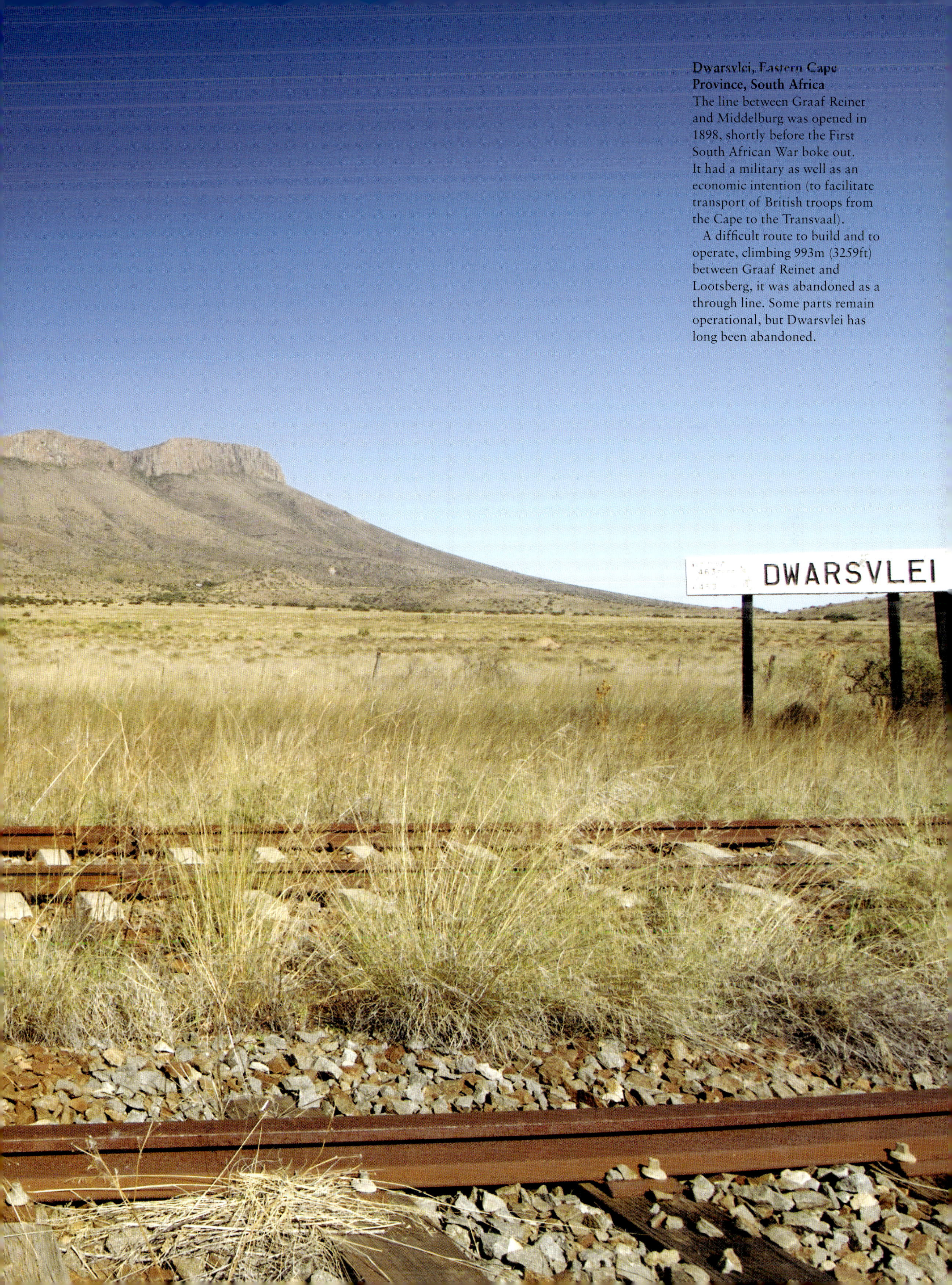

Dwarsvlei, Eastern Cape Province, South Africa
The line between Graaf Reinet and Middelburg was opened in 1898, shortly before the First South African War boke out. It had a military as well as an economic intention (to facilitate transport of British troops from the Cape to the Transvaal).

A difficult route to build and to operate, climbing 993m (3259ft) between Graaf Reinet and Lootsberg, it was abandoned as a through line. Some parts remain operational, but Dwarsvlei has long been abandoned.

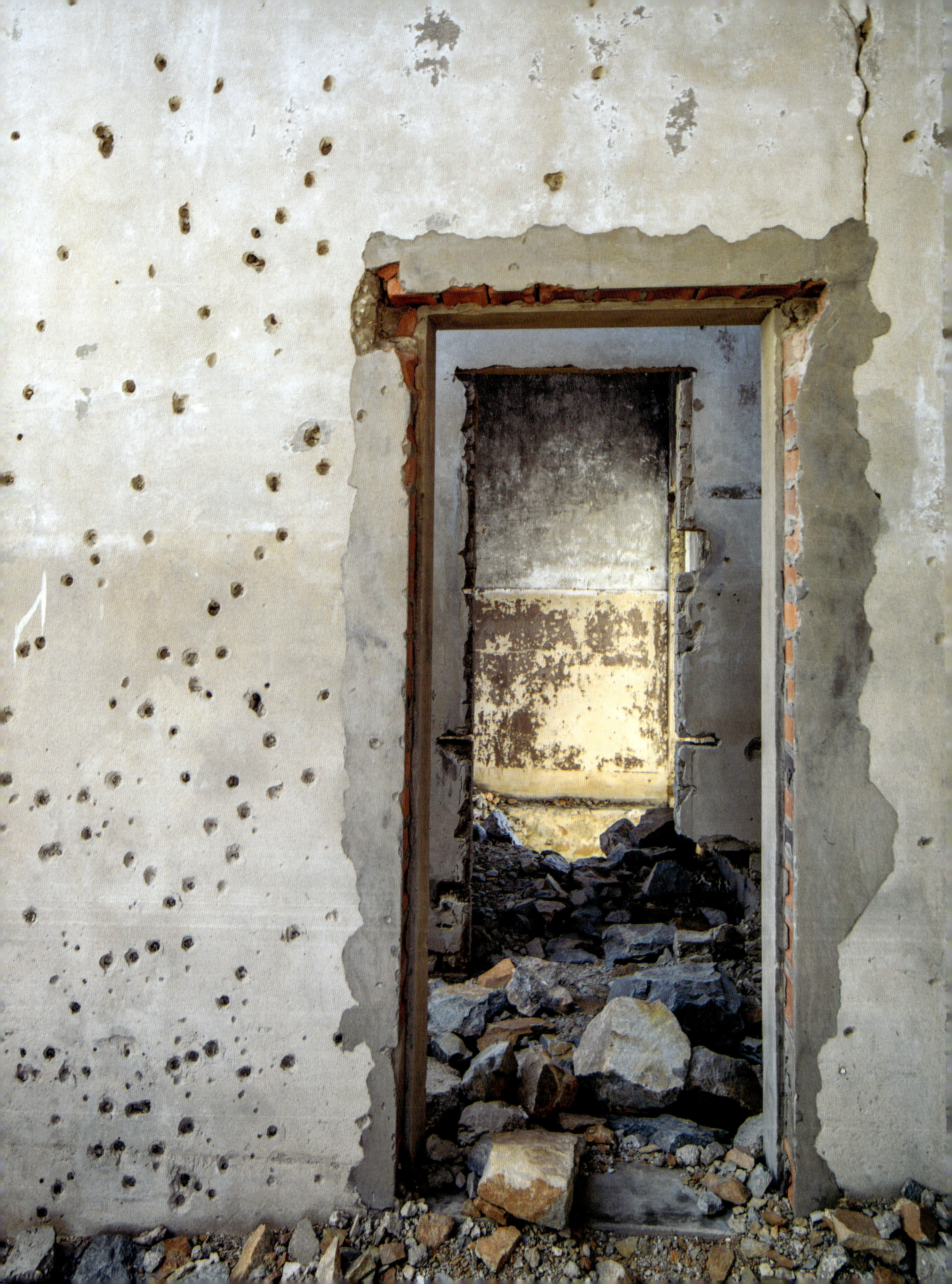

ALL PHOTOGRAPHS:

Grasplatz, Karas Region, Namibia

This was diamond country in the early 1900s, with a brief rush of mining activity. Grasplatz is where the first diamond found in Namibia was unearthed, in 1908. It is on the decommissioned railway from the inland settlement of Aus to the coastal town of Luderitz.

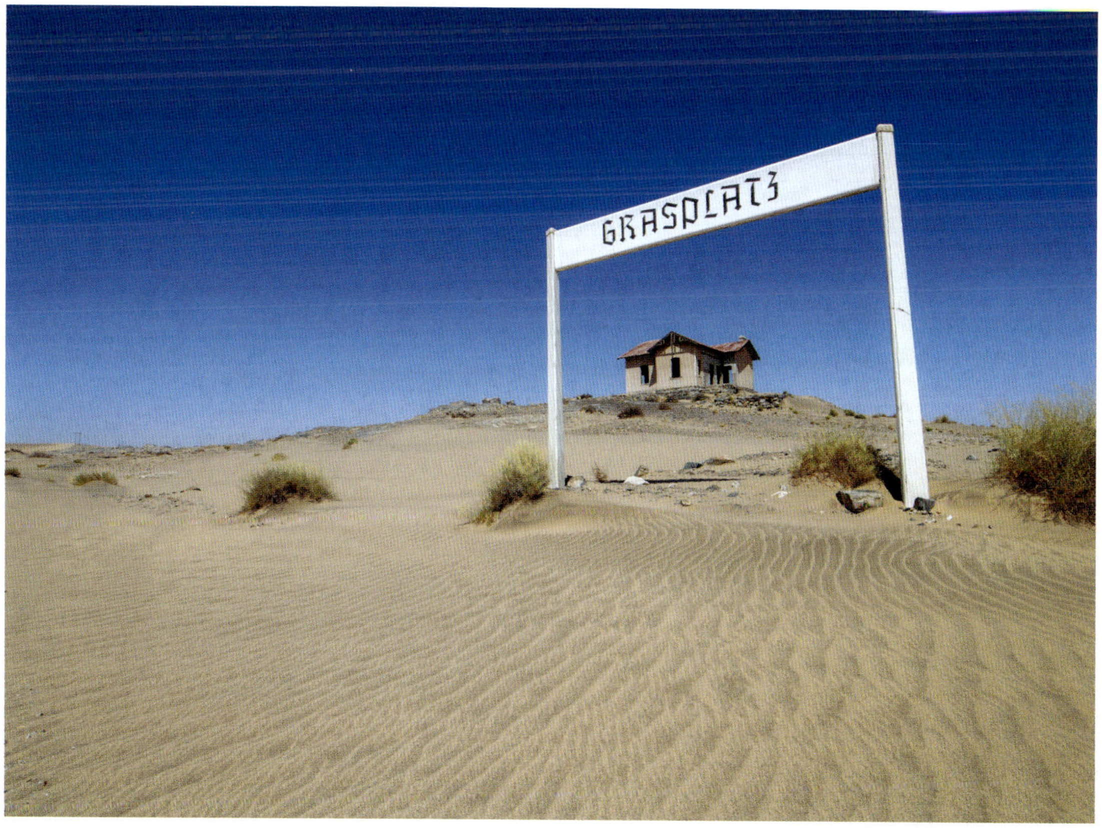

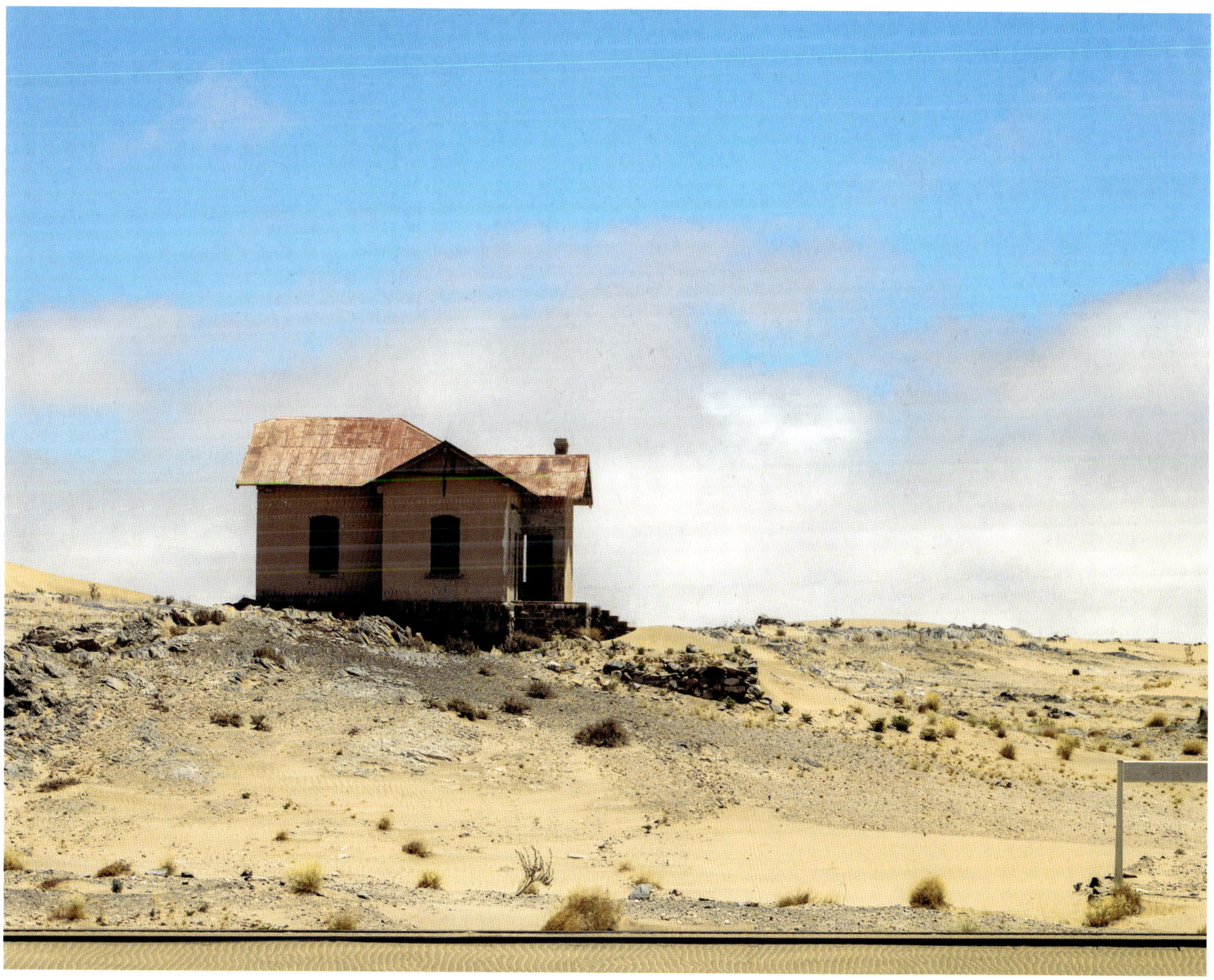

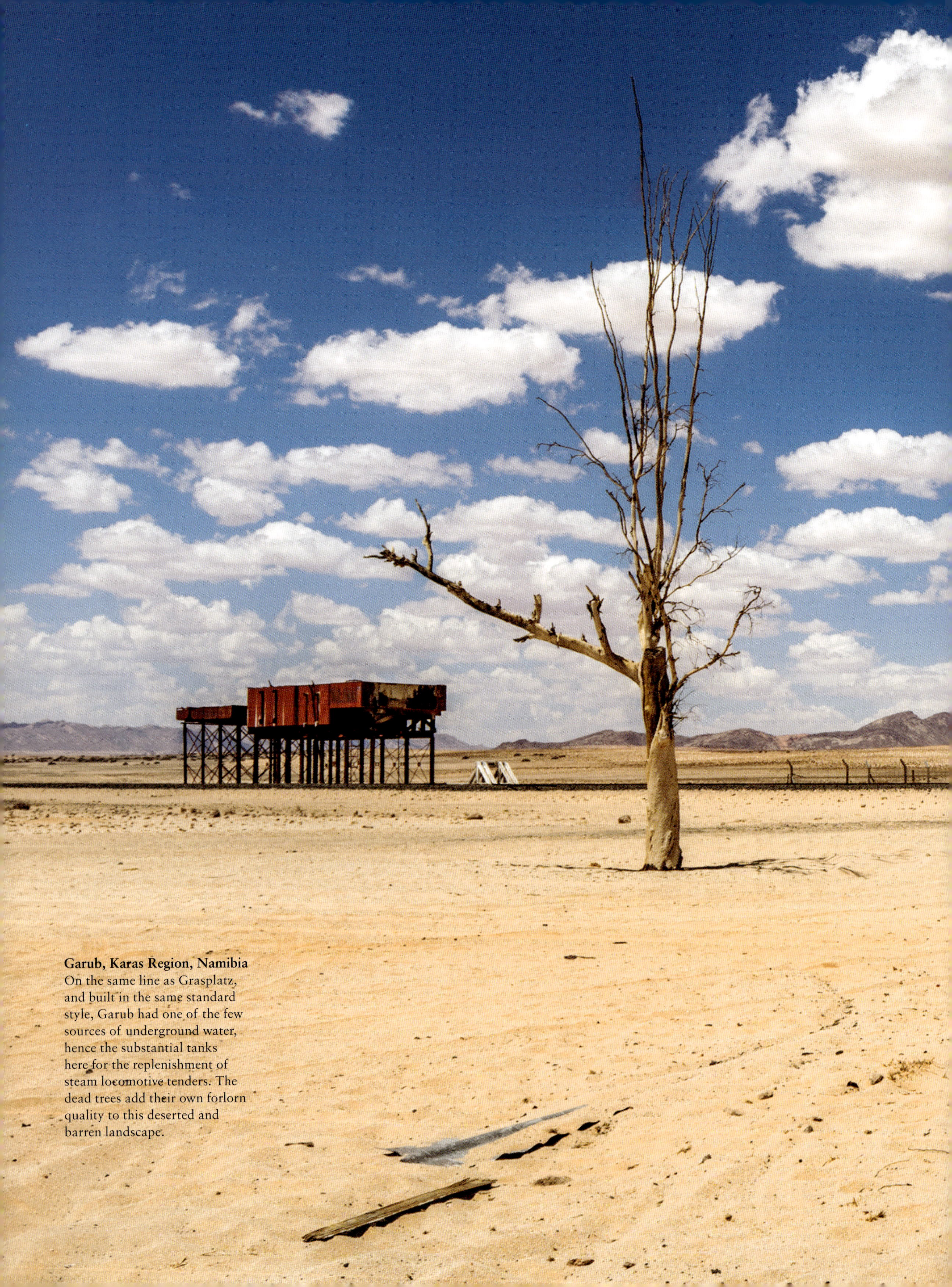

Garub, Karas Region, Namibia
On the same line as Grasplatz, and built in the same standard style, Garub had one of the few sources of underground water, hence the substantial tanks here for the replenishment of steam locomotive tenders. The dead trees add their own forlorn quality to this deserted and barren landscape.

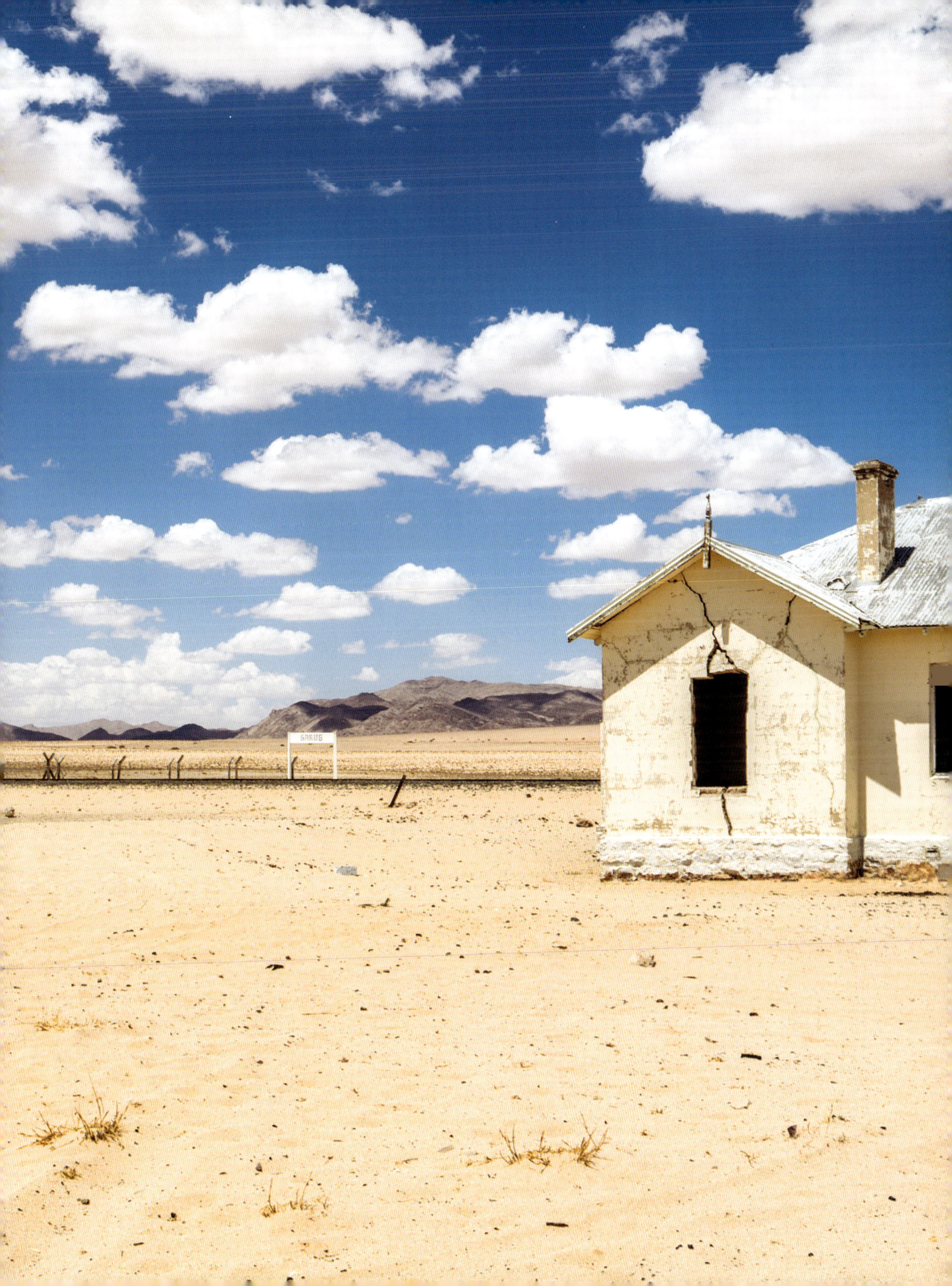
GARUB

RIGHT & OPPOSITE:
Dakar, Senegal
This was the terminus of the Chemin de Fer de Dakar Au Niger, linking Senegal and Niger from 1885 until 2009. While currently out of use, it will be rebuilt as the terminus of a new line linking Dakar city with Blaise Diagne International Airport, with its original façade retained. The first section of this line is forecast to open at the end of 2023.

BELOW:
Thiès, Senegal
Once a busy junction station, now there are no trains at any of its four platforms. Privatized in 2003, the Senegal railway system was in total decline by 2017. Colonial-era stations in Africa had few amenities. Fares had to be low and inadequate maintenance led to low speeds. Such amenities as existed were often racially segregated.

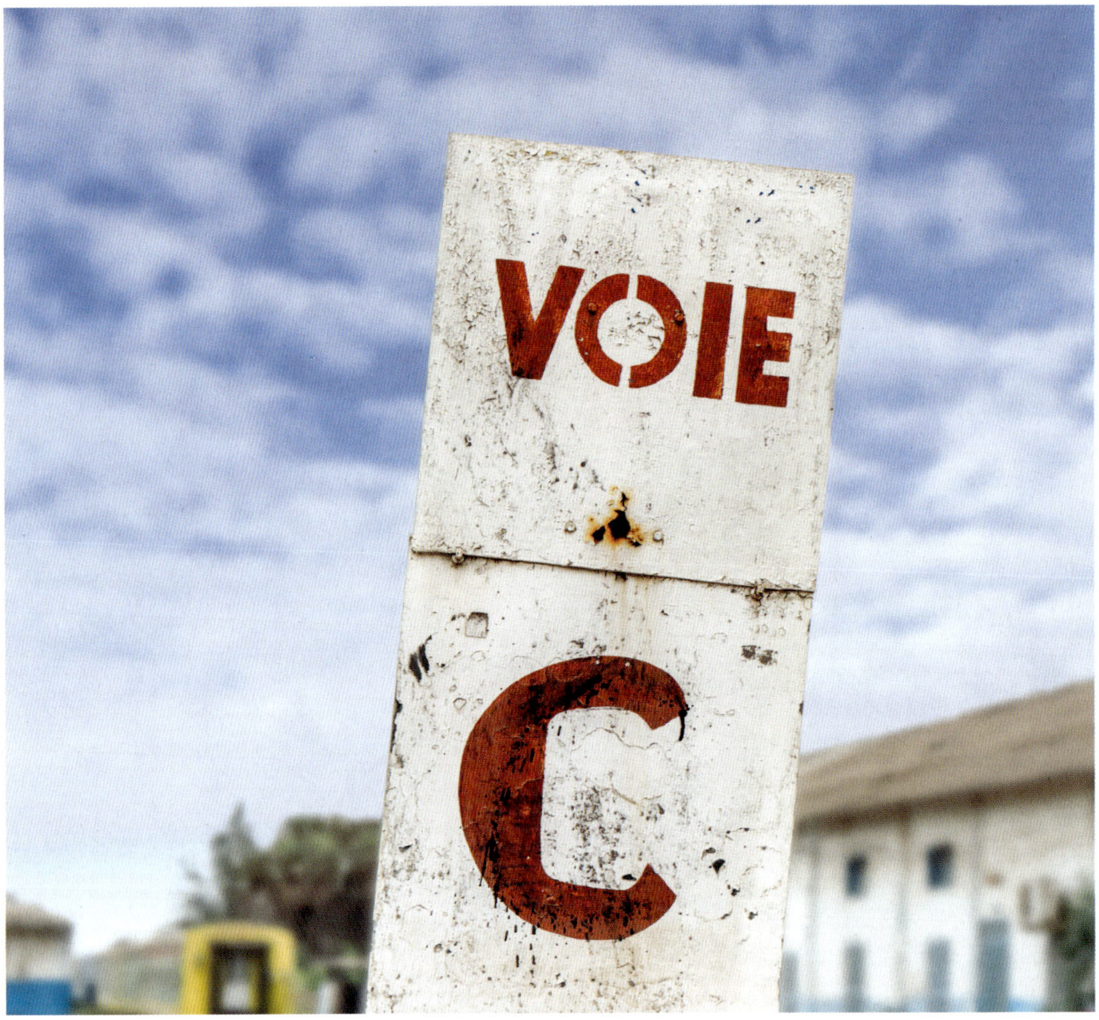

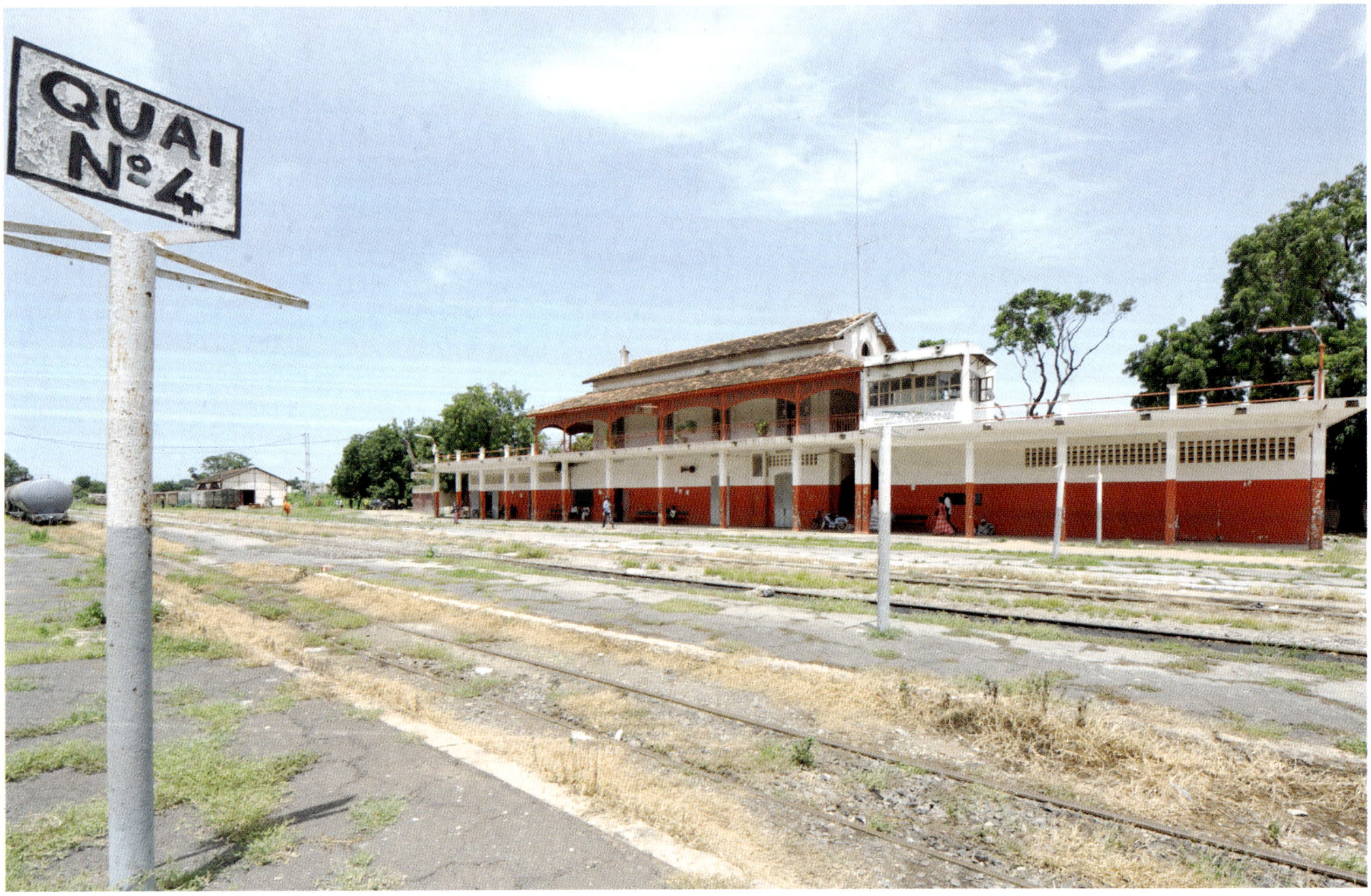

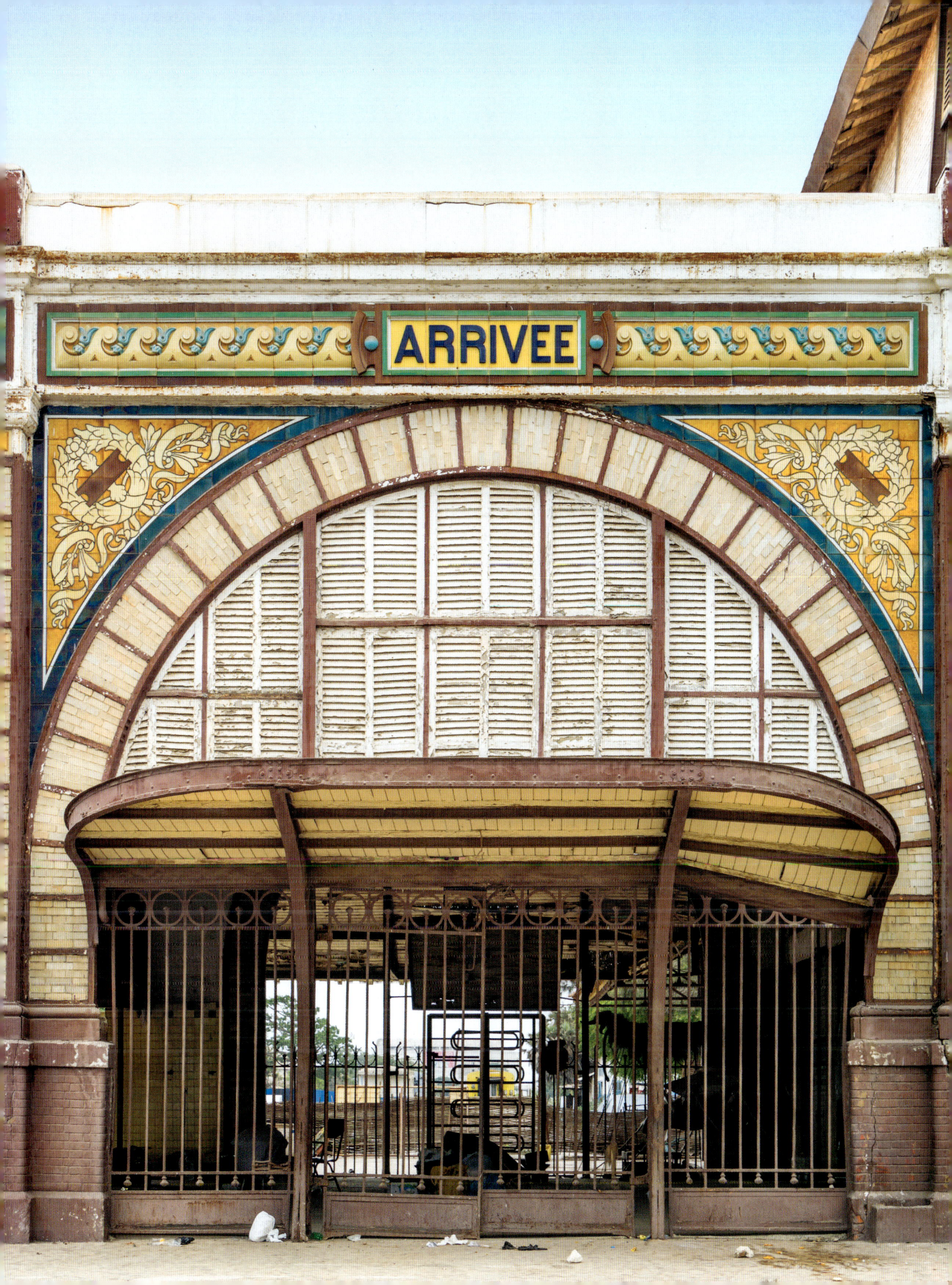
ARRIVEE

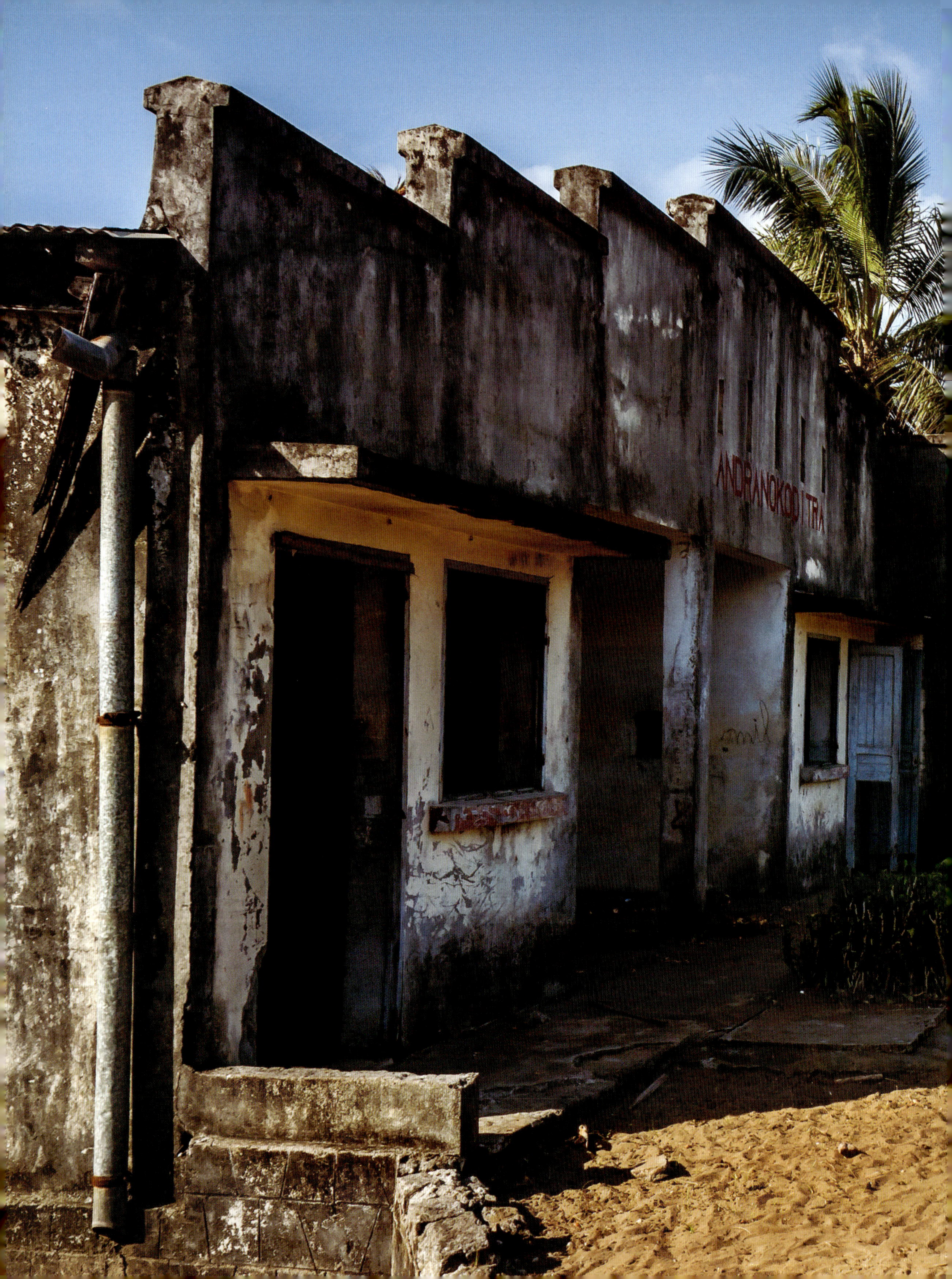
ANDRANOKODITRA

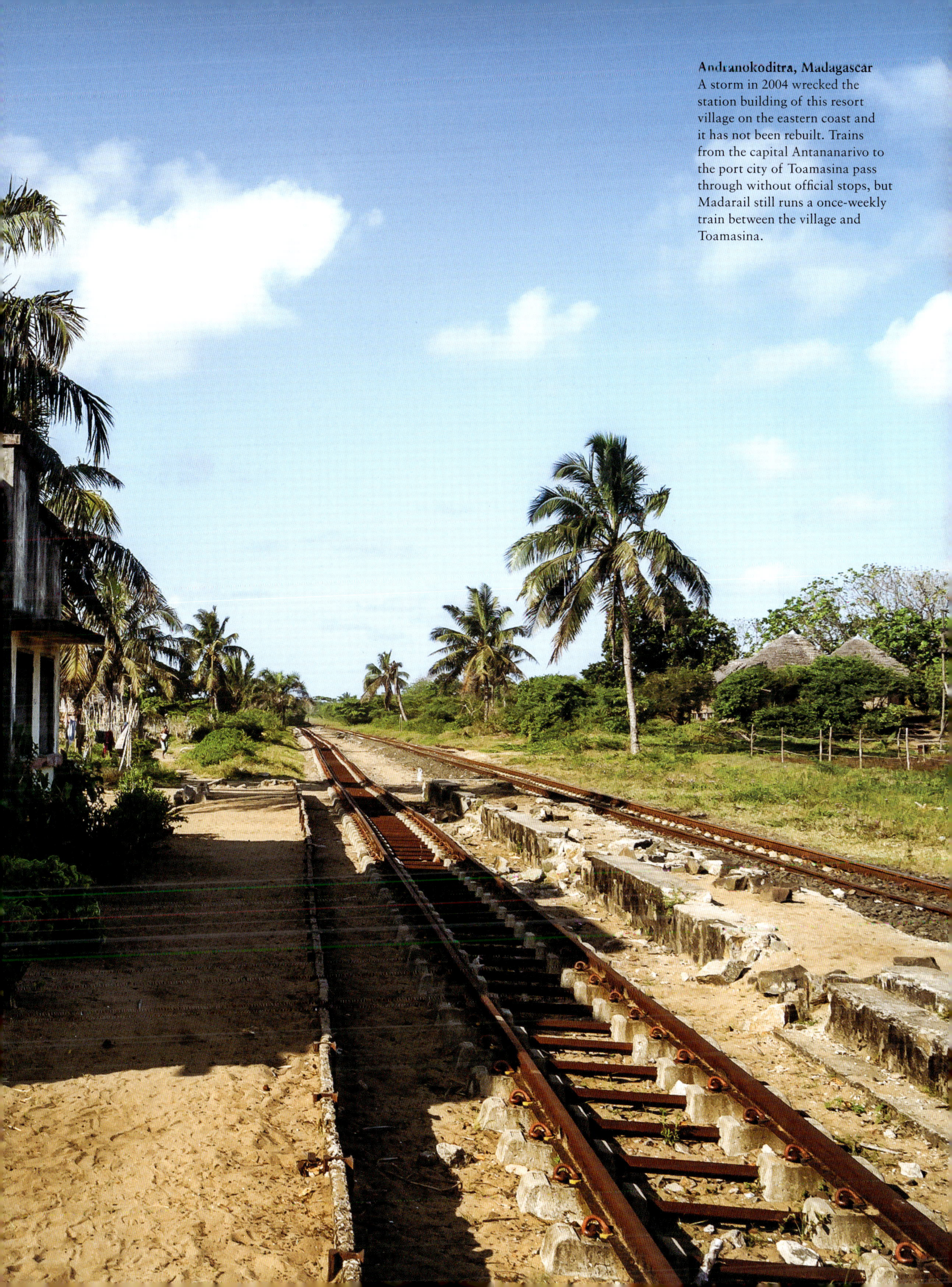

Andranokoditra, Madagascar
A storm in 2004 wrecked the station building of this resort village on the eastern coast and it has not been rebuilt. Trains from the capital Antananarivo to the port city of Toamasina pass through without official stops, but Madarail still runs a once-weekly train between the village and Toamasina.

Massawa, Eritrea
The Eritrean-Ethiopian War, which ended with Eritrean independence in 1991, partly destroyed the fascinating narrow-gauge railway that led from Massawa, on the Red Sea coast, to the inland city of Asmara, 2400m (7872ft) above sea level. Begun in 1887, restored in the 1990s, the line does not run timetabled services but does offer charter services using vintage locomotives and diesel cars.

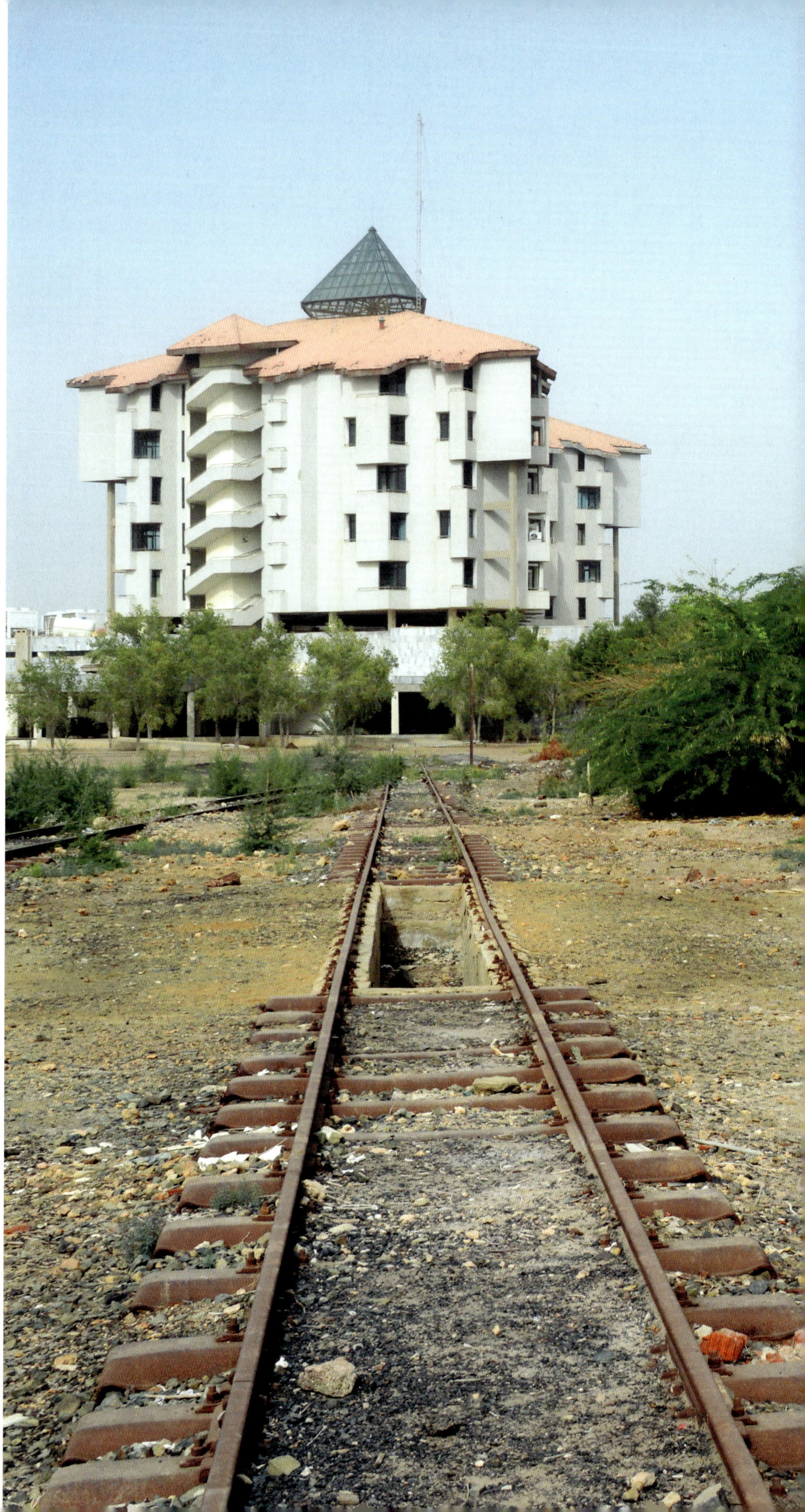

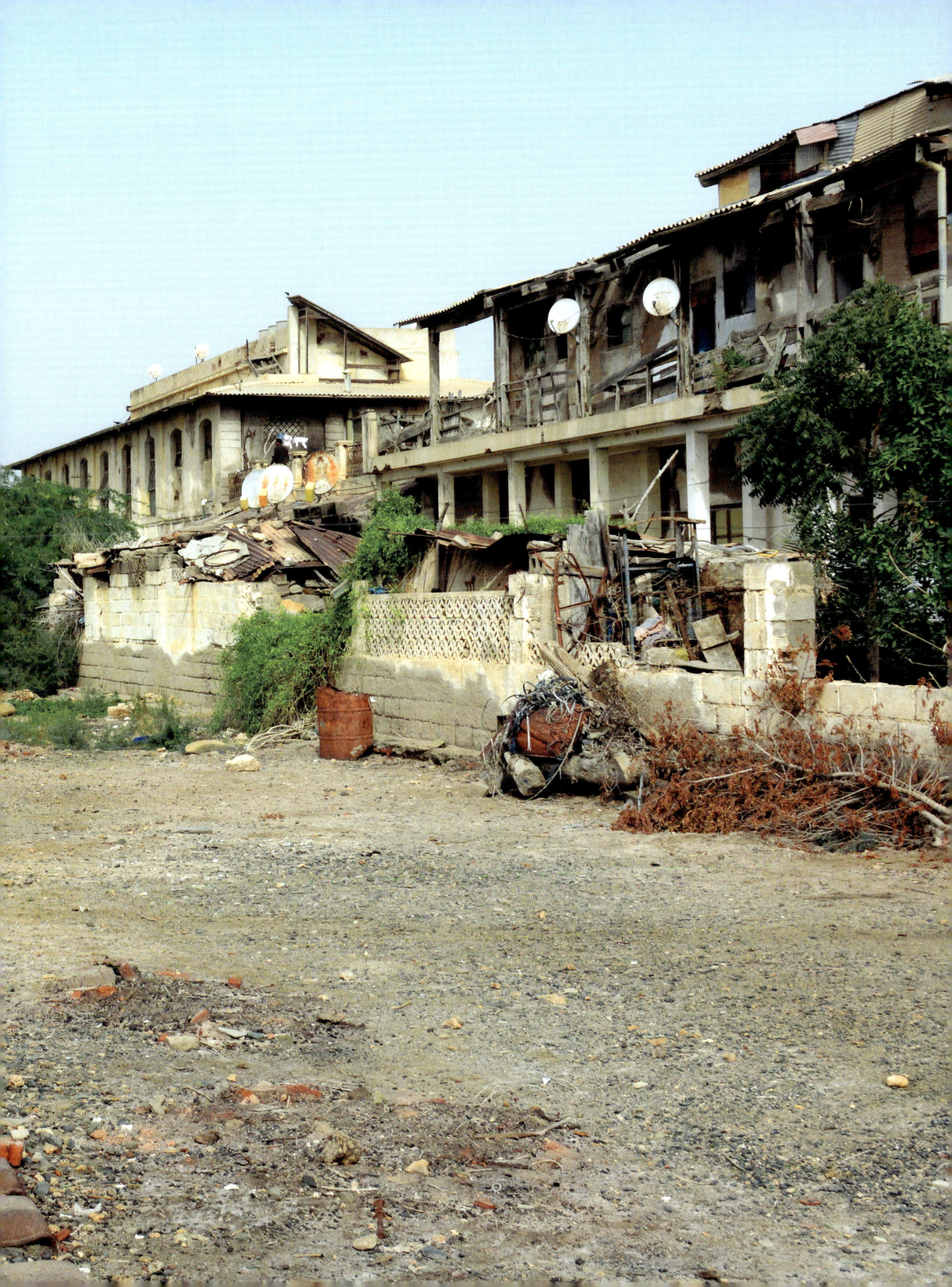

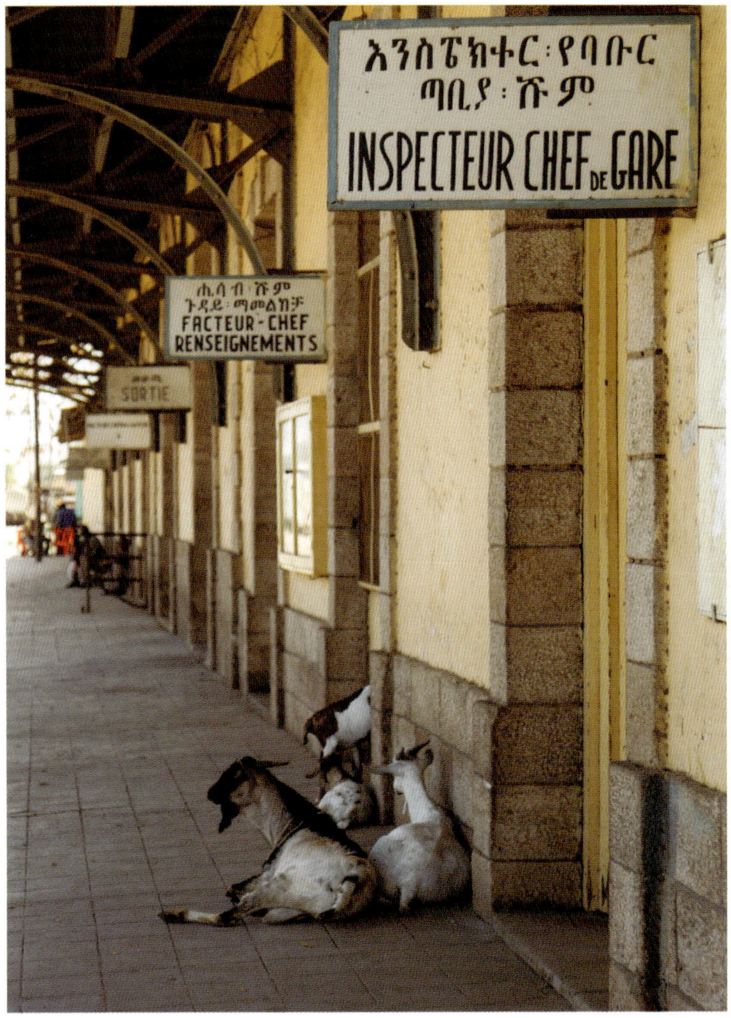

ALL PHOTOGRAPHS:

Dire Dawa, Ethiopia

The other Red Sea railway once beloved of intrepid travellers is the Ethio–Djibouti line, from the port of Djibouti up to Addis Ababa, the Ethiopian capital. This historic metre-gauge line has been supplanted for practical purposes by a new electric line completed in 2017, but a workable section remains around the town of Dire Dawa. Its station still maintains the signage of the old Franco-Ethiopian company.

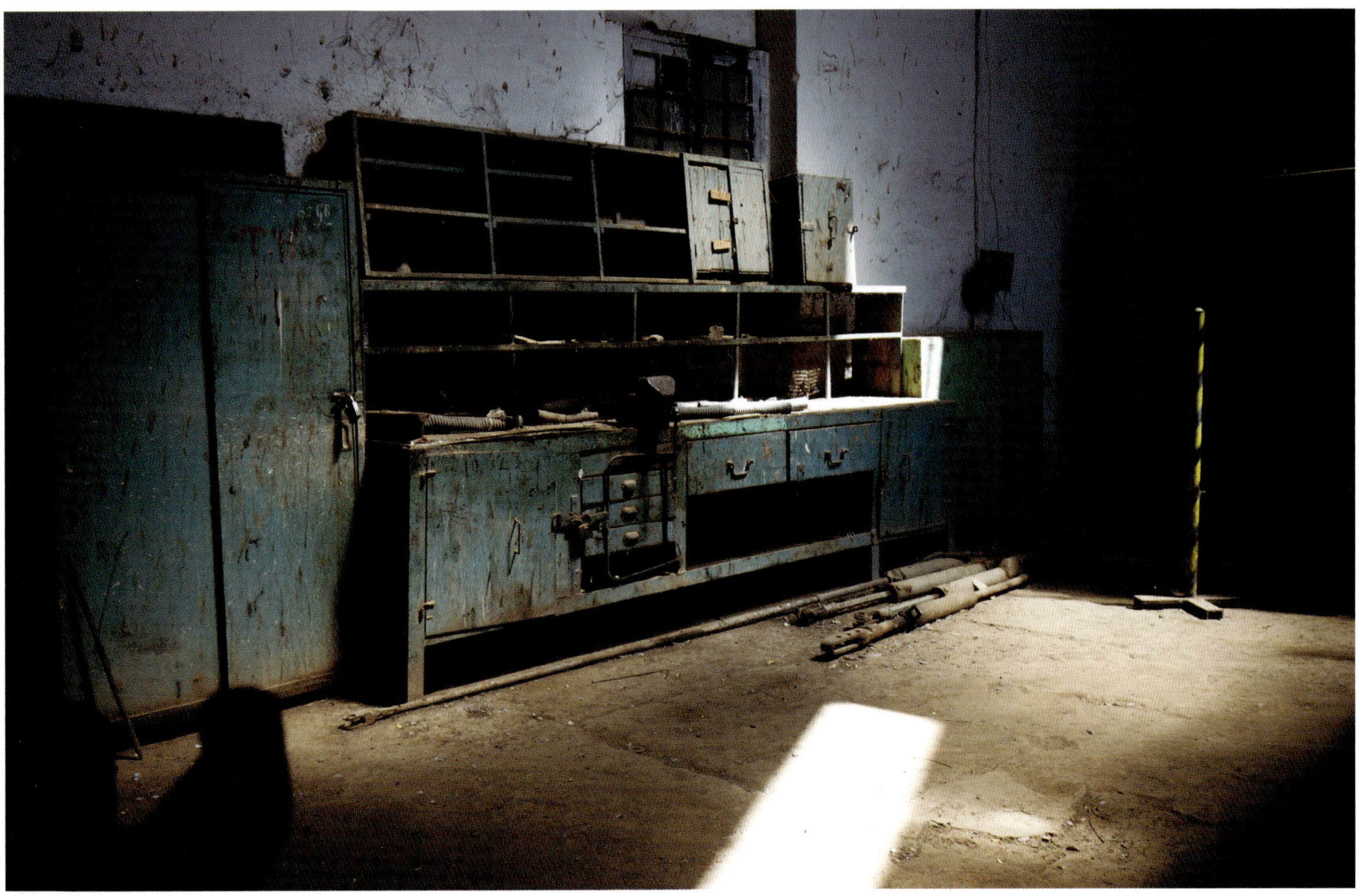

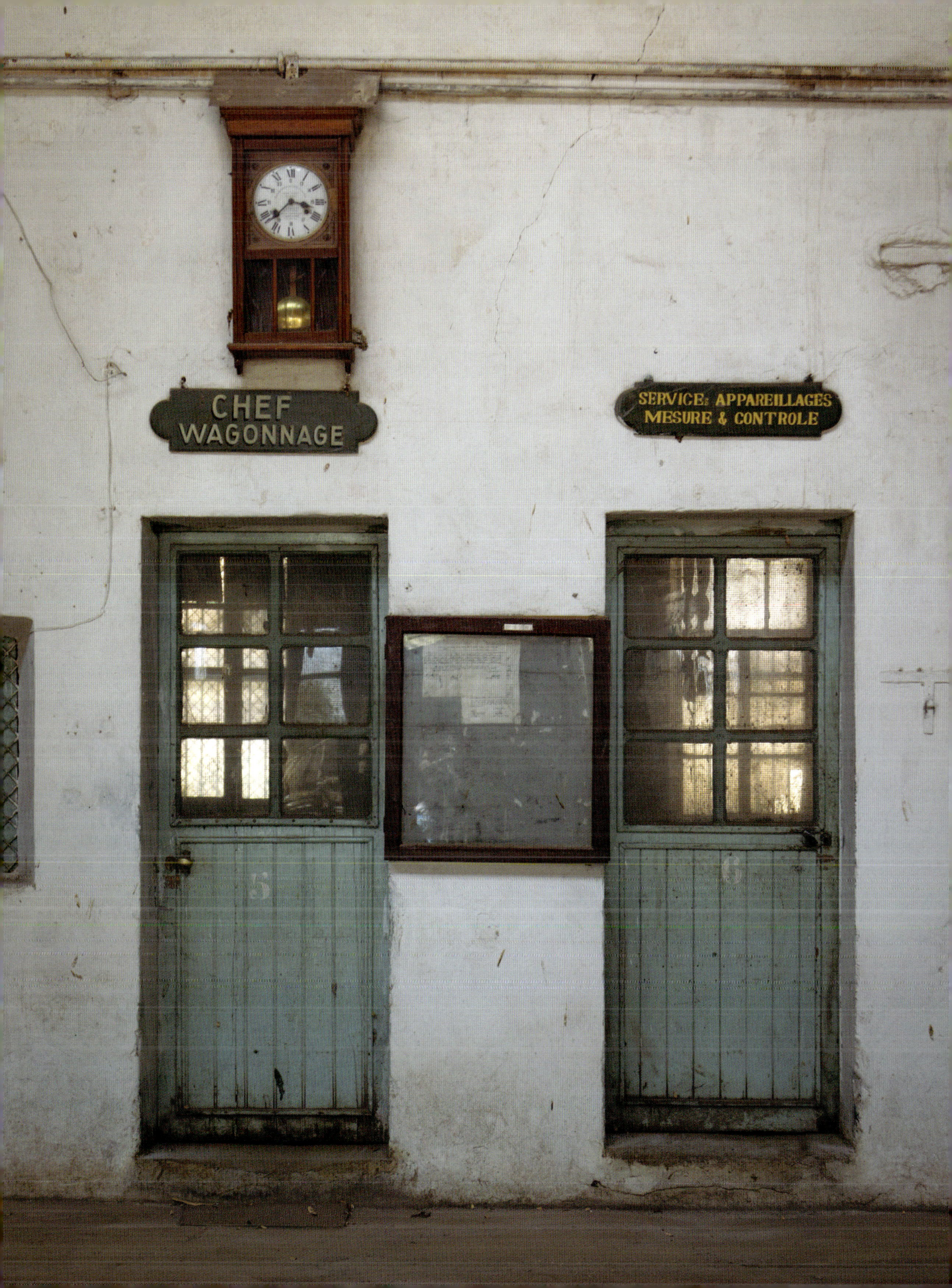
CHEF
WAGONNAGE
SERVICE: APPAREILLAGES
MESURE & CONTROLE
5
6

Station No. 6, Sudan
Stations with no name, just a number, are unusual, but this one is in the middle of the Nubian Desert, between Wadi Halfa on the Egyptian border, and Atbara. The scanty but healthy trees indicate a water supply, and the existence of No. 6 was probably determined by the water and its distance from the next crossing-point. This part of the line was completed in 1897–98. Very sporadic trains pass to and from Wadi Halfa.

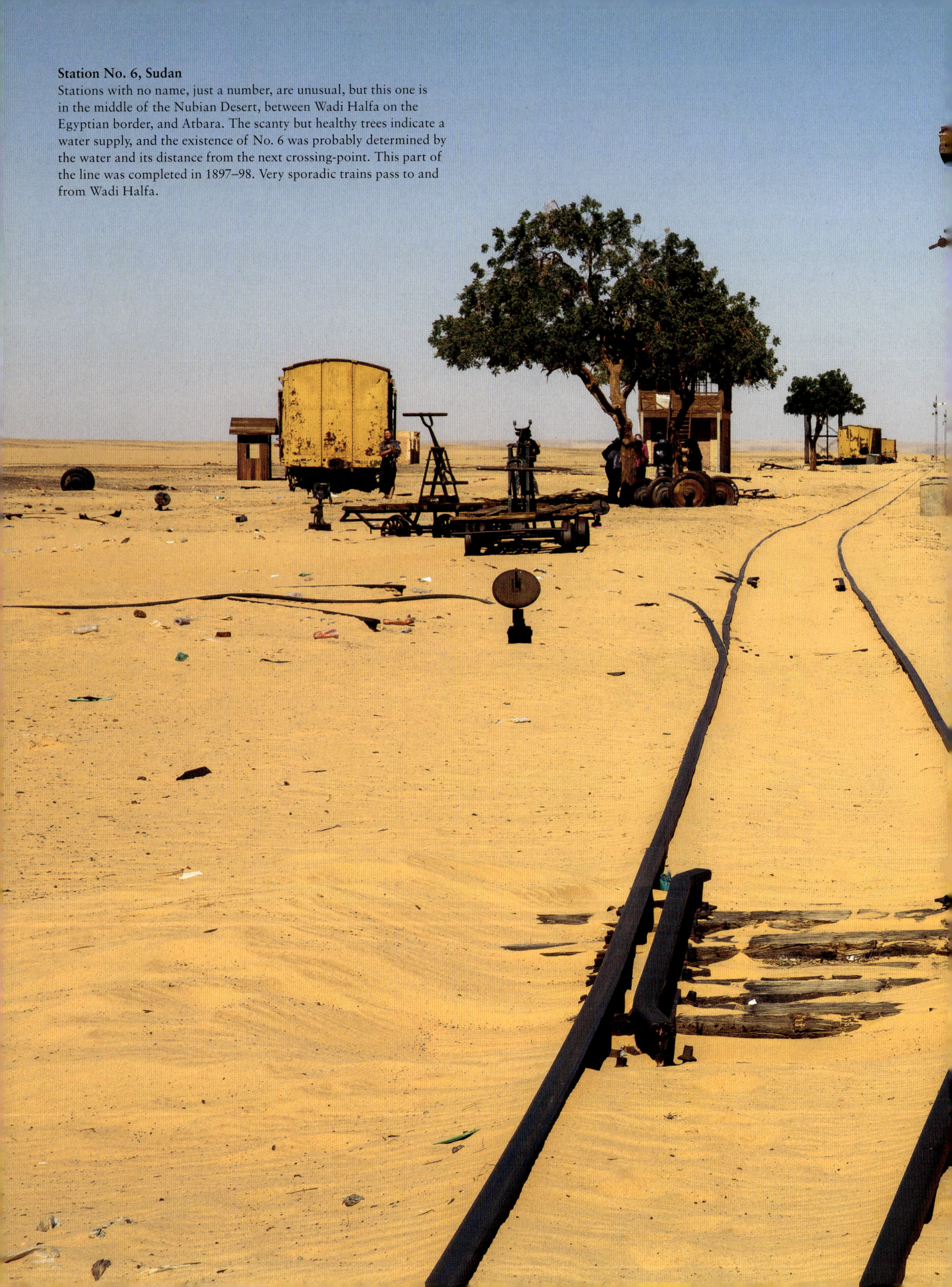

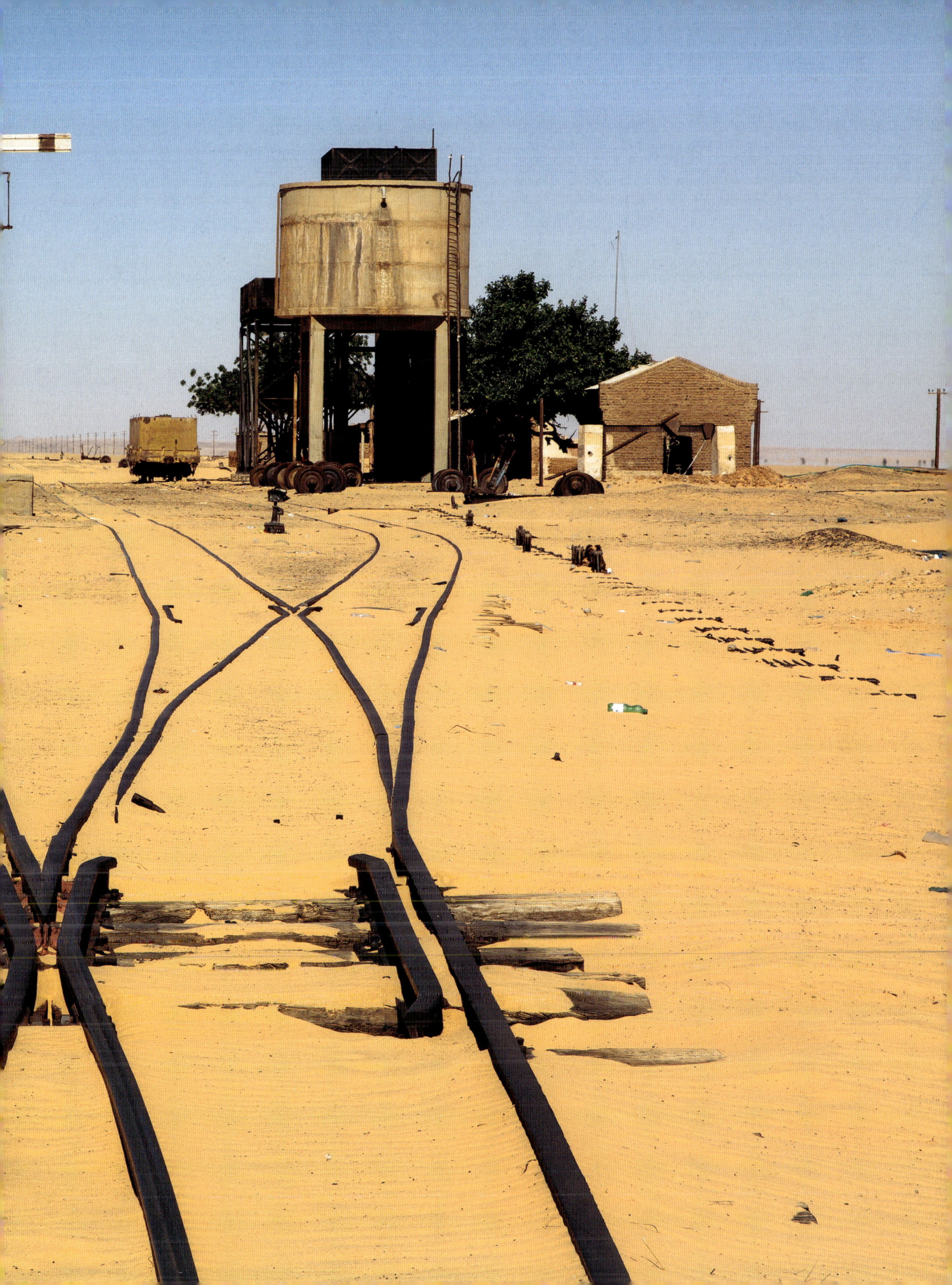

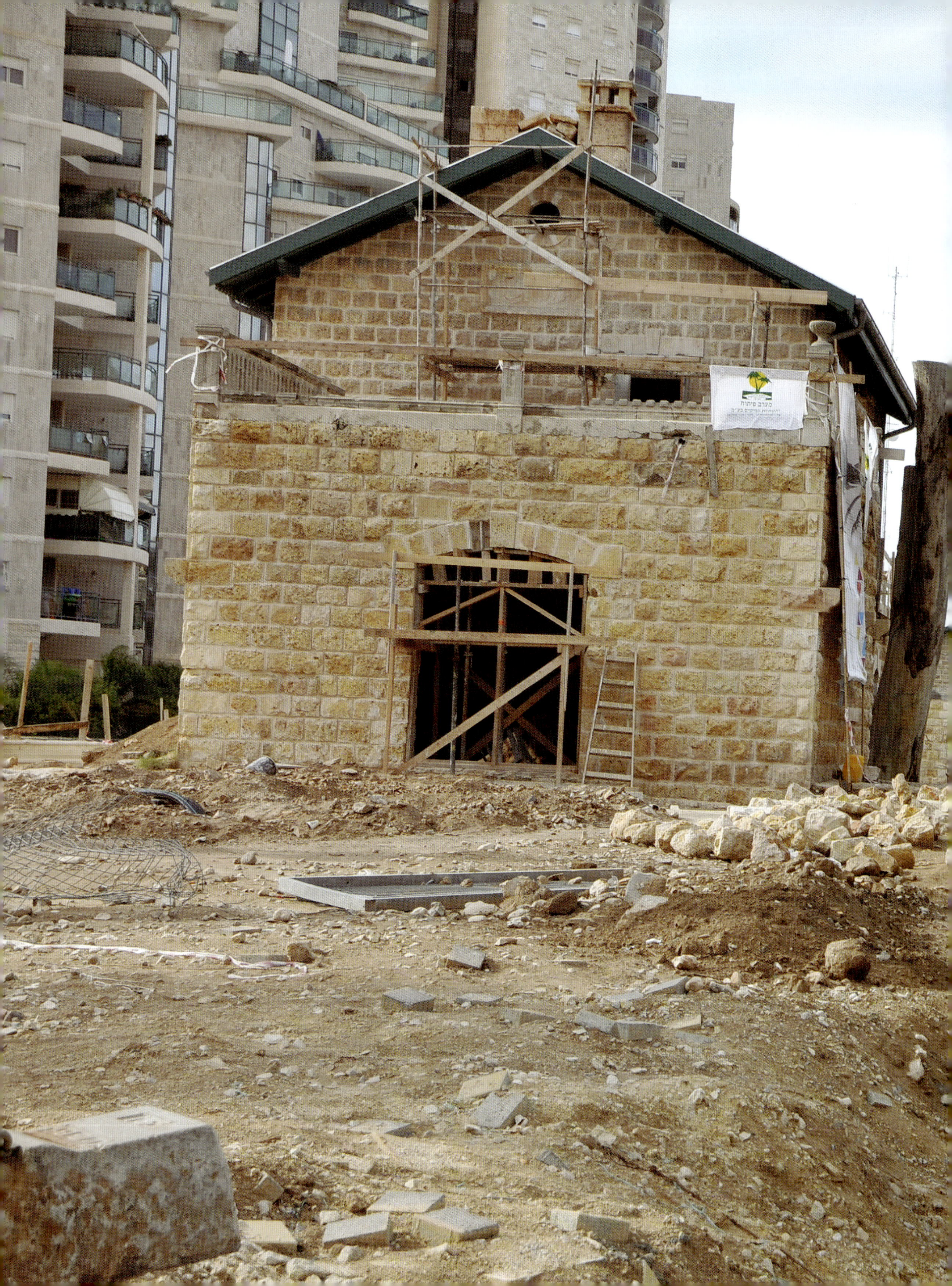
מערב פיתוח

Old Ottoman Station, Beer Sheva, Israel
Now a railway museum, the station was built in 1915 as part of a German-Turkish plan to extend military communication lines for an attack on the British-held Suez Canal. The British army took Beer Sheva in 1917 and the system was adapted for British colonial purposes, with closure of the station in 1927. Under Israeli rule, new rail connections to Beer Sheva were made. The locomotive is a British-built 2-8-0, one of many built for the War Department between 1940 and 1945.

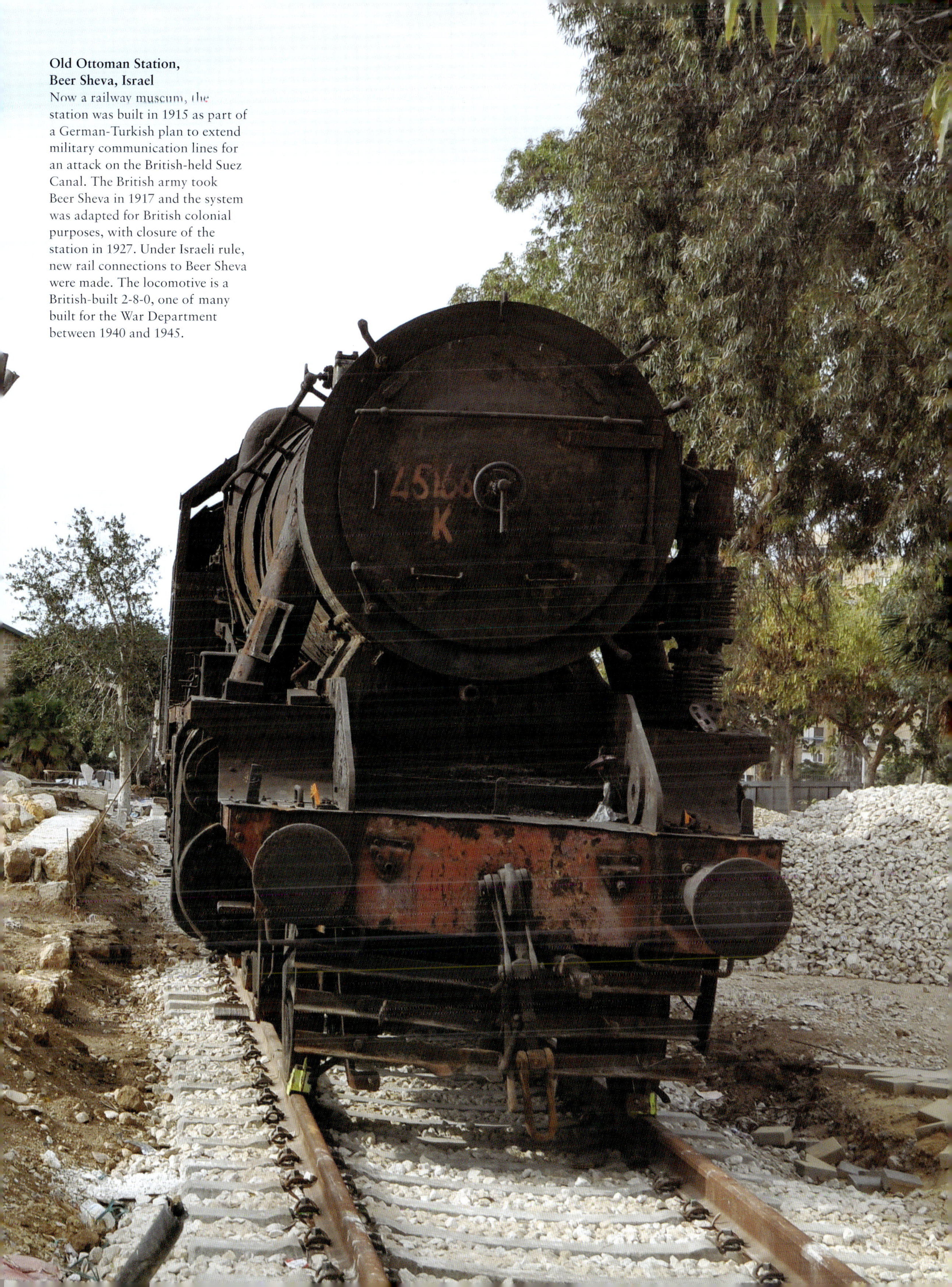

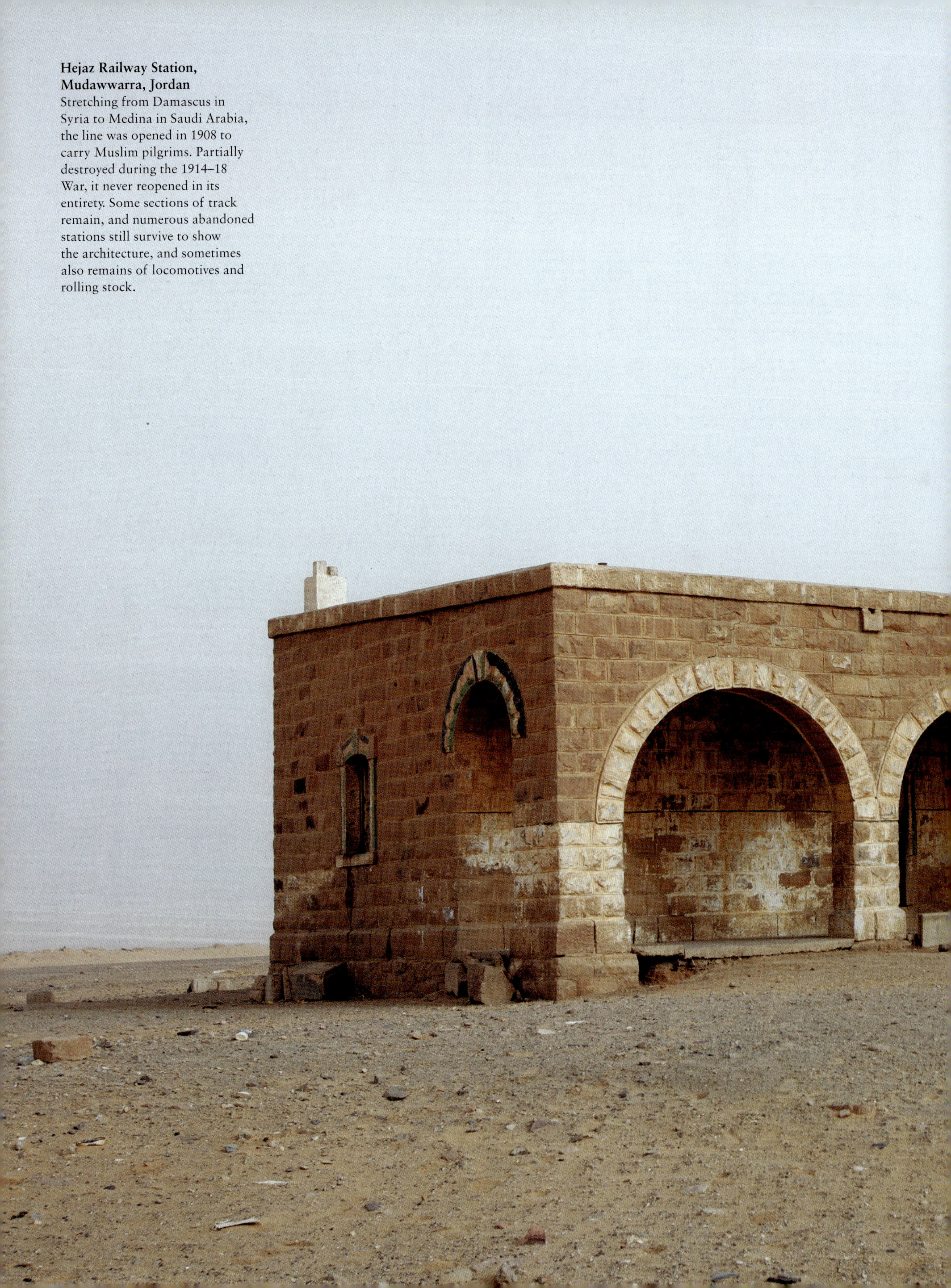

Hejaz Railway Station, Mudawwarra, Jordan
Stretching from Damascus in Syria to Medina in Saudi Arabia, the line was opened in 1908 to carry Muslim pilgrims. Partially destroyed during the 1914–18 War, it never reopened in its entirety. Some sections of track remain, and numerous abandoned stations still survive to show the architecture, and sometimes also remains of locomotives and rolling stock.

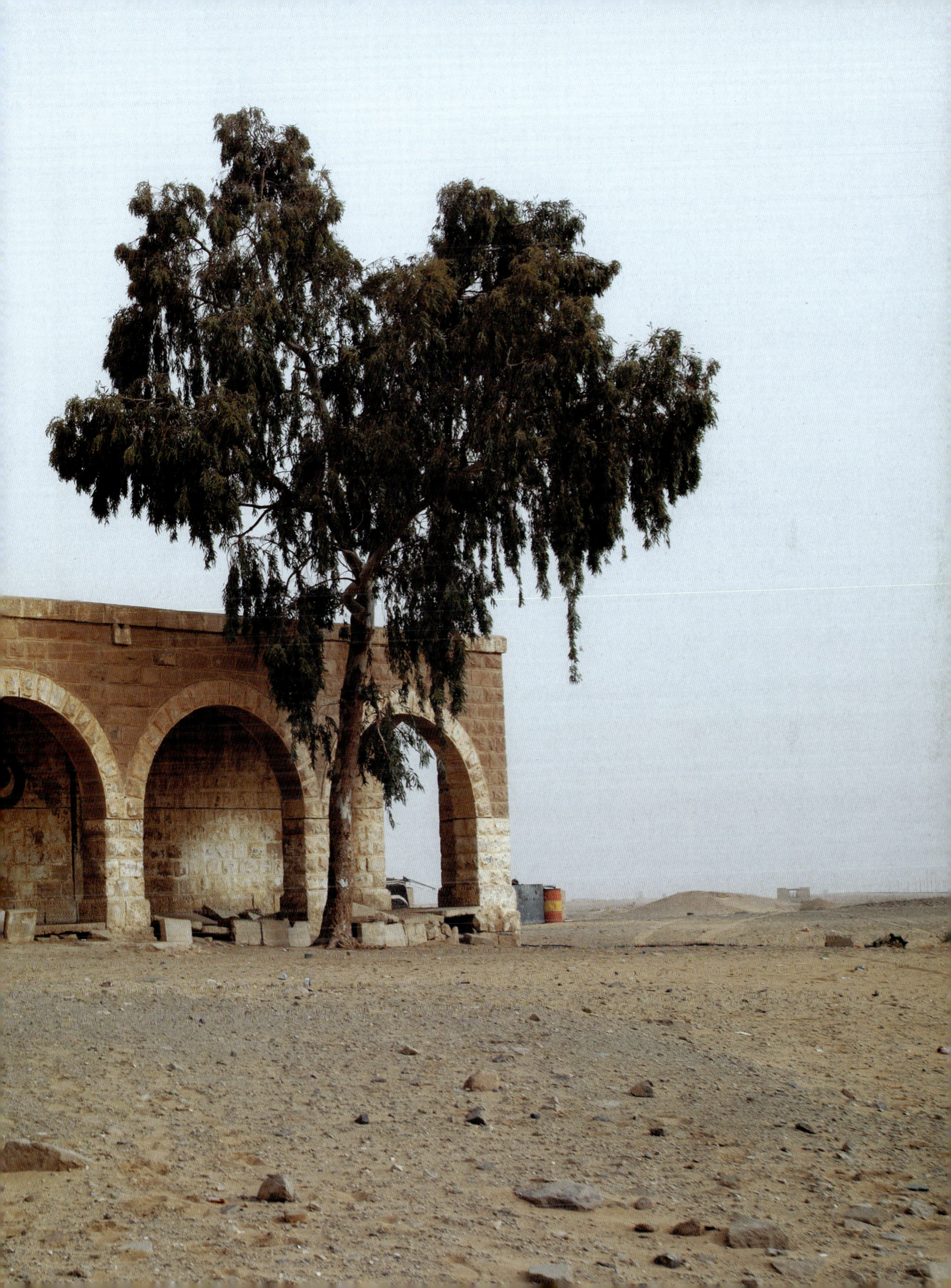

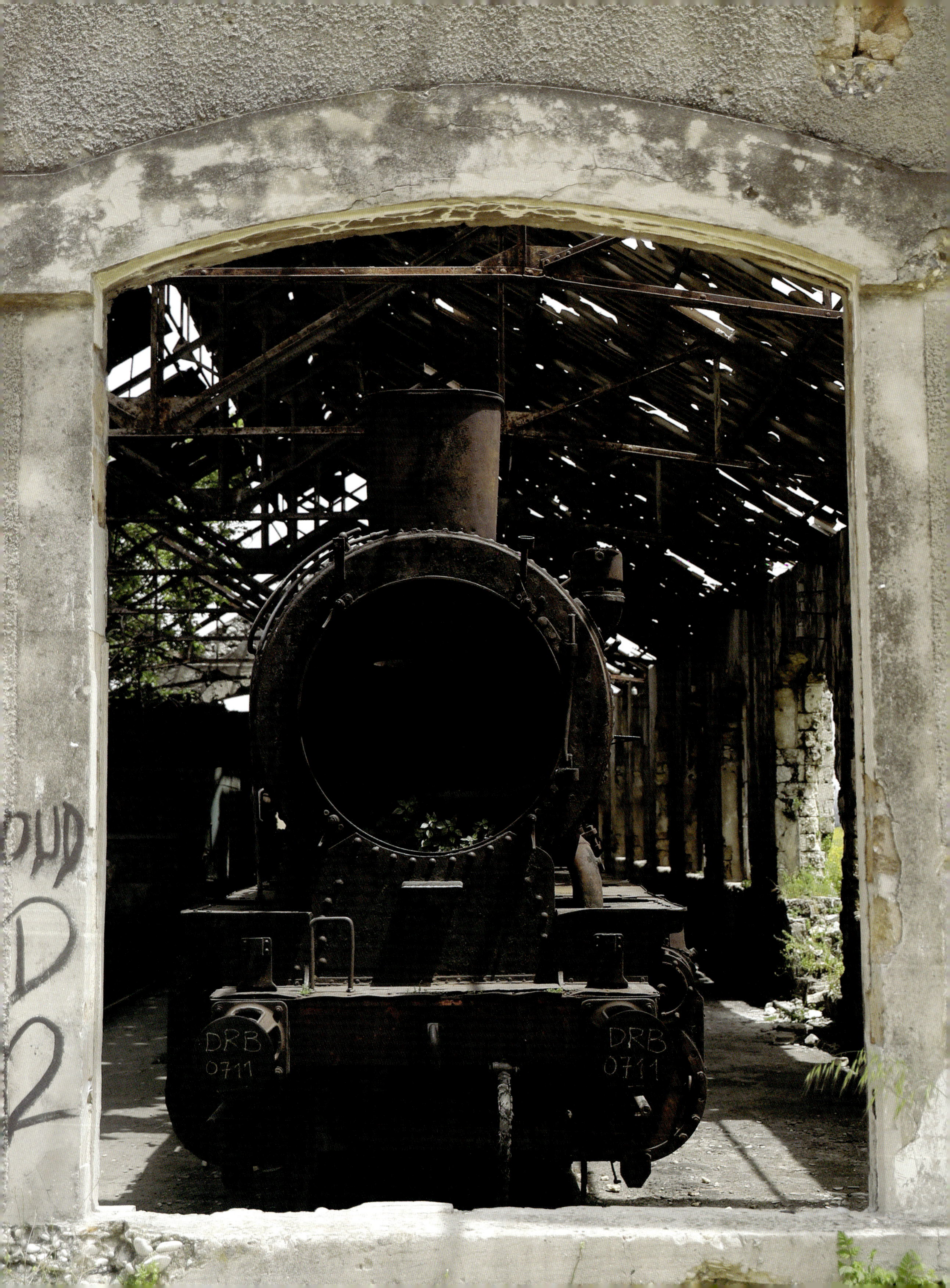
DRB 0711
DRB 0711

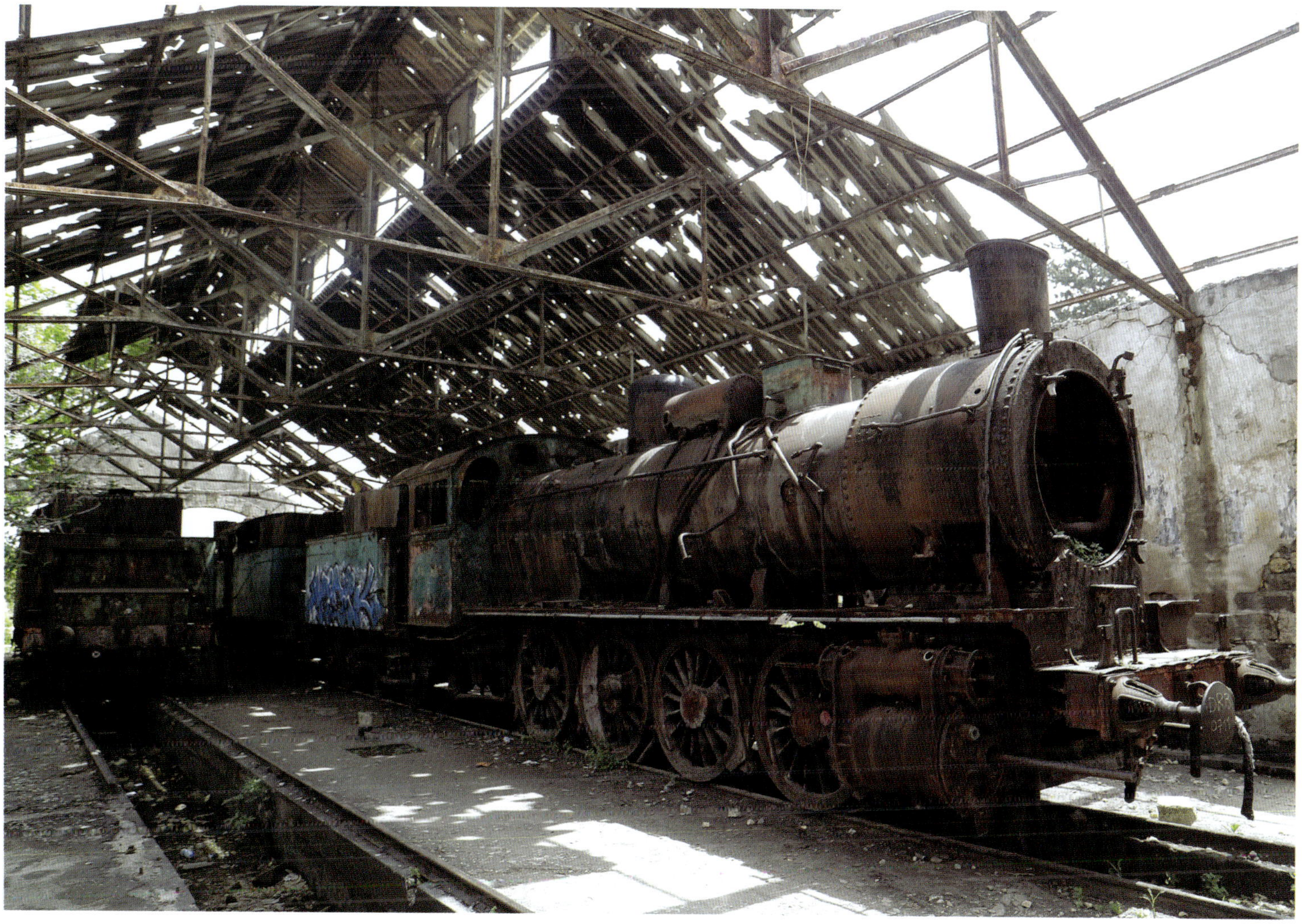

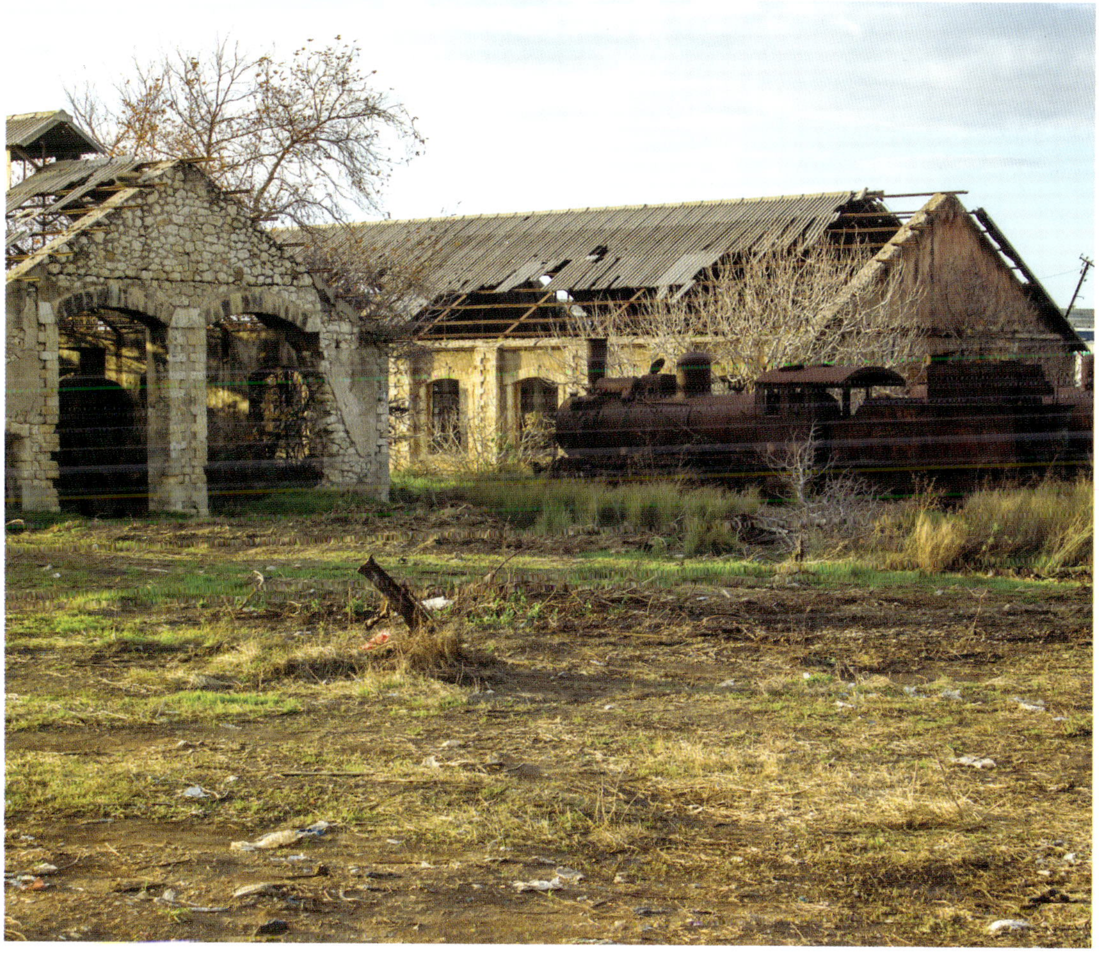

ALL PHOTOGRAPHS:

Locomotive shed, Tripoli, Lebanon

A rugged German-built G7 0-8-0 locomotive, vintage 1895, stands inside the ruins of the locomotive shed at Tripoli station. The lower image shows a G-8 from 1901. Gifted to Syria by the Allies in 1918, they are older than the station, which opened in 1911. Civil war has destroyed much of the buildings. Once a terminus of the Orient Express, Tripoli's old station is due to be converted into a museum to commemorate Lebanon's railway past (it has no railways today). These ruins, evocative of a past still within living memories, are a popular background for wedding photos.

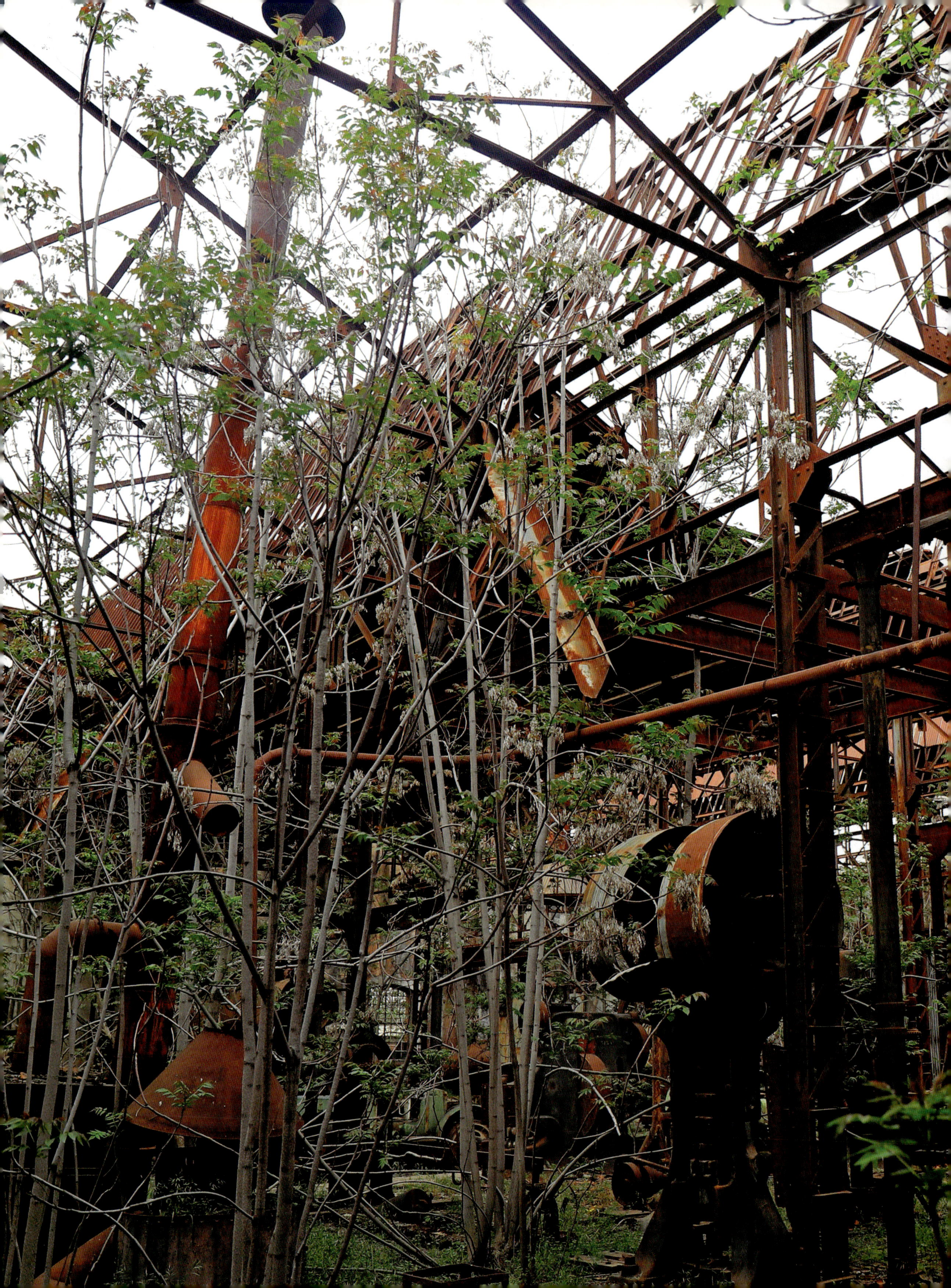

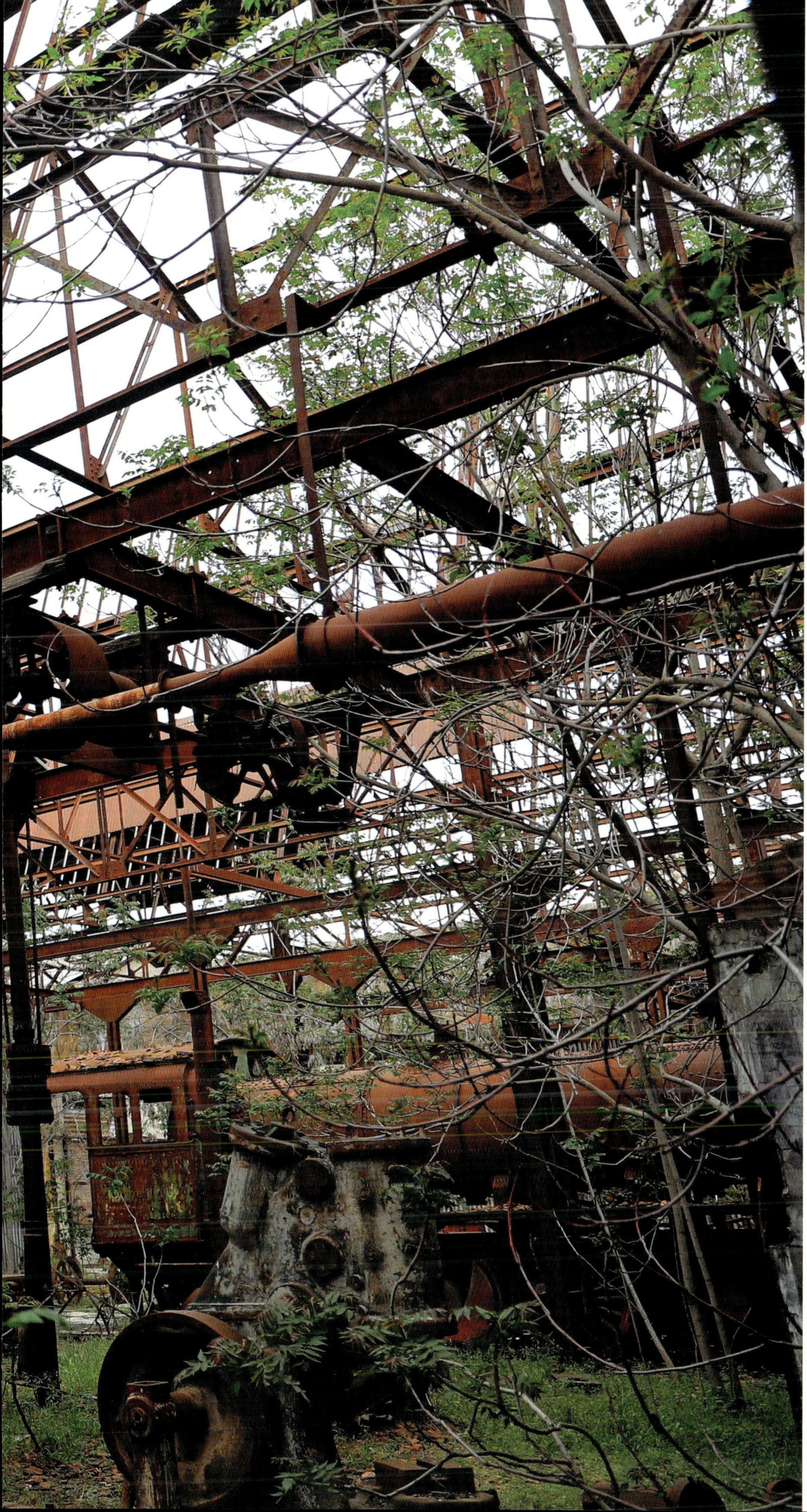

Riyaq Station Locomotive Workshops, Lebanon
The shattered walls reveal the ruins of the wheelwrights' shop in what was once a busy junction with a repair and maintenance depot. At one time Riyaq's railway connections ran through to Syria, Palestine and Iraq.

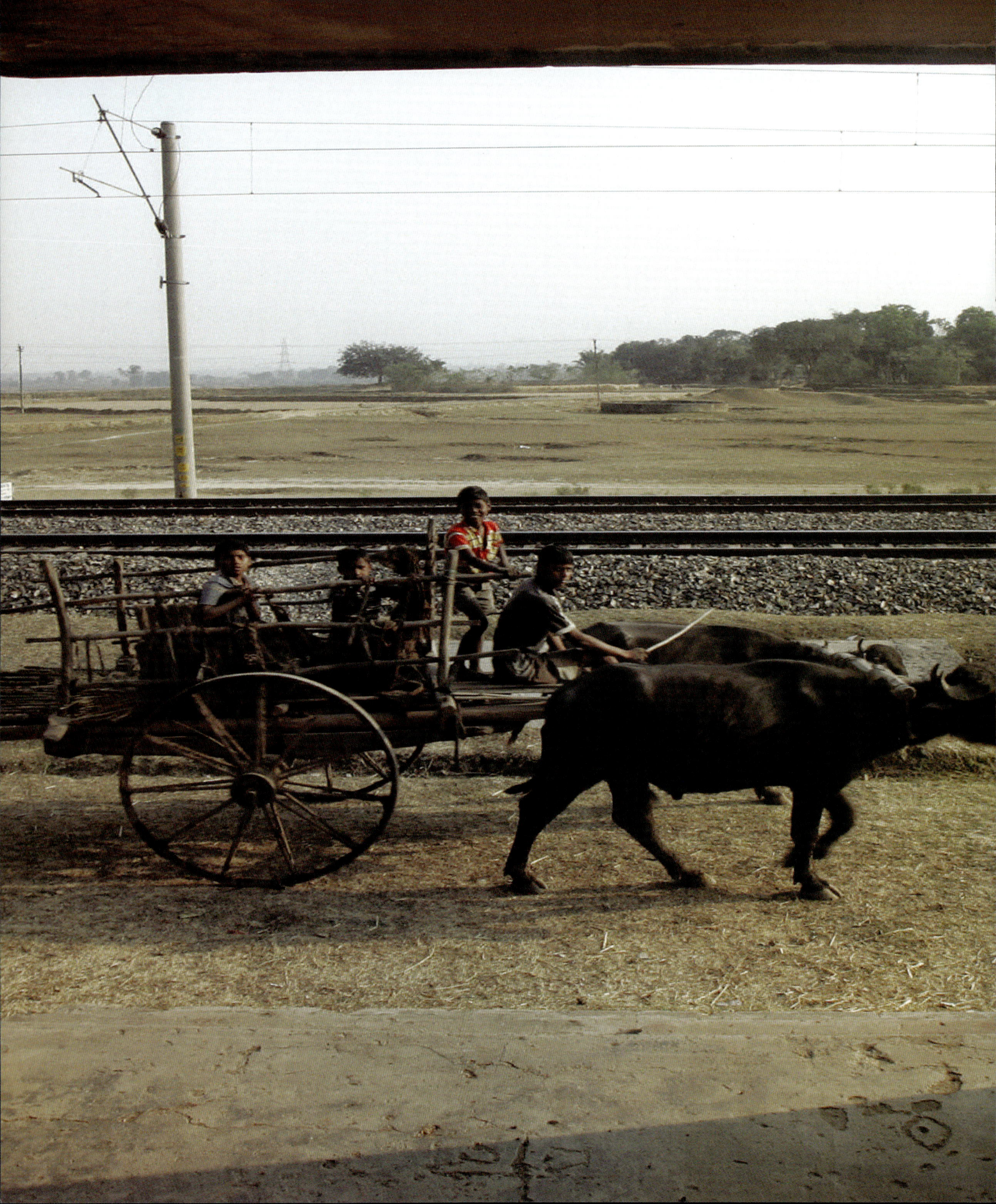

Asia and the Pacific

In the 19th century, railway development in this vast region largely followed British and European practice. Now, railways underpin and sustain the enormous growth of Chinese industrial power, while Japan led the way with high-speed trains and has some of the fastest intercity services in the world. Completion of new high-speed lines has led to the closure of some old cross-country lines along with all their stations. But the pace of modernization has left even some very recent stations unused and abandoned. In many places there is a fascinating juxtaposition of old and new; this is especially true of India, with its vast and long-established railway network.

Some truly historic former stations are shown in this chapter. They include the first station to be opened on the island of Java, and the private station of an Indian prince. Others give insight into the history of one-time coalmining districts of Japan and Australia. South Korea provides a station with literary and cinema associations. There is even an Indian station whose story is abandonment, not by the railway company but by passengers and train crews. Though New Zealand no longer runs regular intercity passenger trains, some of the country's most picturesque branch lines still see heritage services passing by long-abandoned stations.

OPPOSITE:
Begunkodar, West Bengal, India
A traditional form of local transport ambles alongside the electrified track of the South Eastern Railway's Purulia–Muri line at Begunkodar. The station here is a relatively recent one, having been opened in 1960 to serve villages in the neighbourhood. But it has an unusual history, said to have been 'abandoned' not by the railway company but by railway staff and passengers.

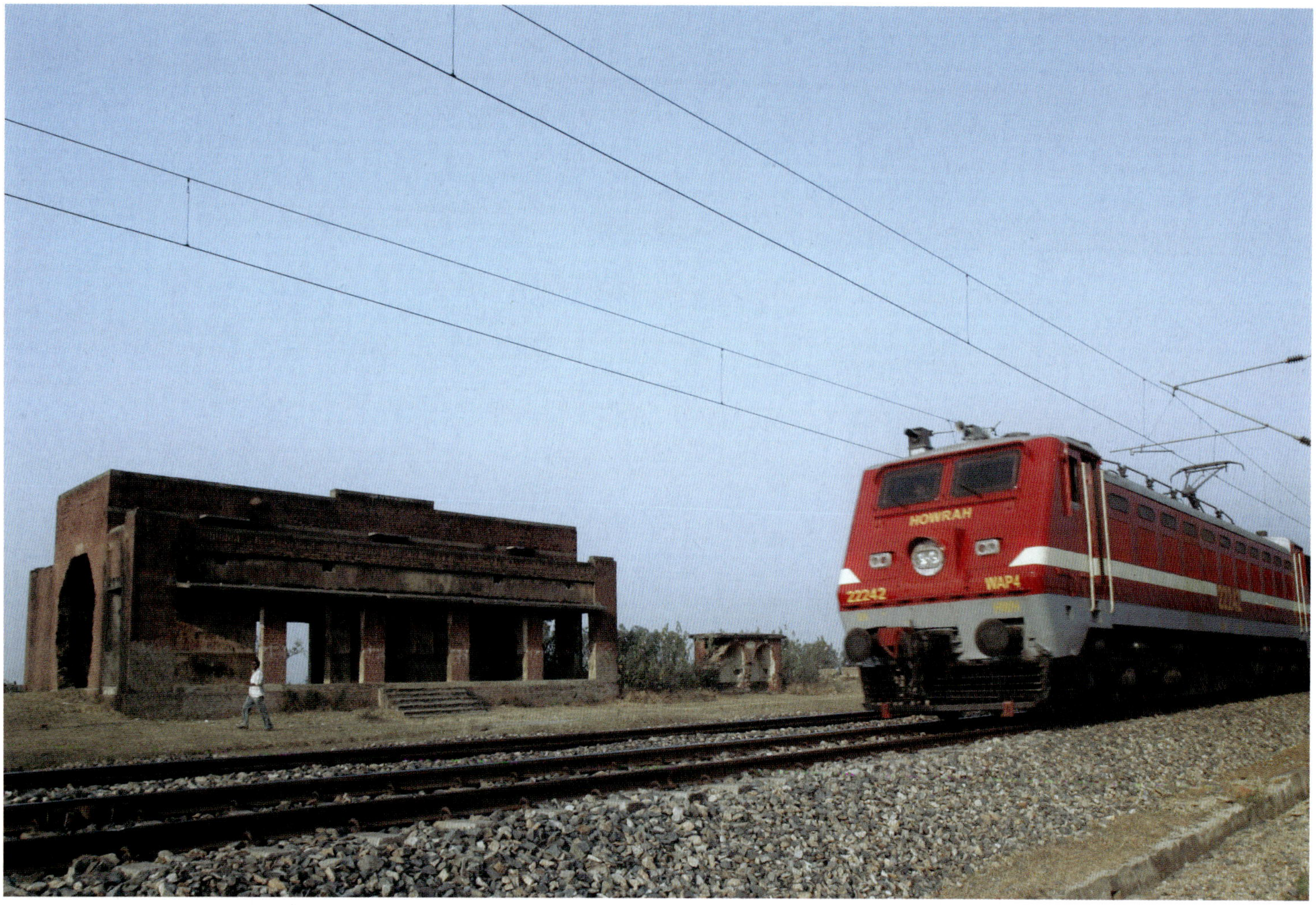

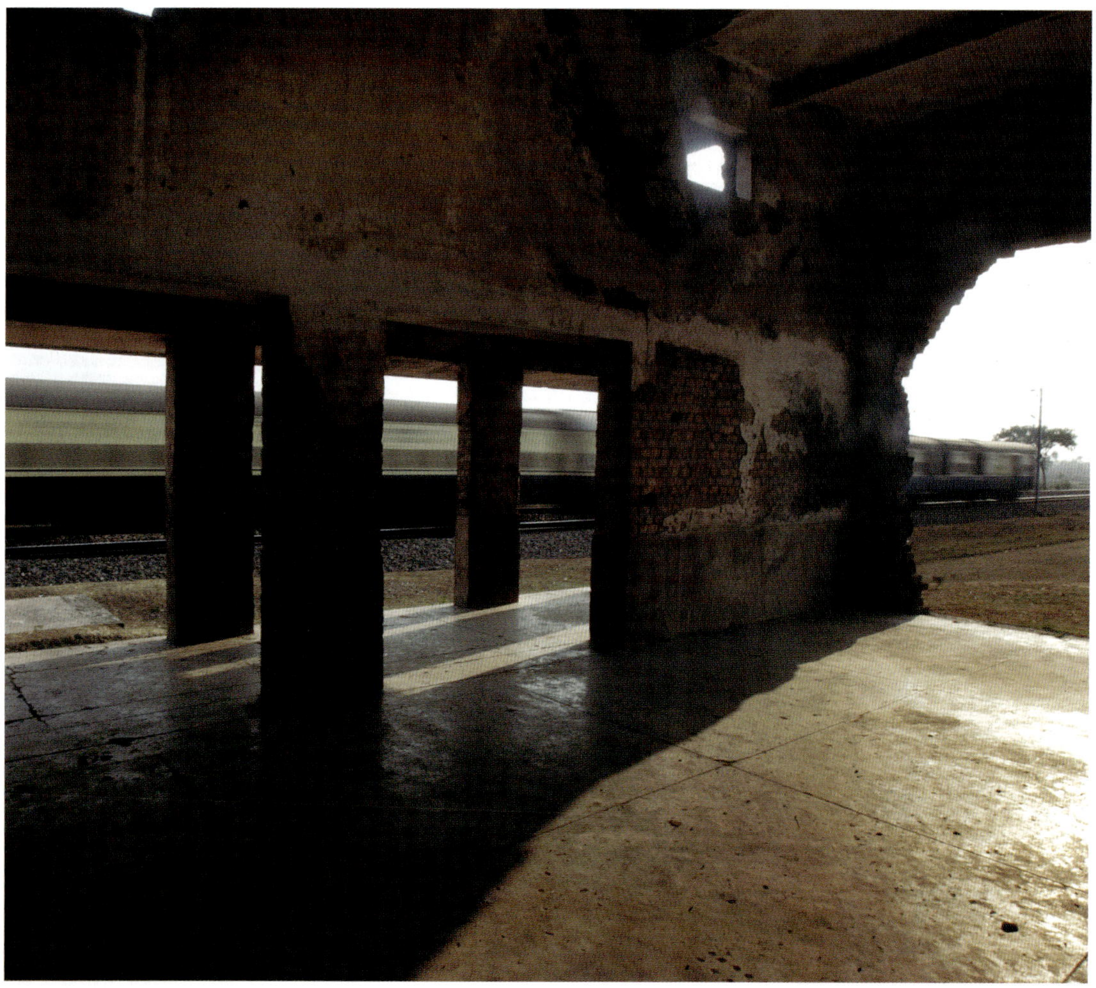

ABOVE & LEFT:

Begunkodar, West Bengal, India

Never formally closed, this wayside station has acquired a reputation for being haunted and being 'deserted' by passengers and railway staff. A stationmaster and his wife were allegedly murdered here in the 1960s. More probable explanations for its reputation centre on its remoteness from local communities, and on possible local efforts to generate 'ghost tourism'. Despite rumours to the contrary, trains stop here and the station is in regular use.

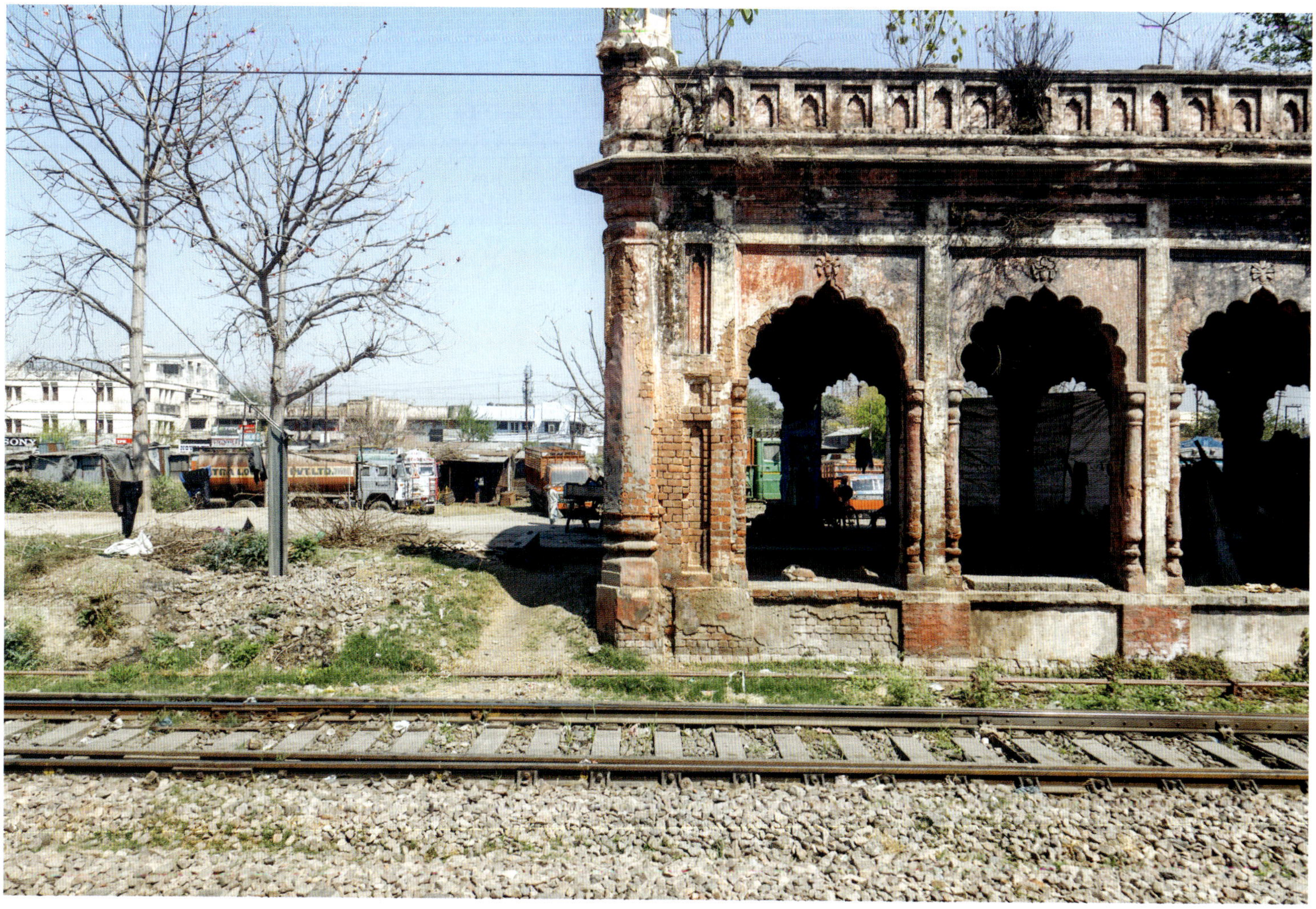

ABOVE:
Rampur, Uttar Pradesh, India
Ornate but now tree-grown and dilapidated, this open-sided building at Rampur Junction was the private station of the Nawab of Rampur. With a single broad platform, it was also where his train's carriages were stabled. Following the death of Nawab Raza Ali Khan in 1966, a protracted legal battle over inheritance rights has meant that the building and carriages have been left for decades in deteriorating conditions.

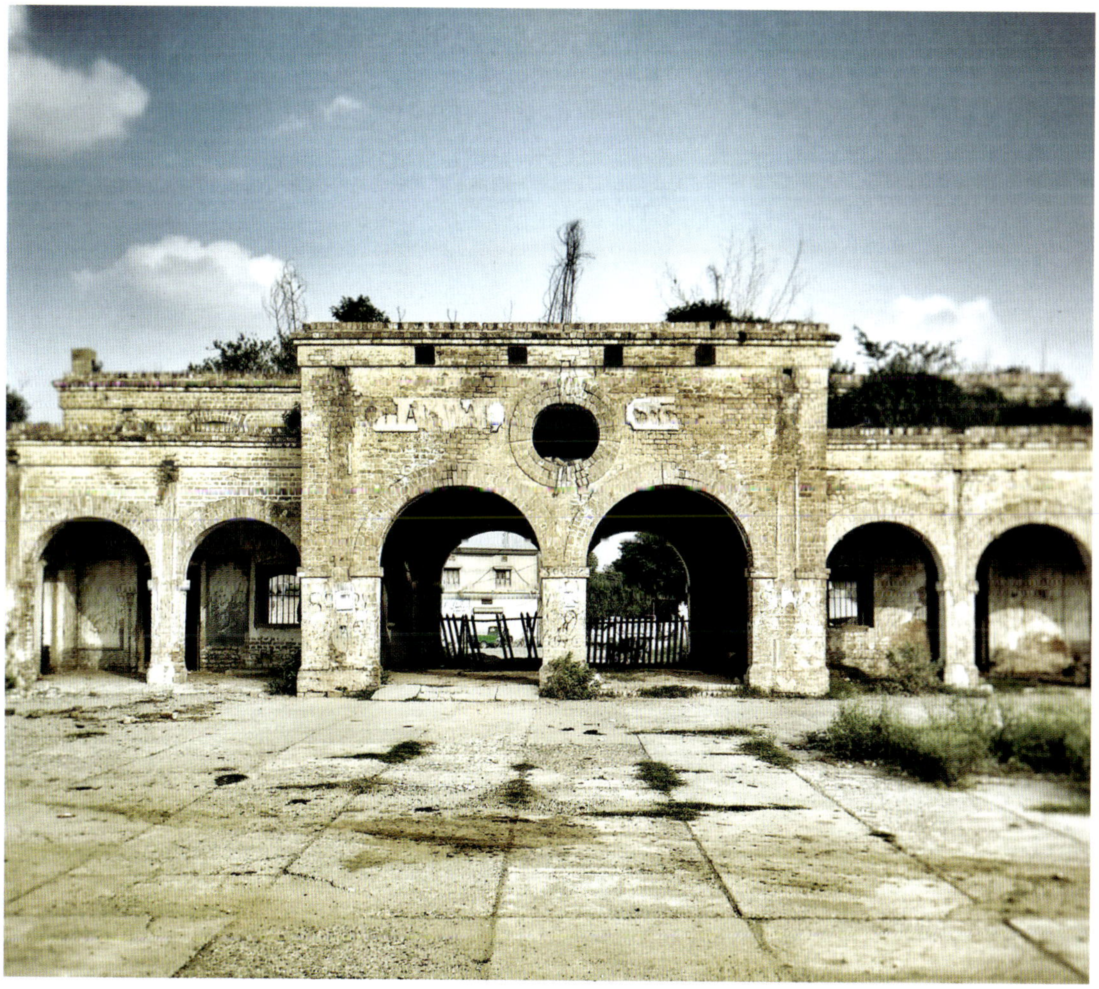

RIGHT:
Bhoun Station, Chakwal, Potohar District, Pakistan
The 75km (47 miles) railway from Mandra to Bhoun, completed in 1916, was closed down in 1993 and subsequently dismantled. The imposing terminal building, 12km (7 miles) south of Chakwal city, remains stripped of all fittings, in an increasingly ruinous state. With the 're-discovery' of the social and economic value of railways, this line, like many others, has been proposed for reinstatement.

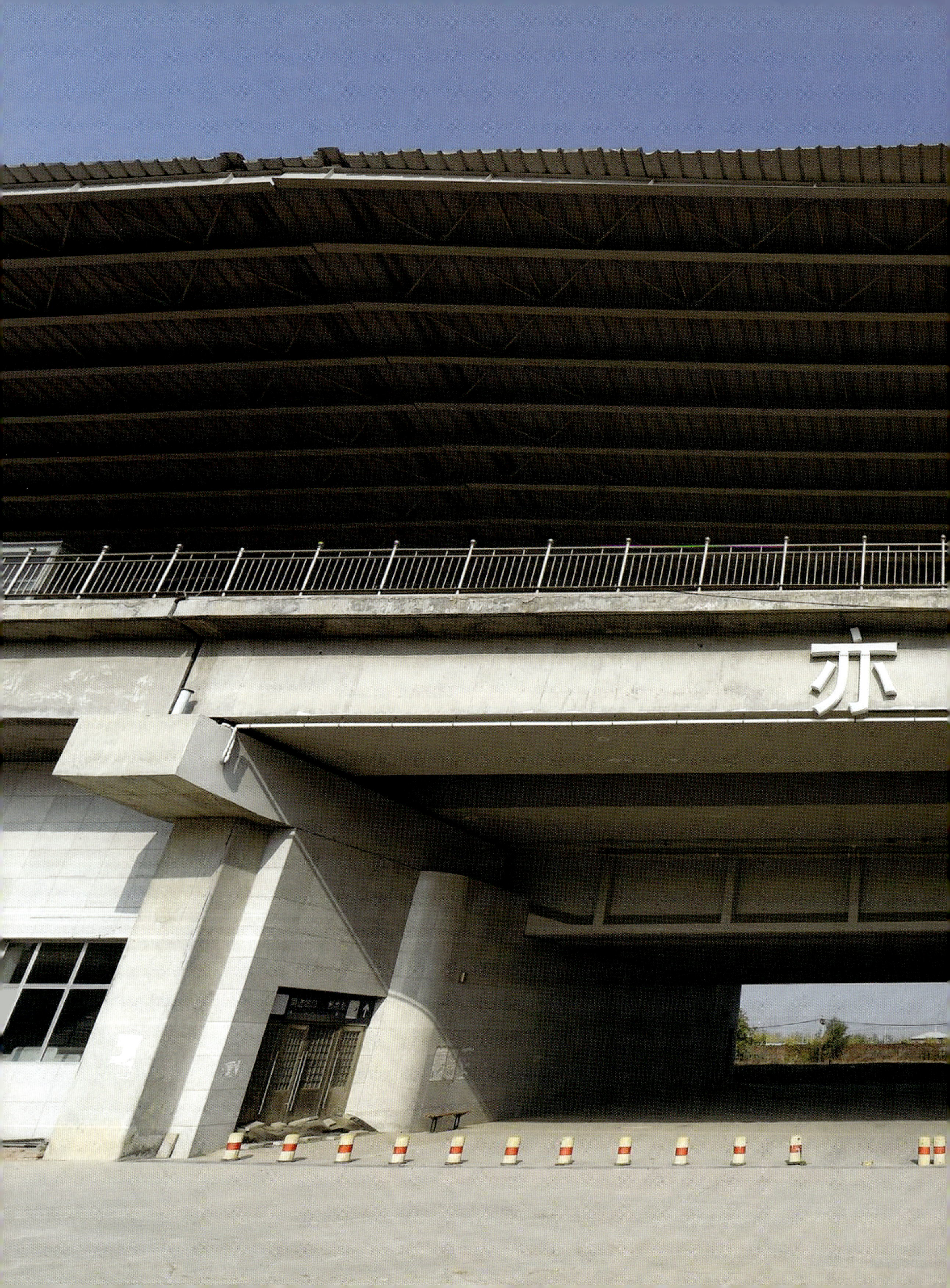

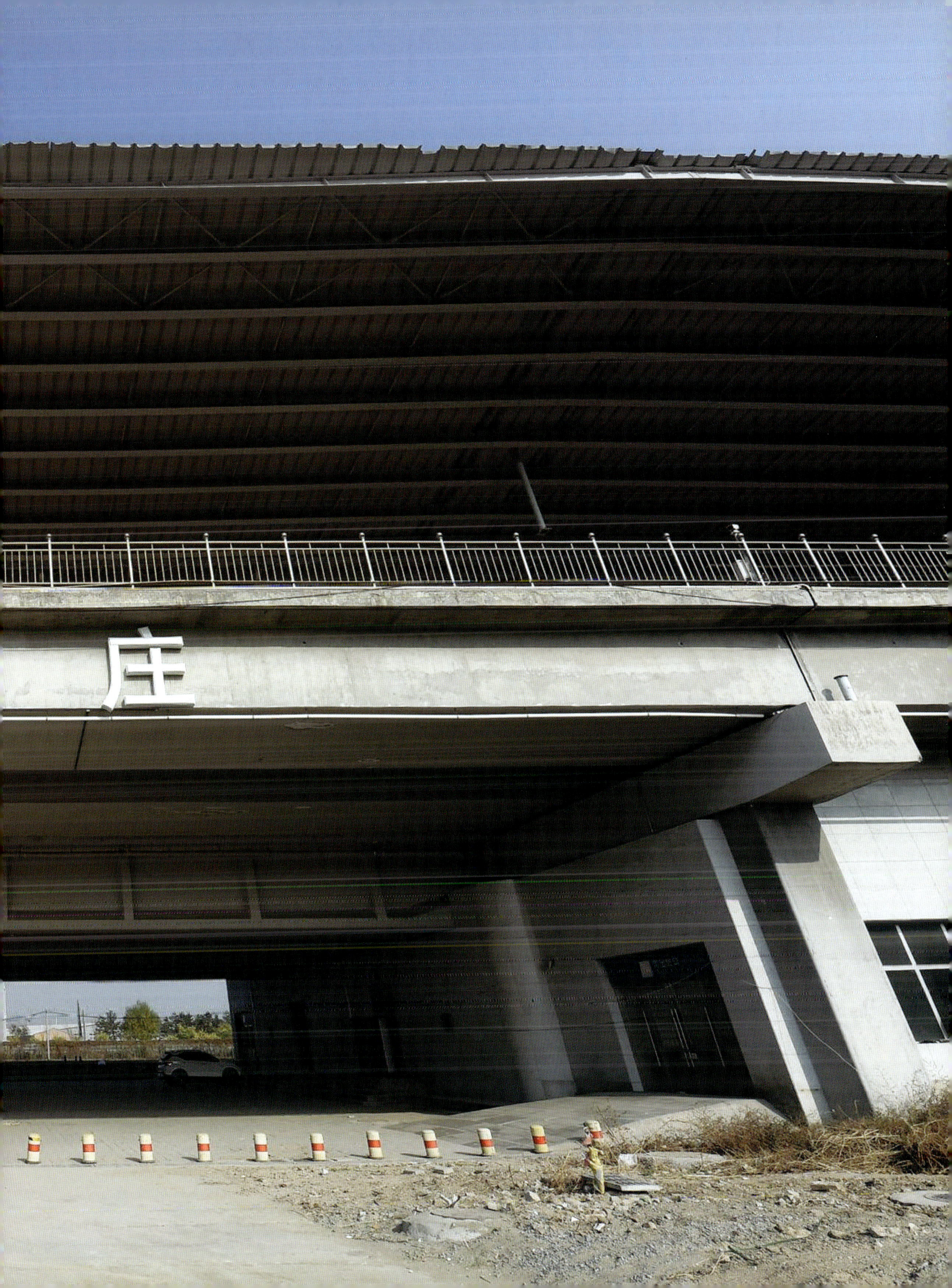
庄

PREVIOUS PAGES, RIGHT, BELOW & OPPOSITE BELOW:

Yizhuang Station, Tongzhou District, Beijing, China

Although Beijing's Yizhuang subway line, south-east of the city, opened in 2010 – mostly overground – its terminus, intended to form an interchange with the Beijing–Tianjin intercity railway, remained unopened in early 2022. The station structure appears complete, with interior signage already installed.

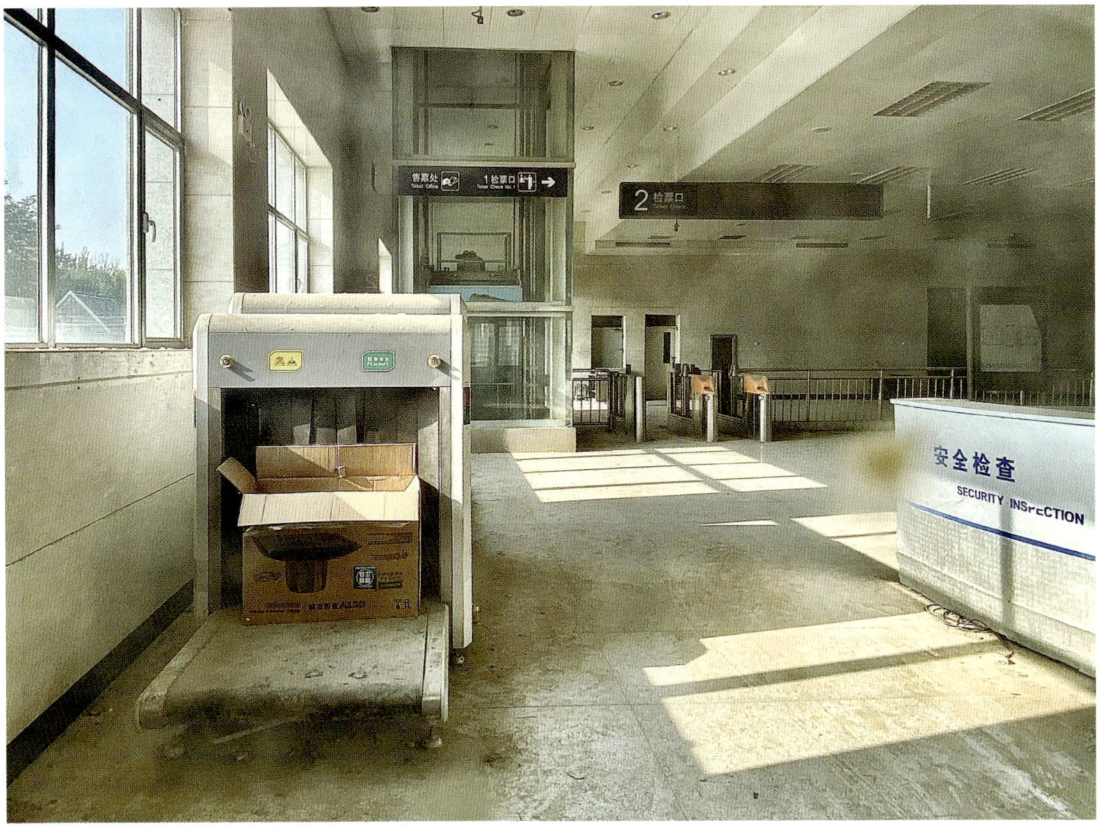

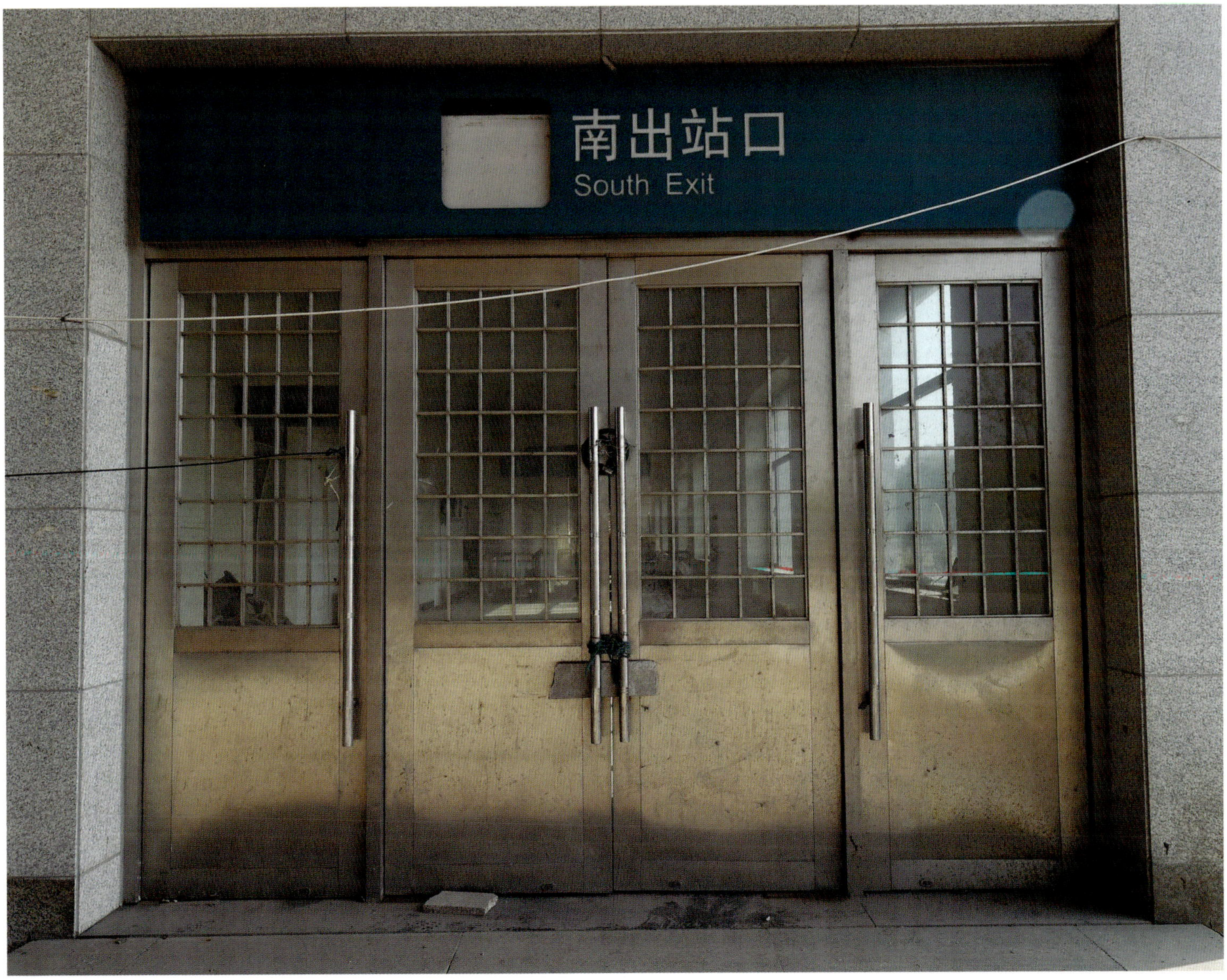

RIGHT:

Tianjin West, China

Tianjin is China's third-largest city, located about 130km (81 miles) south of Beijing, with around 14 million inhabitants. The original Tianjin West station was built between 1908 and 1910, designed in the style of a French chateau, with mansard roofs, an ornamental main stairway and a cupola with a clock. Today it is replaced by the international-style 24-platform new Tianjin West. Both stations were designed by German architects.

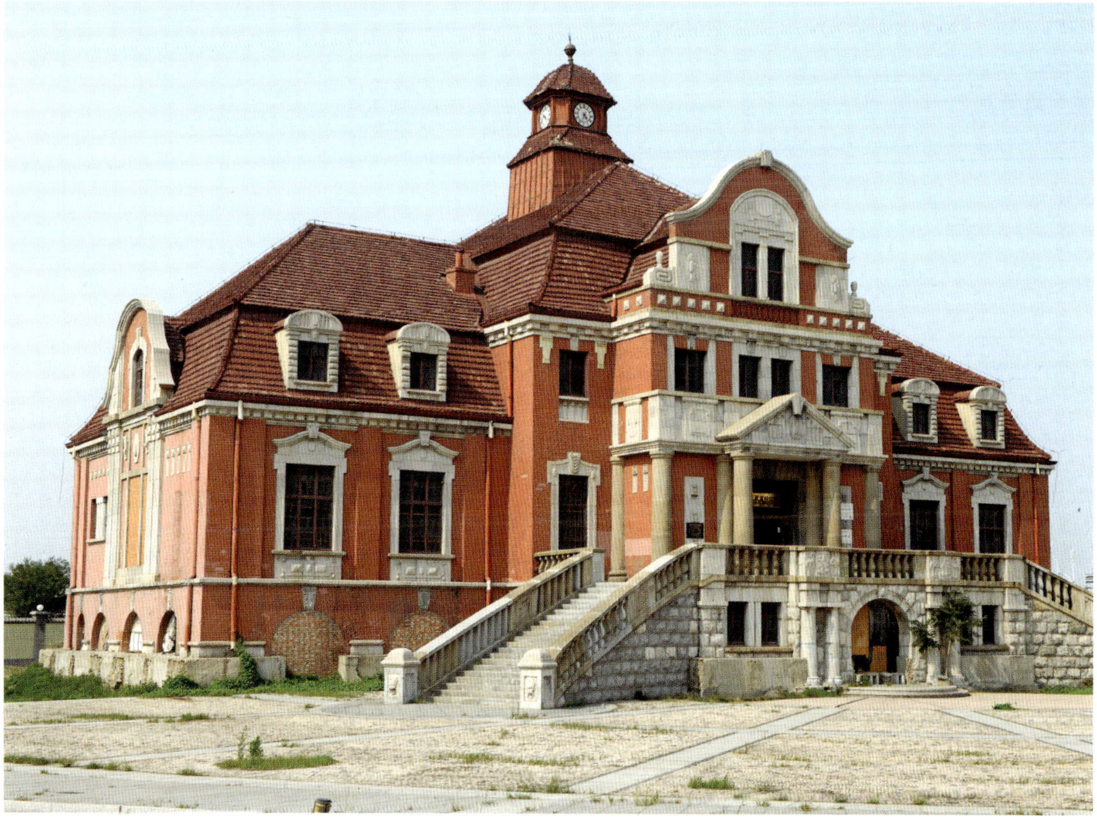

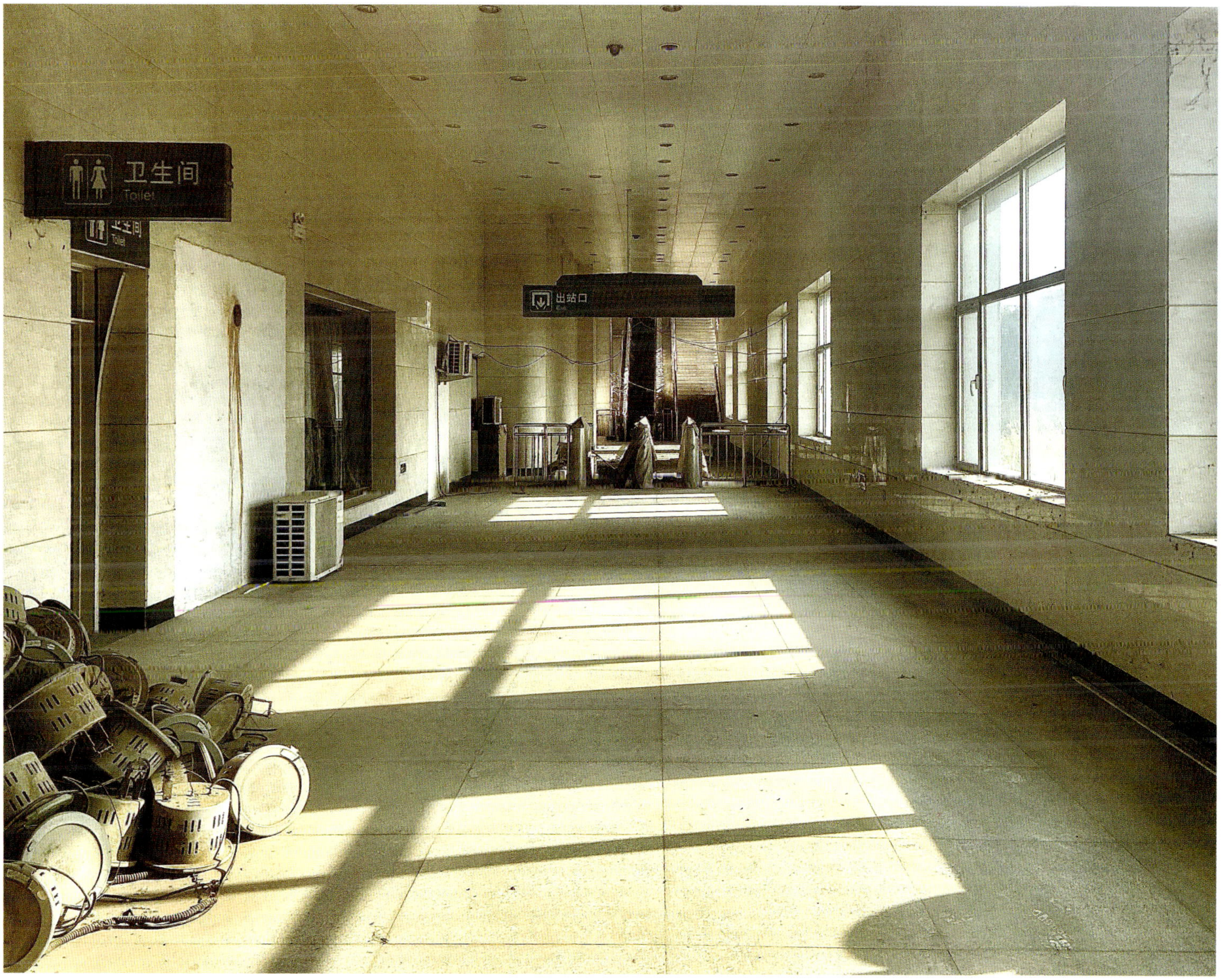

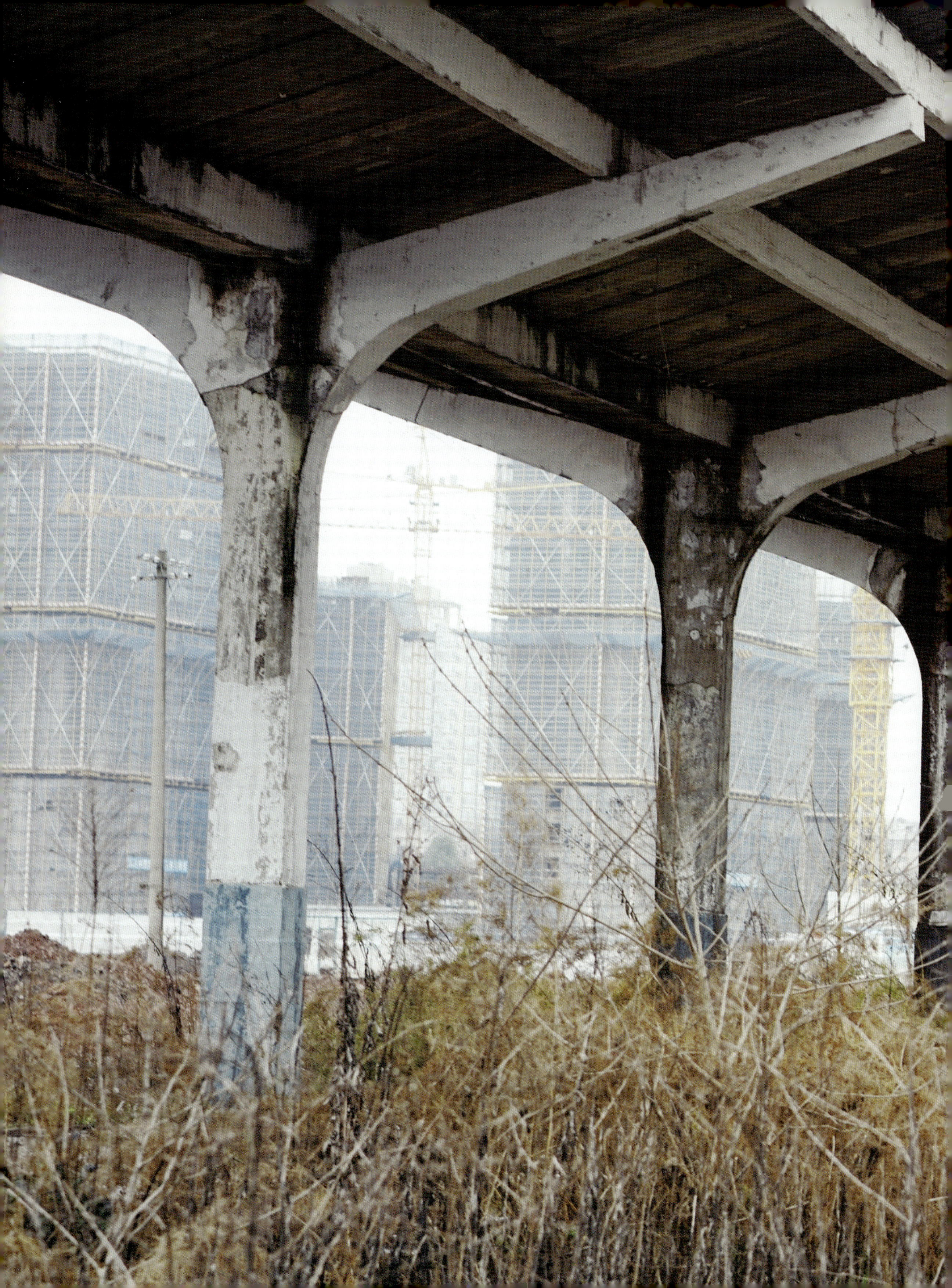

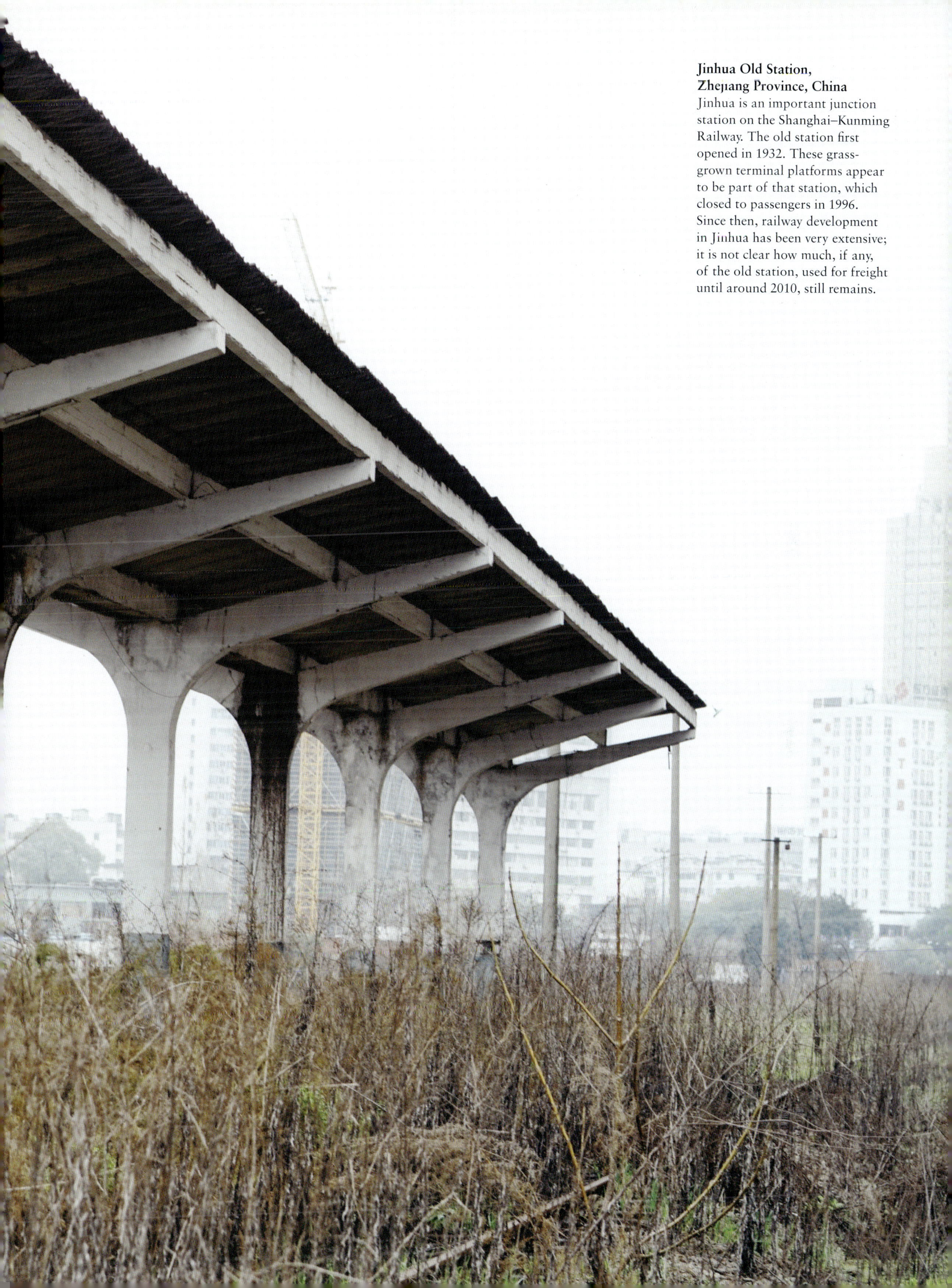

Jinhua Old Station, Zhejiang Province, China
Jinhua is an important junction station on the Shanghai–Kunming Railway. The old station first opened in 1932. These grass-grown terminal platforms appear to be part of that station, which closed to passengers in 1996. Since then, railway development in Jinhua has been very extensive; it is not clear how much, if any, of the old station, used for freight until around 2010, still remains.

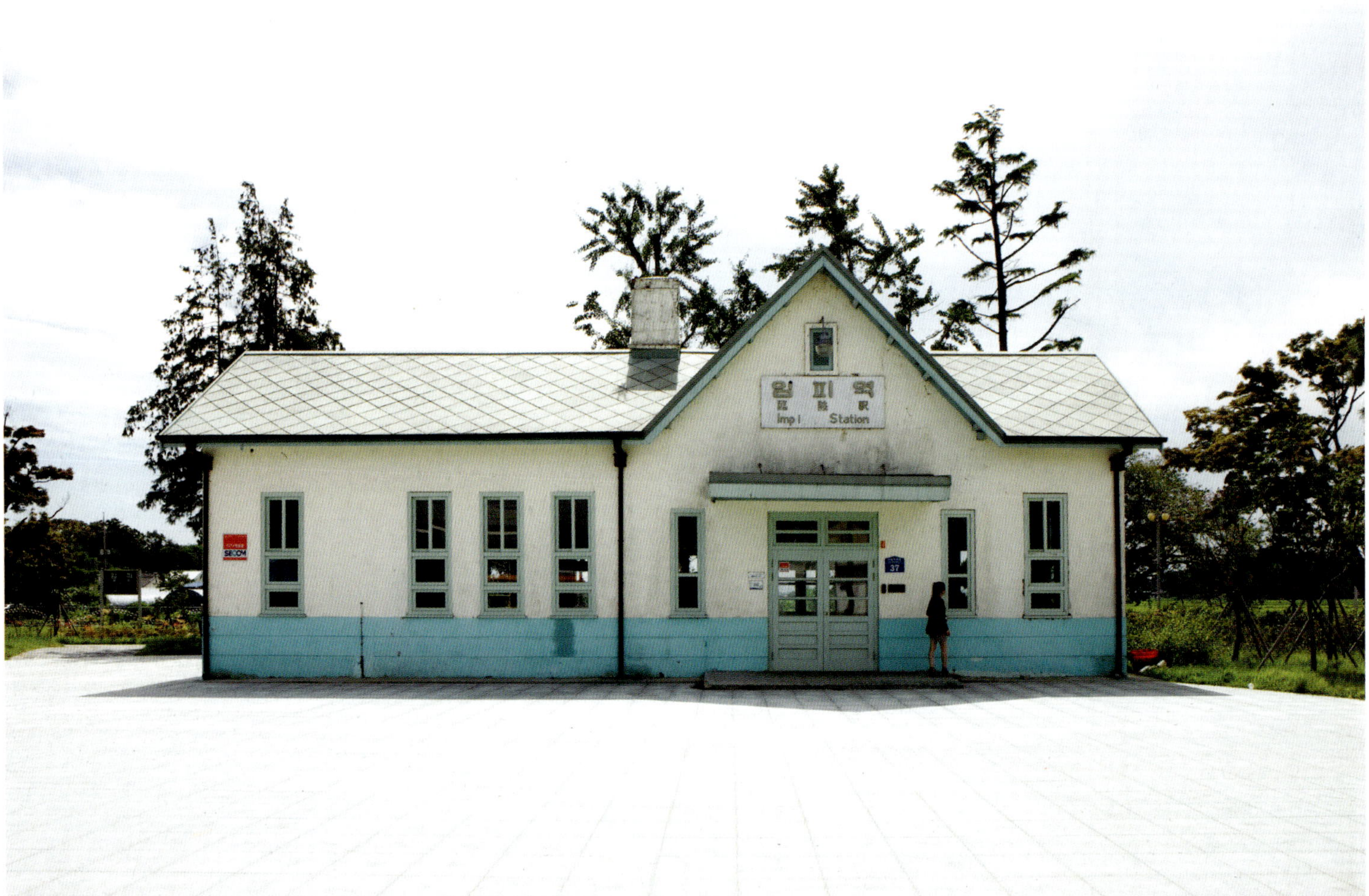
임피역
Impi Station

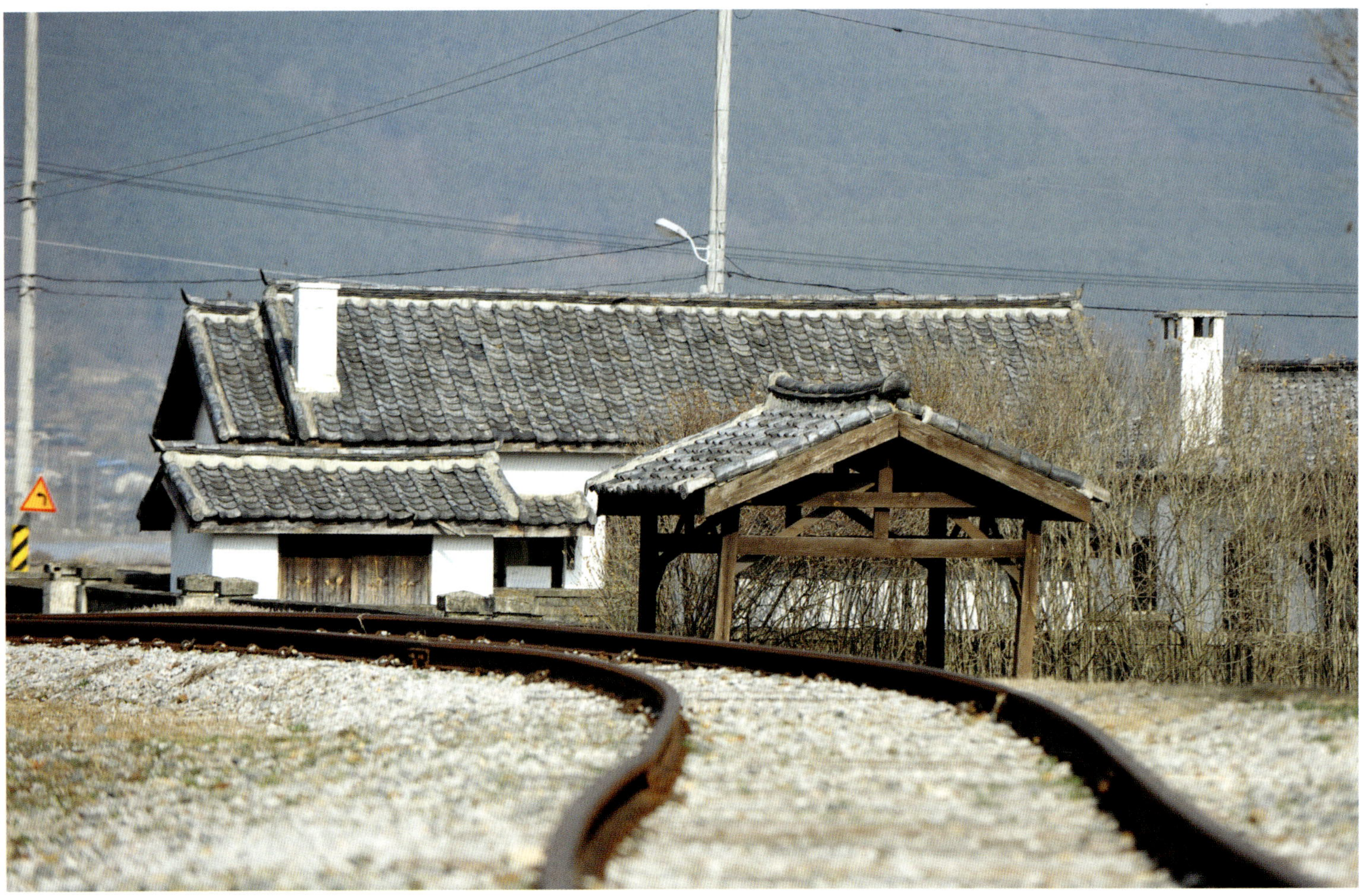

OPPOSITE ABOVE:

Impi Station, Gunsan-si Jeollabuk-do Province, South Korea

On the original Gunsan line, the station opened in 1912 to link the harbour of Gunsan with main railways in the interior of the country. In 2008, a new line was completed by South Korean Railways and the old line, with its attractive wayside stations like Impi, was closed. The building is restored, with waxwork-style figures inside, to show how an old-style village station functioned.

OPPOSITE BELOW:

Seodo Old Station, Namwon, Jeollabuk-do Province, South Korea

The traditional-style tiled and whitewashed building was constructed in 1932, and is well-known for featuring in the famous novel *Honbul* (*The Ghost Fire*) by Choi Myunghee, and the Netflix film *Mr Sunshine*. The station closed in 2002 and was replaced by a new Seodo Station, when the railways of the district were reconstructed.

BELOW & OVERLEAF:

Tomatsu, Hokkaido, Japan

No rails remain at Tomatsu Station but it does not lack visitors. Built in 1929 and closed in 1987 when the Horonai line was shut down, it has been reinvented by local volunteers as a resource centre for the history of railways and coal mines in the fomerly industrial Sorachi region. The barn-type roof, shaped like a Japanese shogi tile, is distinctive.

Inside, the ticket window is framed by memorabilia of bygone industry and community life: workers' groups, family photos, landscapes and advertisements for events. The fluoresent light tubes and bare decor all make clear that this is very much a local initiative.

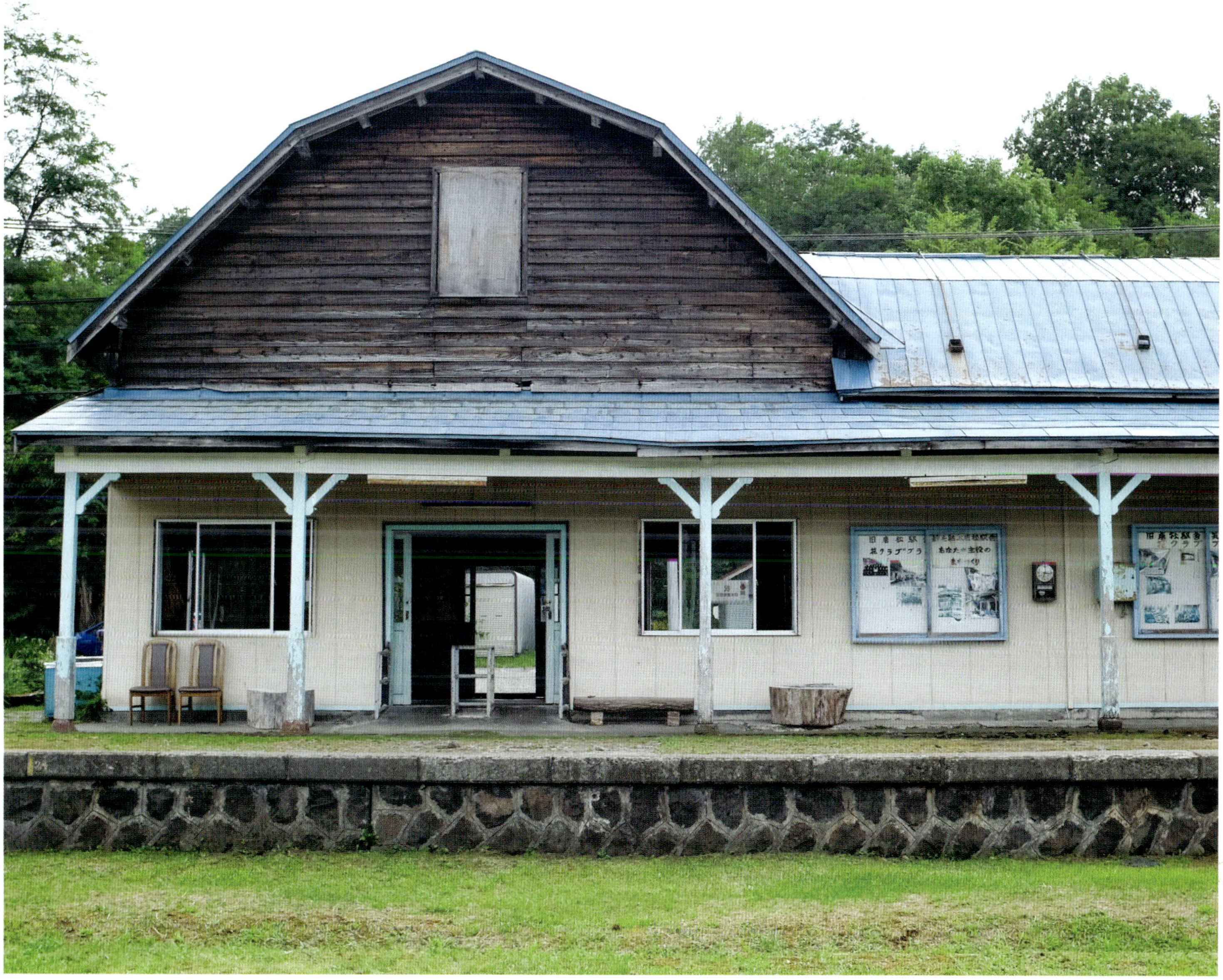

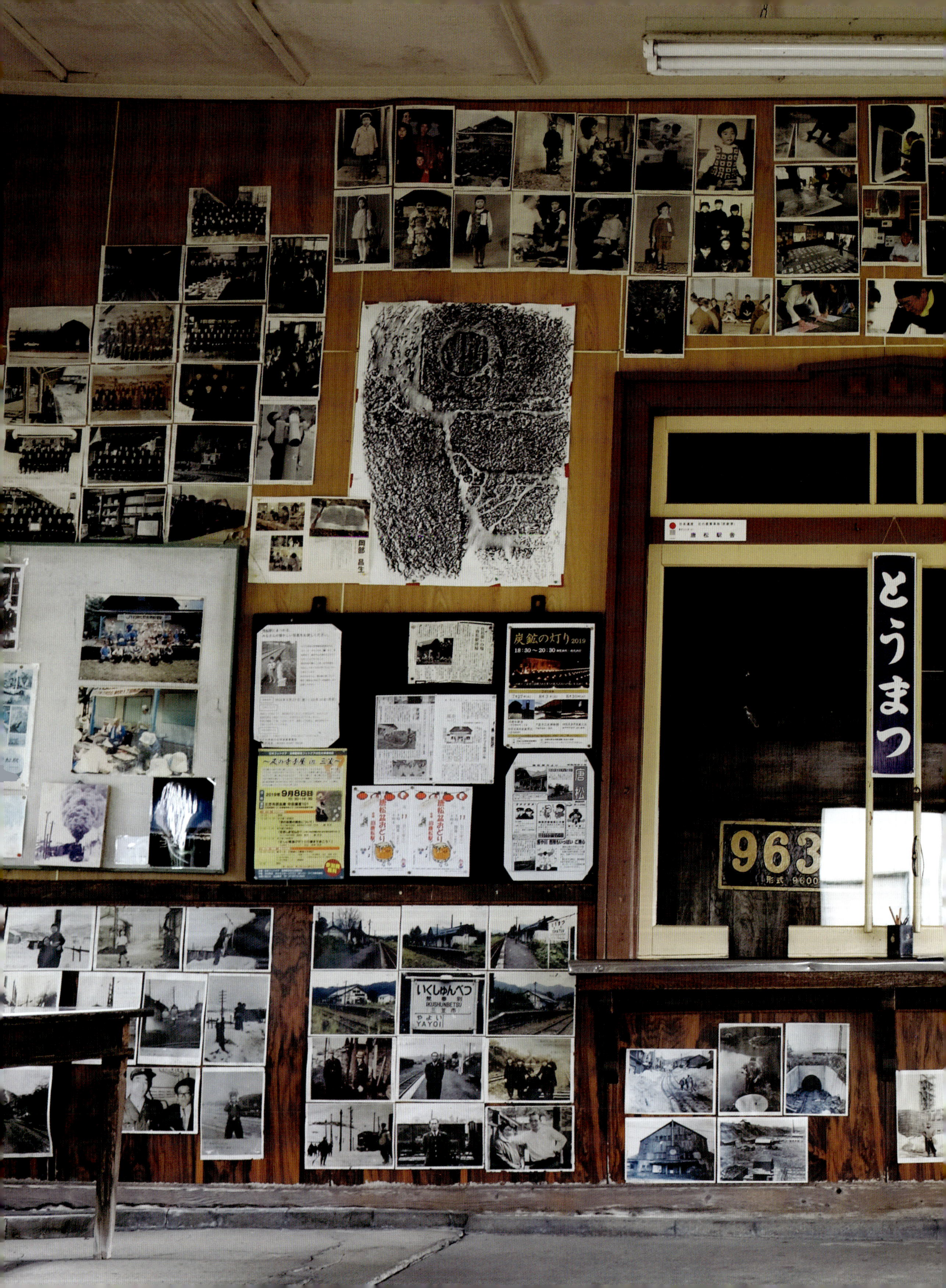

炭鉱の灯り 2019
唐松
とうまつ
963
形式 9600
いくしゅんべつ
IKUSHUNBETSU
やよい
YAYOI

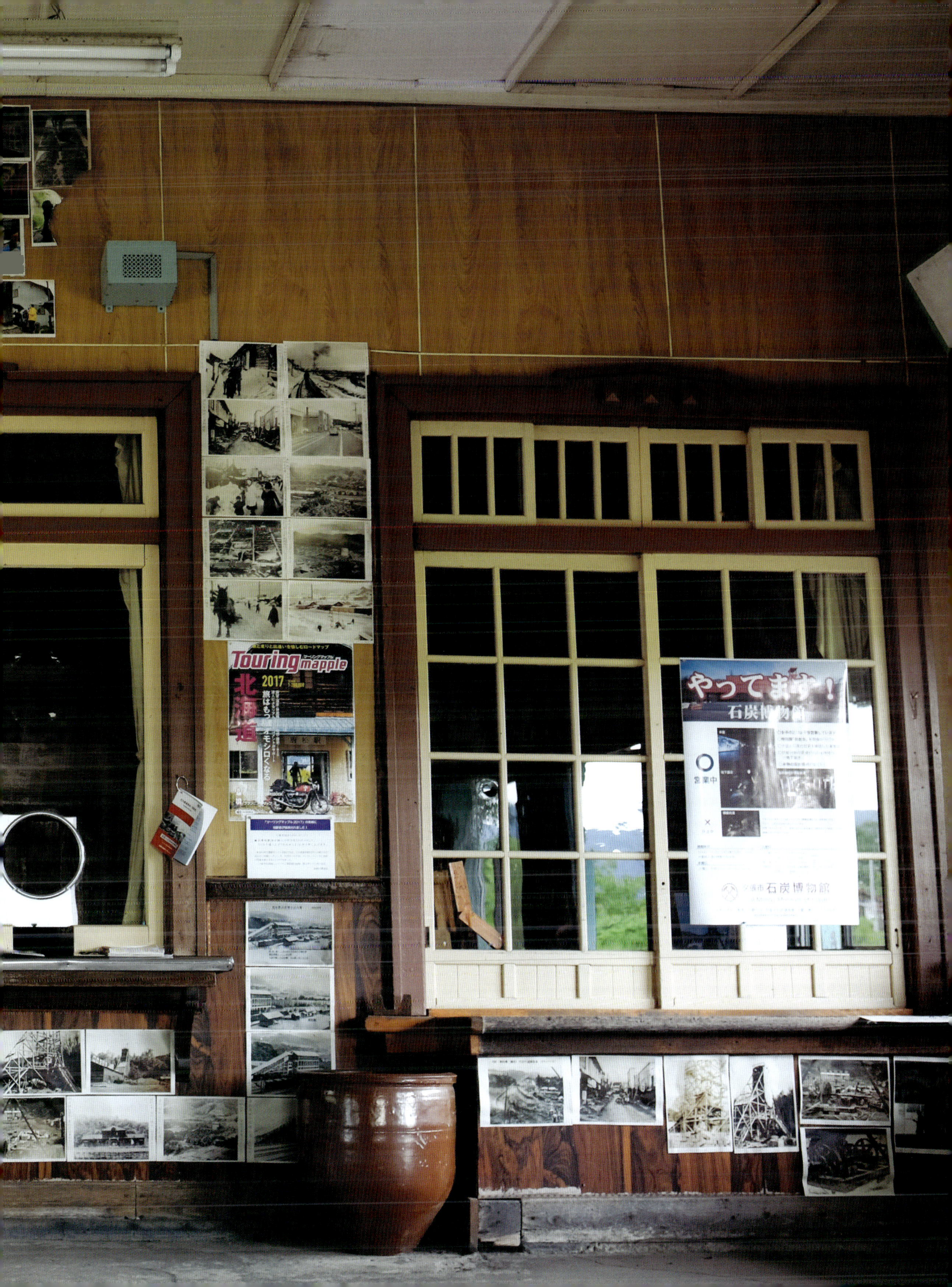
Touring mapple
2017
北海道
石炭博物館

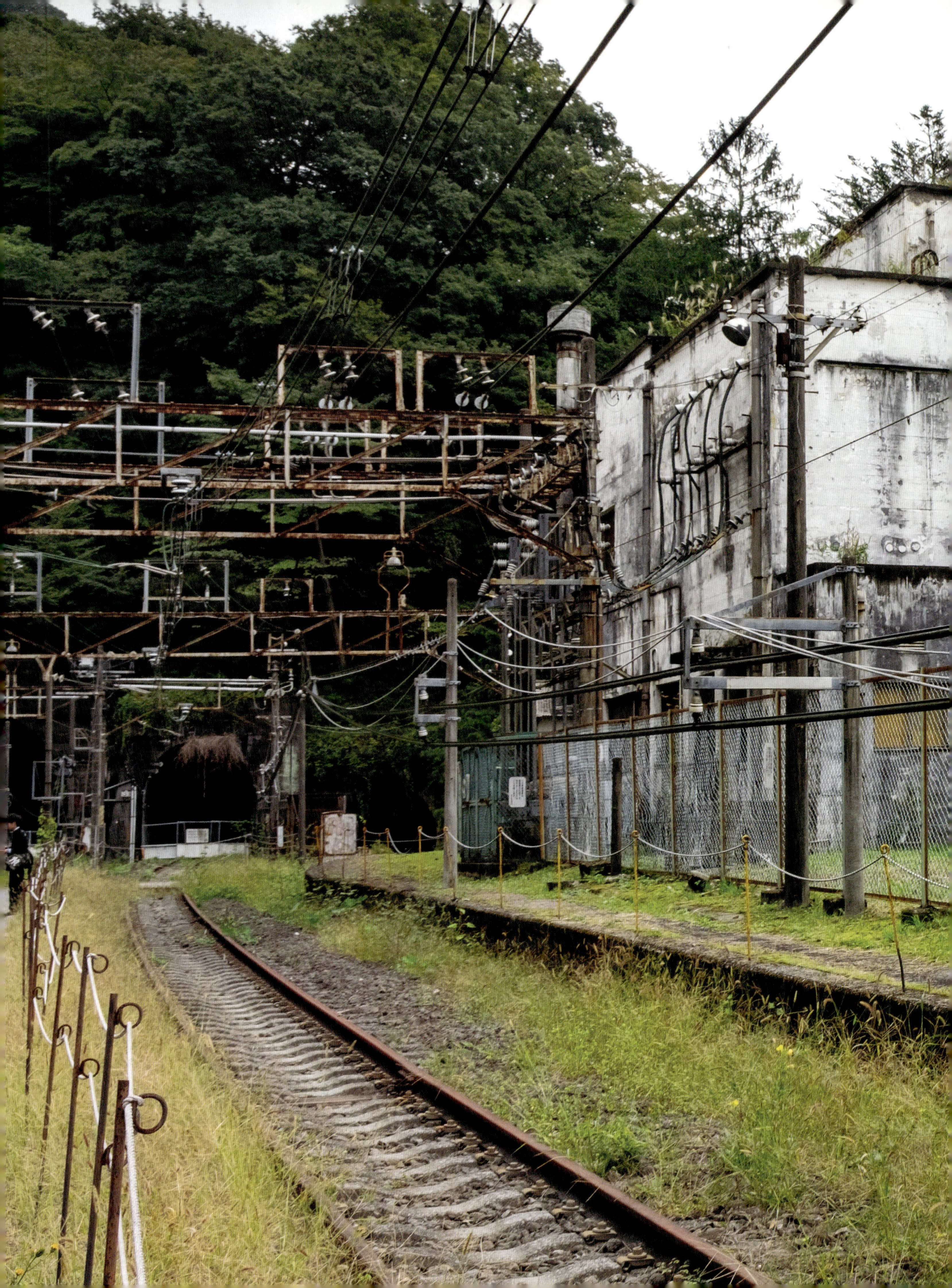

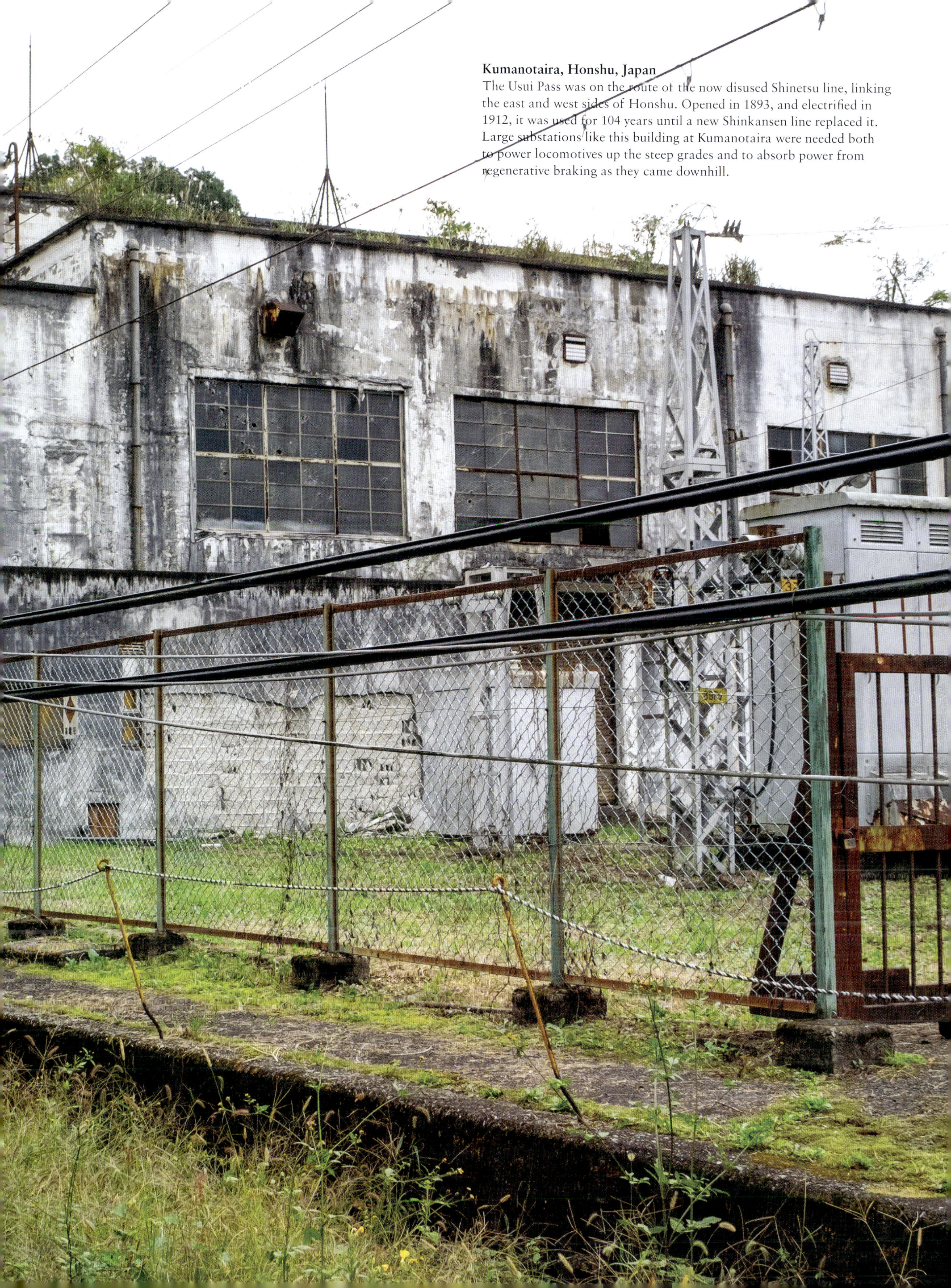

Kumanotaira, Honshu, Japan
The Usui Pass was on the route of the now disused Shinetsu line, linking the east and west sides of Honshu. Opened in 1893, and electrified in 1912, it was used for 104 years until a new Shinkansen line replaced it. Large substations like this building at Kumanotaira were needed both to power locomotives up the steep grades and to absorb power from regenerative braking as they came downhill.

Testudo Ogami station, Itsukaichi Railway, Tokyo, Japan

In Tokyo's Akishima City area a mounted carriage bogie denotes the site of a former suburban station on the Itsukaichi Railway. Opened in 1925, it closed in 1944 when trains were routed over a different section of line.

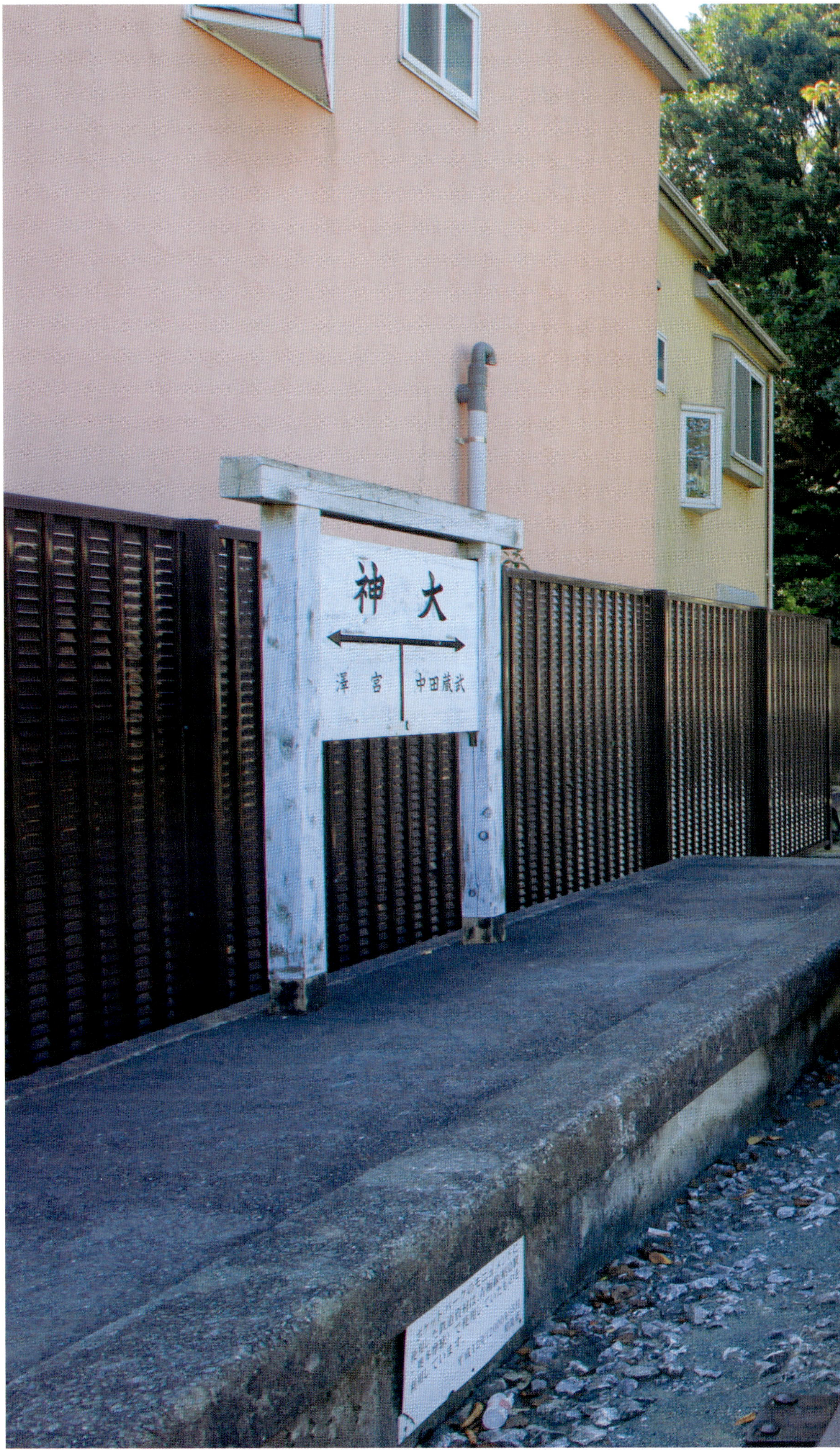

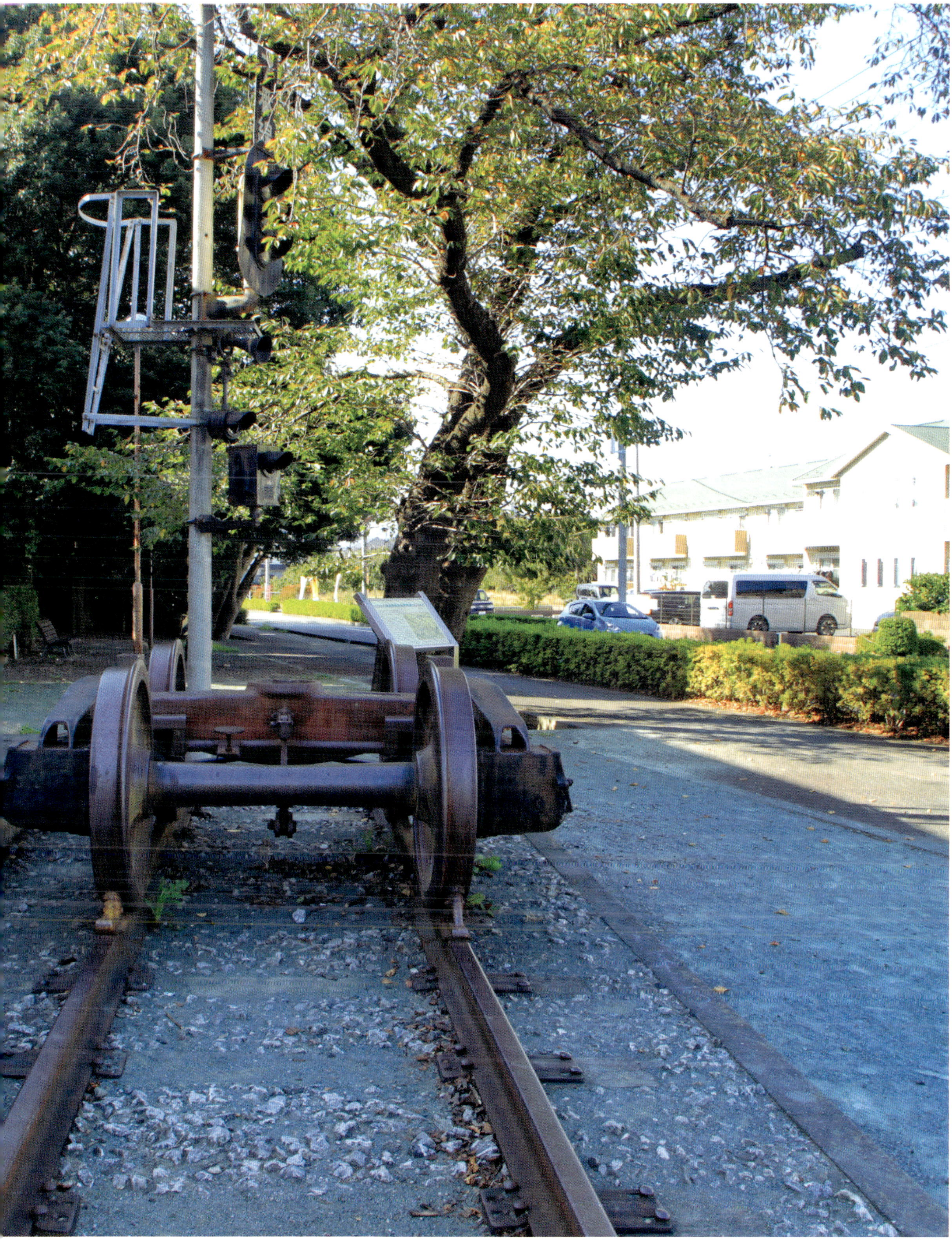

ALL PHOTOGRAPHS:

Gua Musang, Kelantan, Malaysia

Shaded by mango trees and banana palms, signage and buildings from around the old station at Gua Musang, a junction and crossing-point on Malaysia's East Coast line in dramatic limestone country. The old station is boarded up, but the yard sees some freight usage. The new station is 500m (547yds) to the south, part of a substantial urban development.

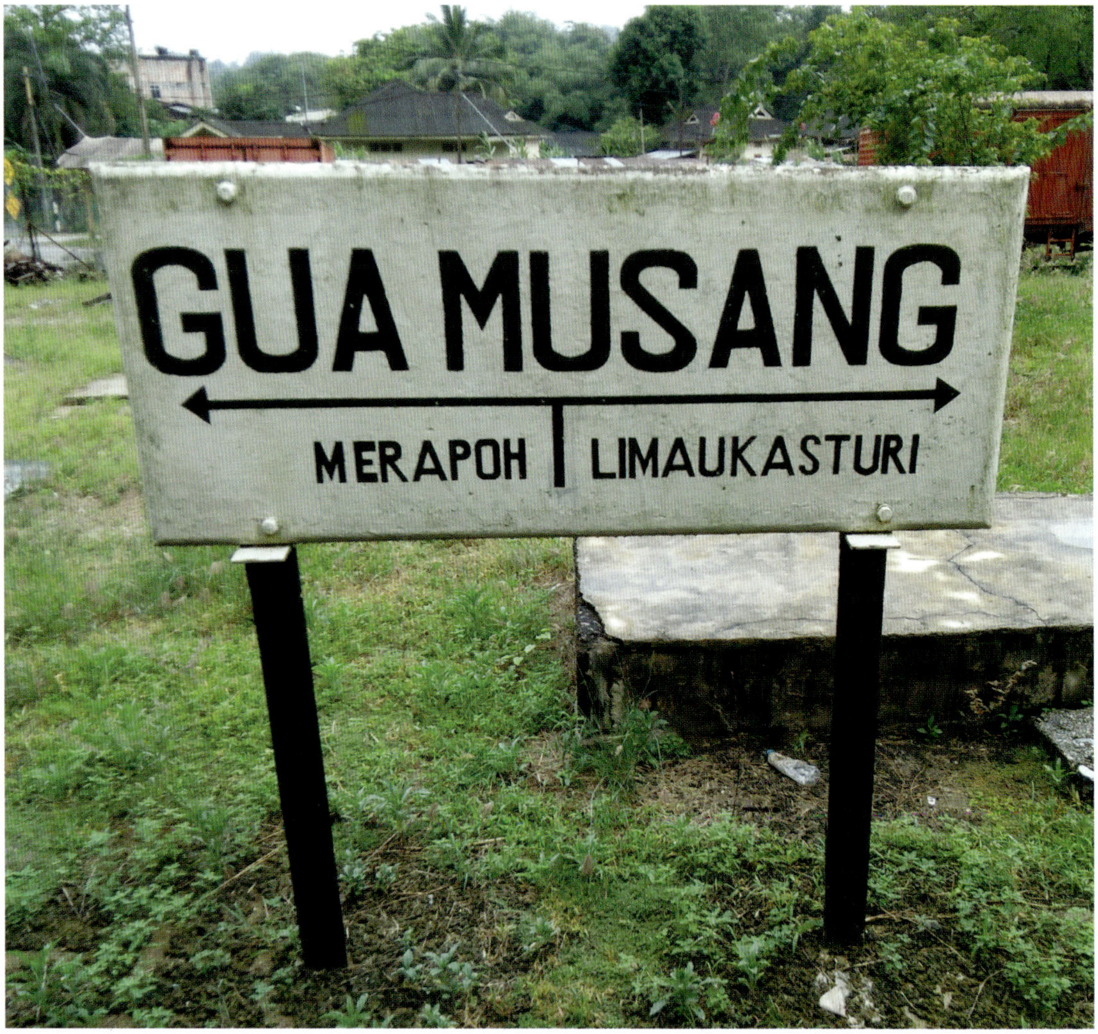

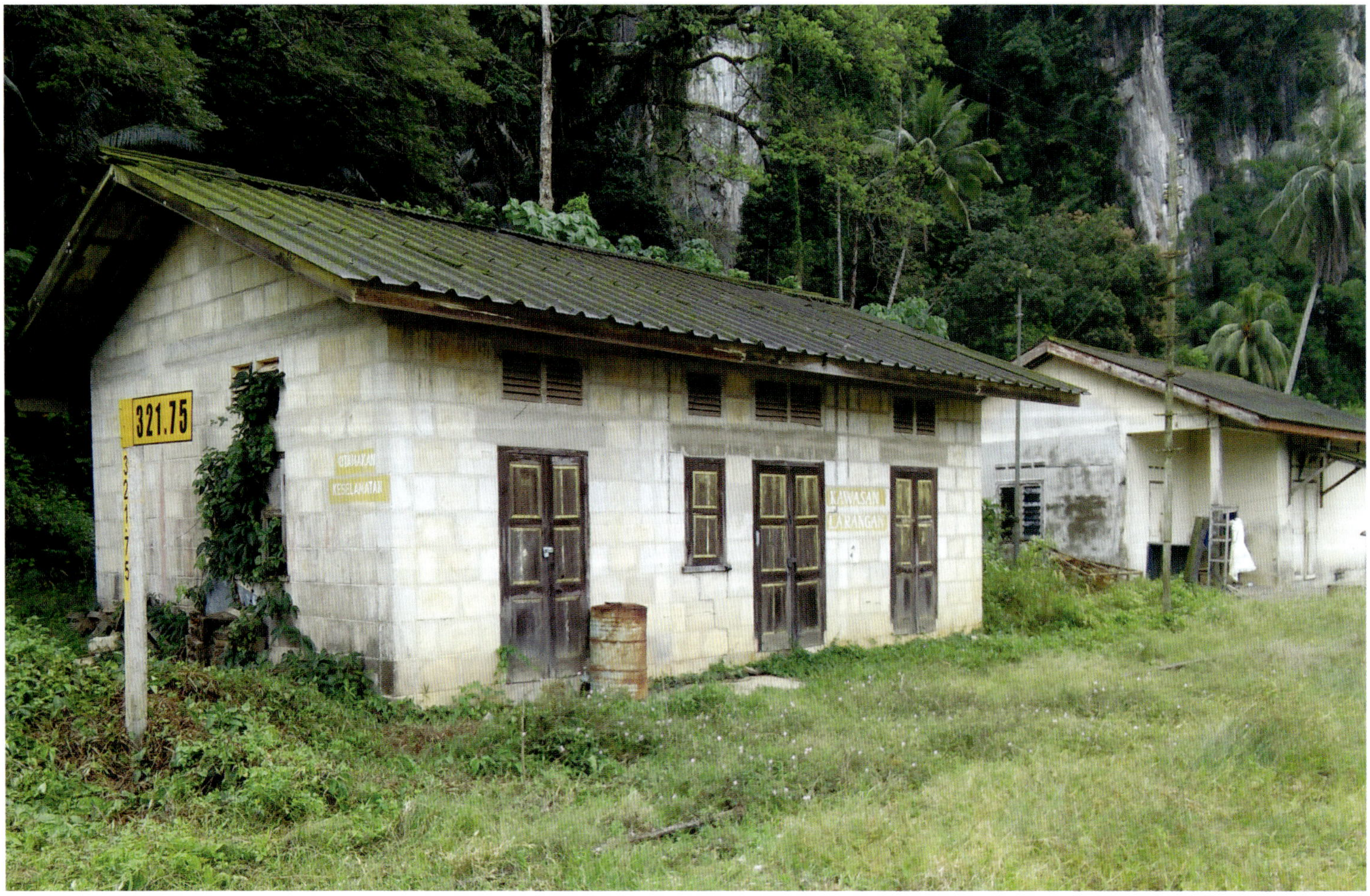

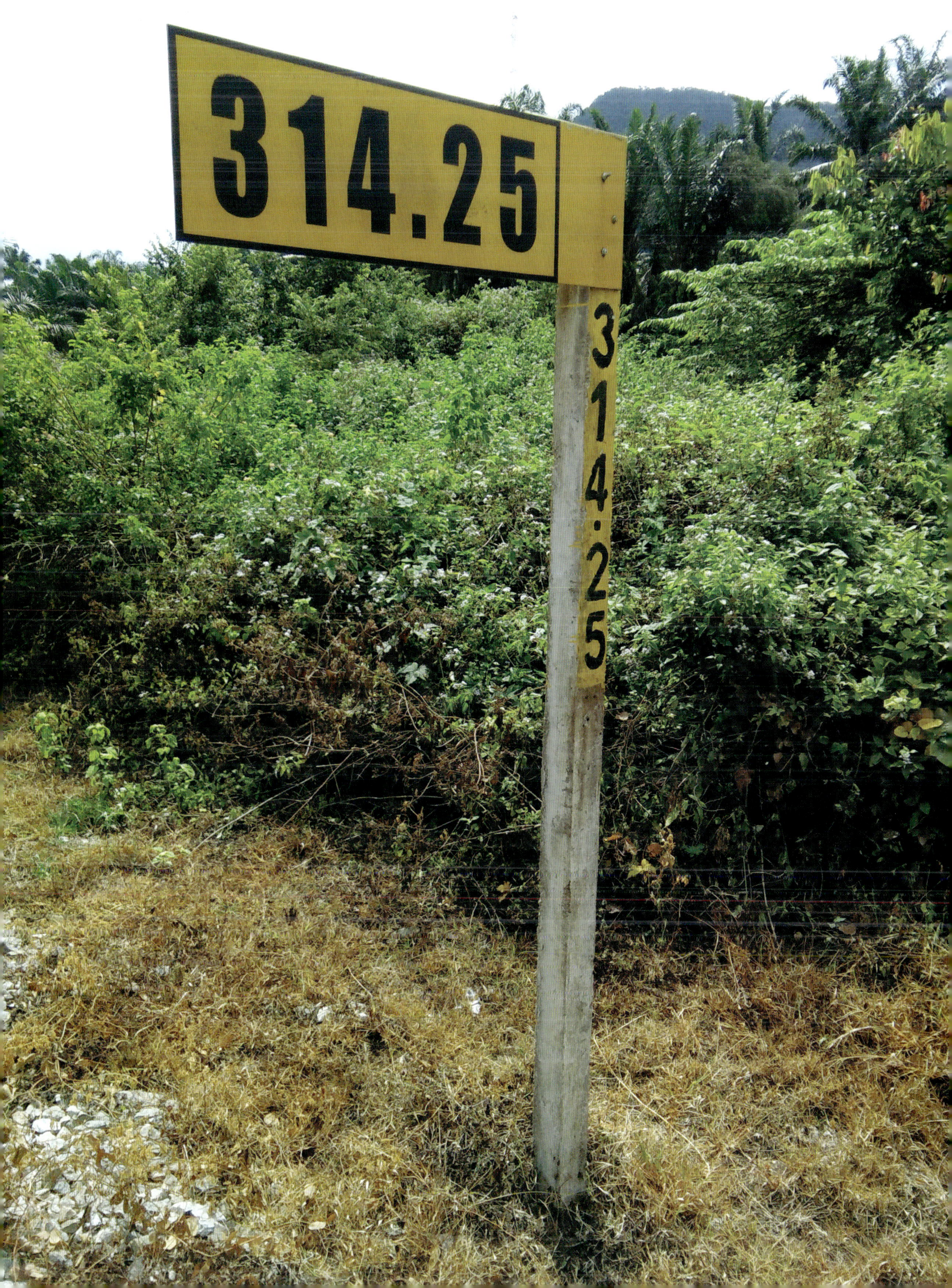
314.25
314.25

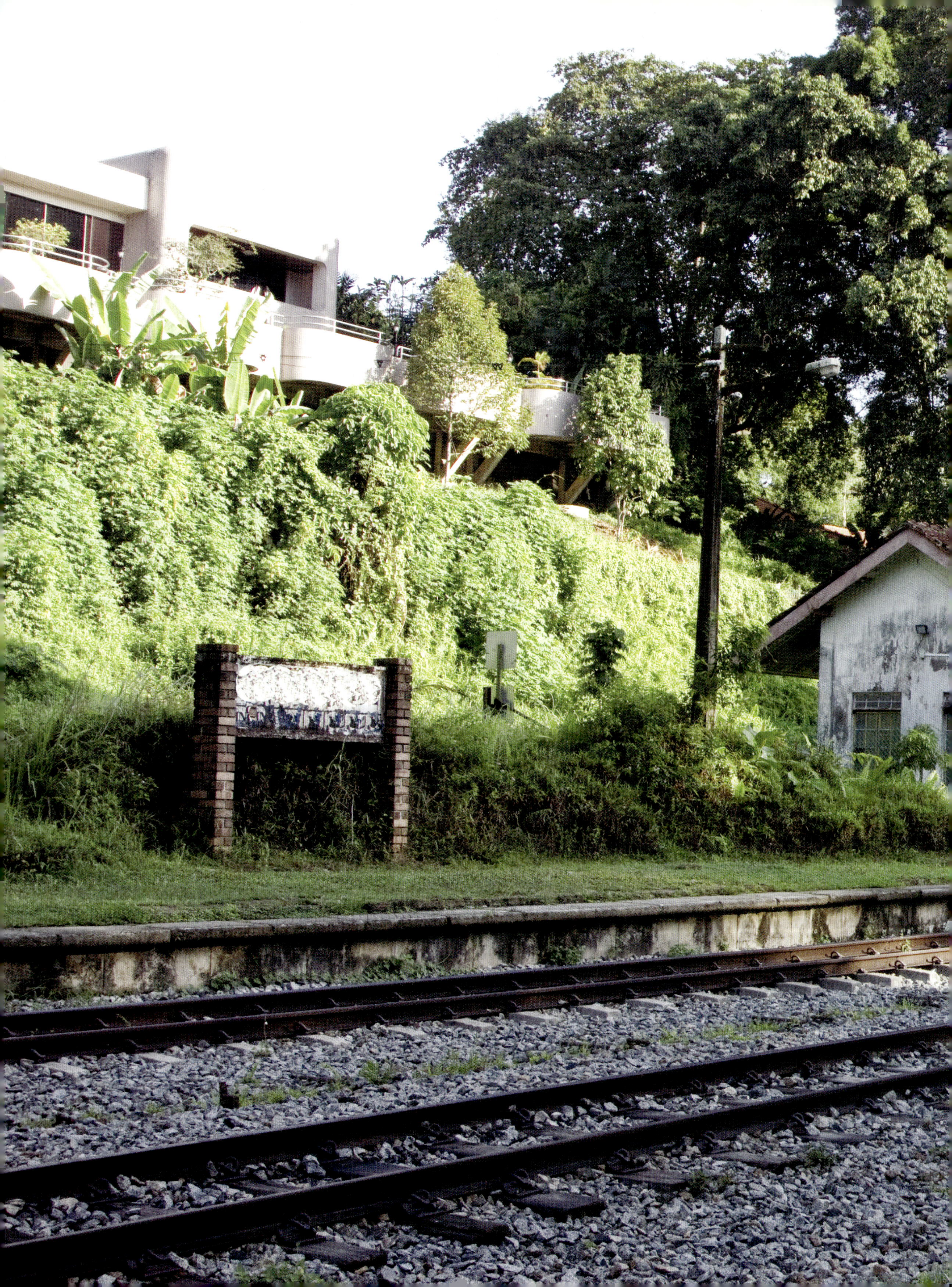

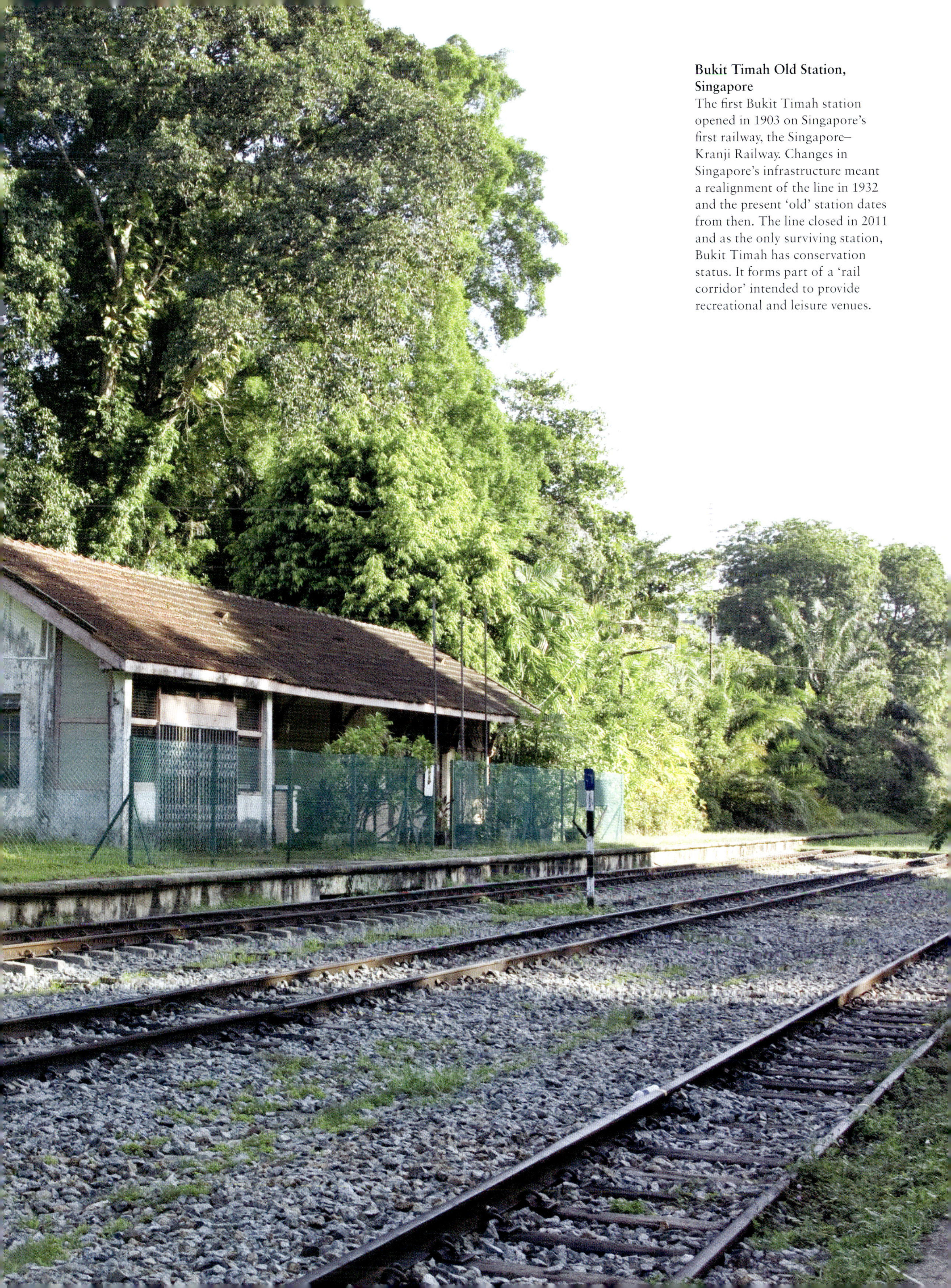

Bukit Timah Old Station, Singapore
The first Bukit Timah station opened in 1903 on Singapore's first railway, the Singapore–Kranji Railway. Changes in Singapore's infrastructure meant a realignment of the line in 1932 and the present 'old' station dates from then. The line closed in 2011 and as the only surviving station, Bukit Timah has conservation status. It forms part of a 'rail corridor' intended to provide recreational and leisure venues.

ALL PHOTOGRAPHS:

Semarang Gudang Station, Java, Indonesia

The first railway in Indonesia, built under Dutch rule in 1864 as the *Nederlandische-Indische Spoorweg*, ran from Semarang to Tanggung. The original terminus, half-forgotten and derelict, frequently flooded, ceased to serve passengers after 1914 but remained in use as a goods station, and was named Semarang Gudang to distinguish it from the city's later stations. Efforts to restore it are hampered by the waterlogged site.

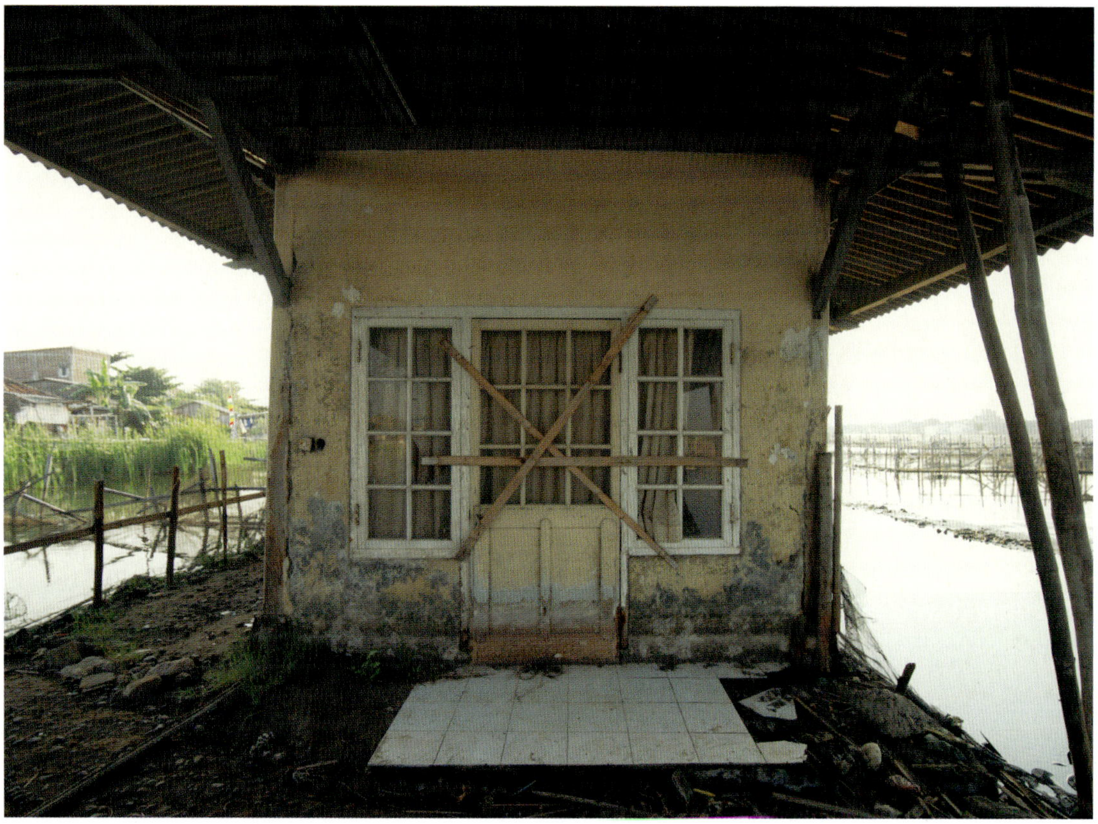

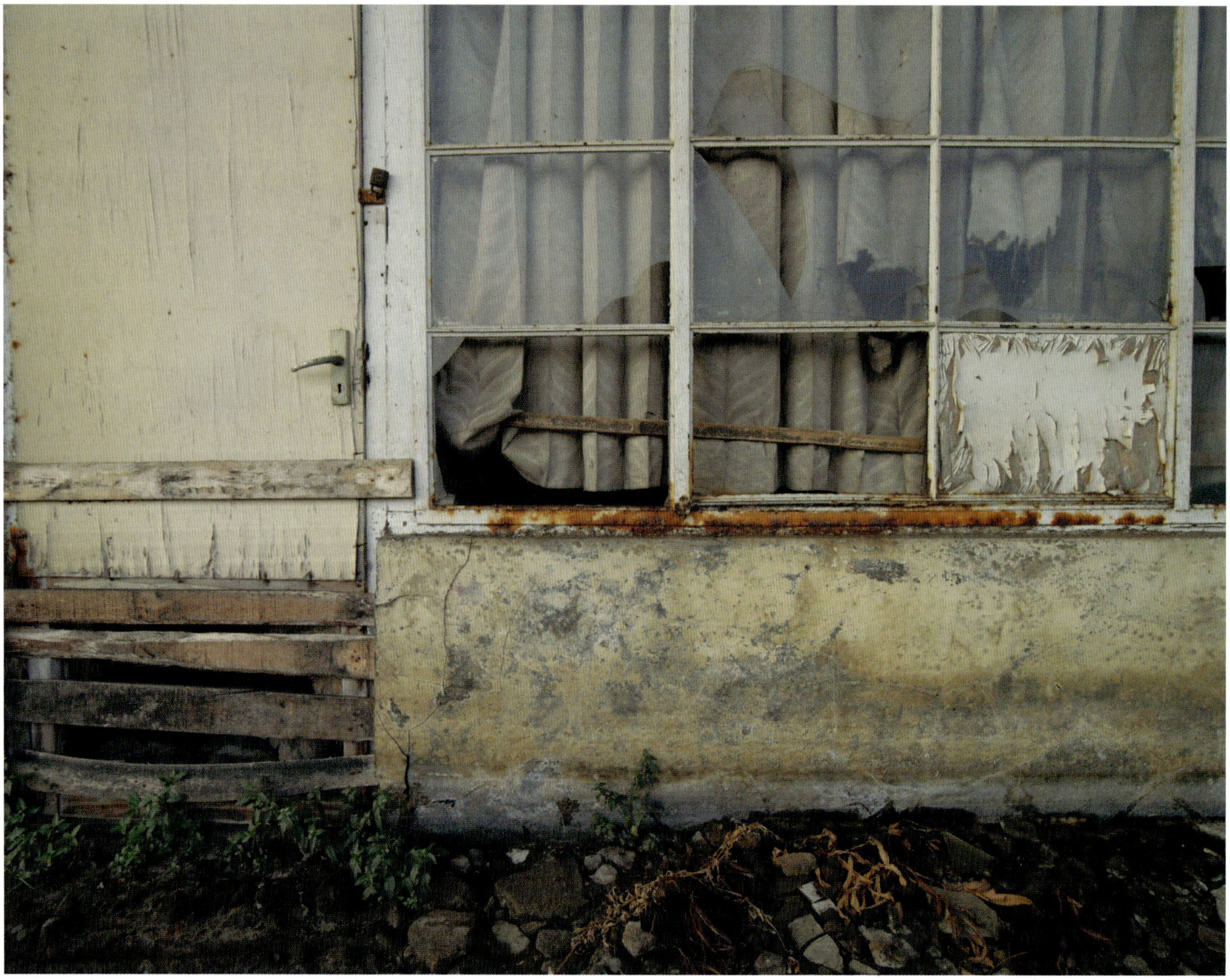

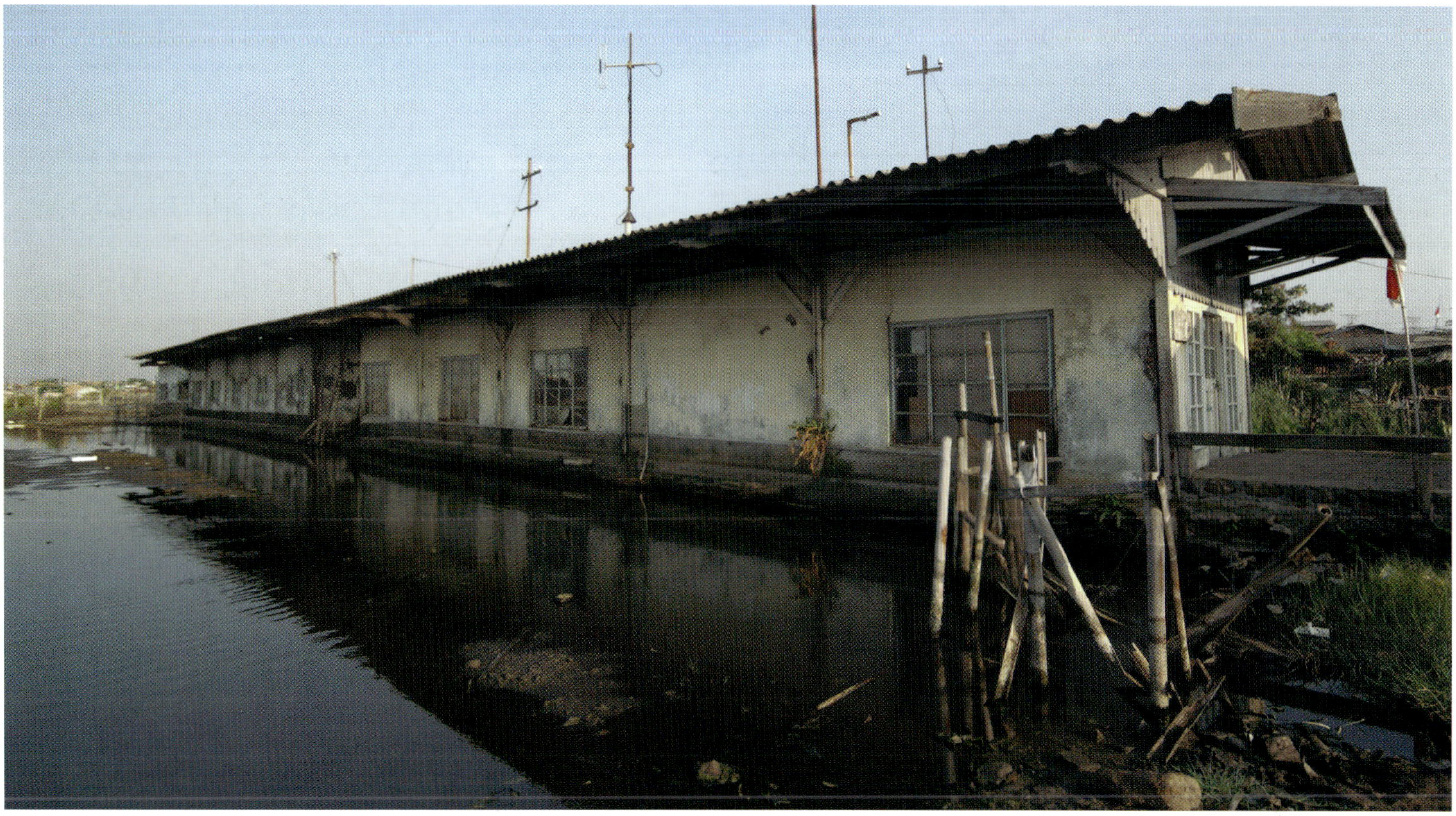

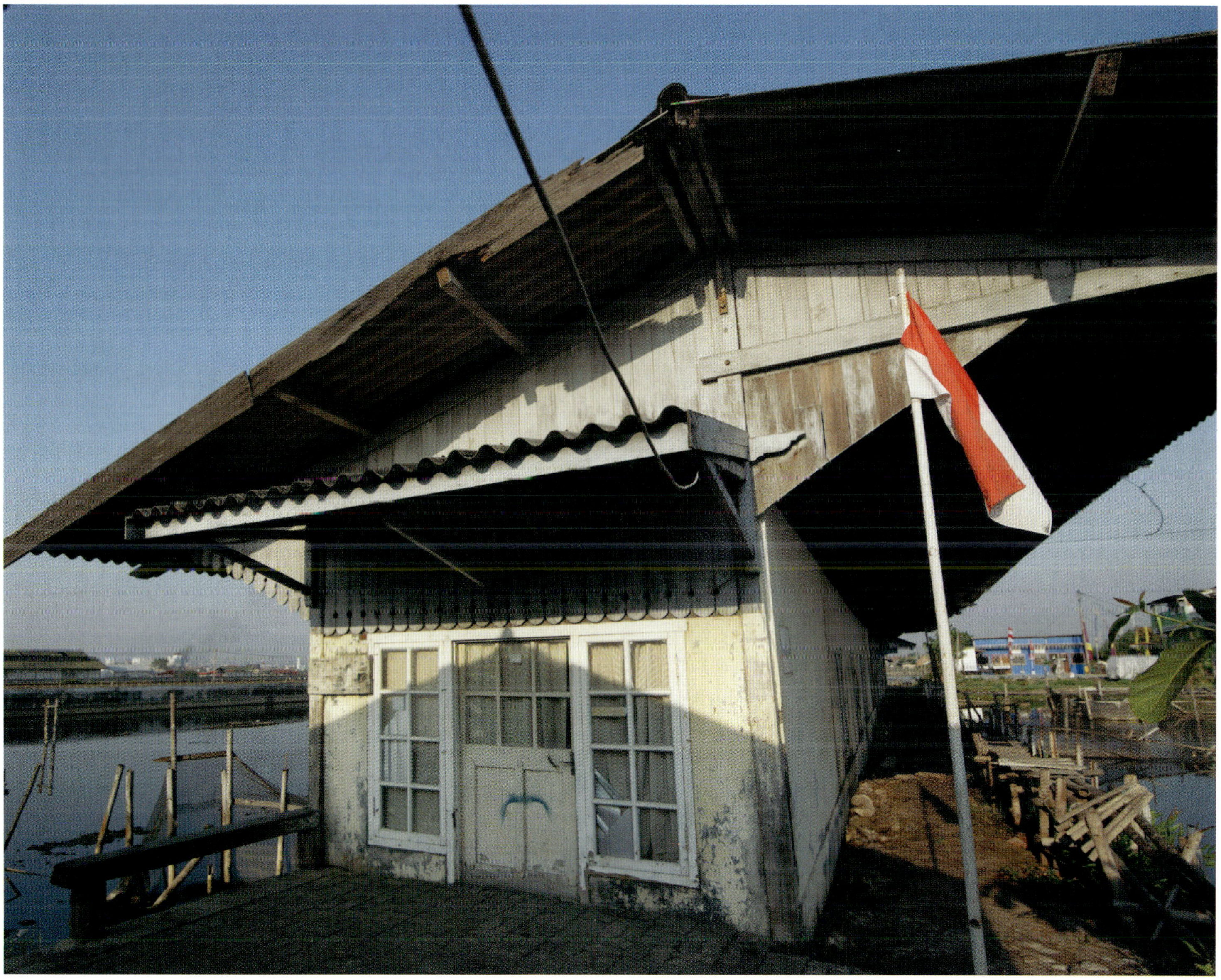

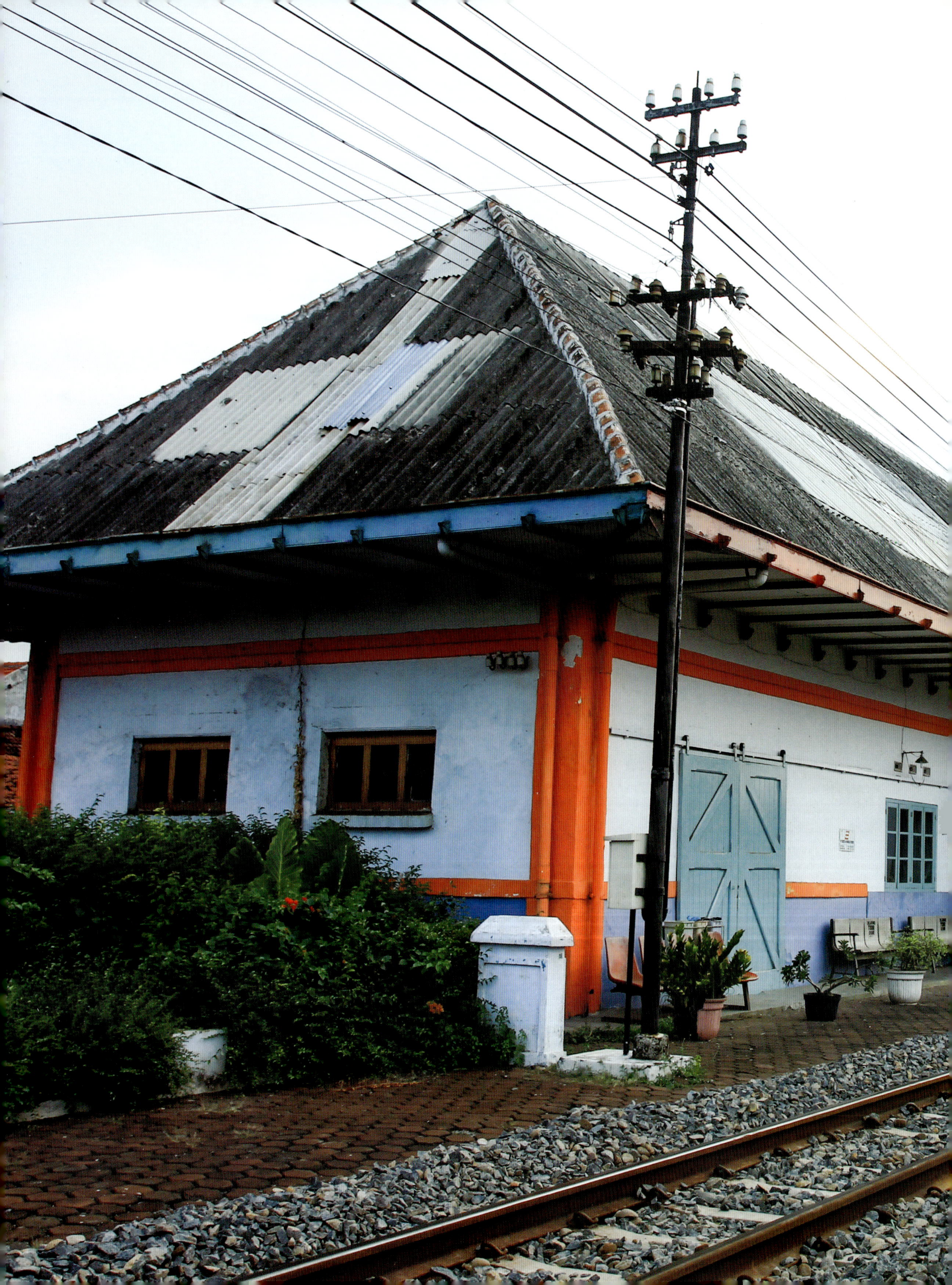

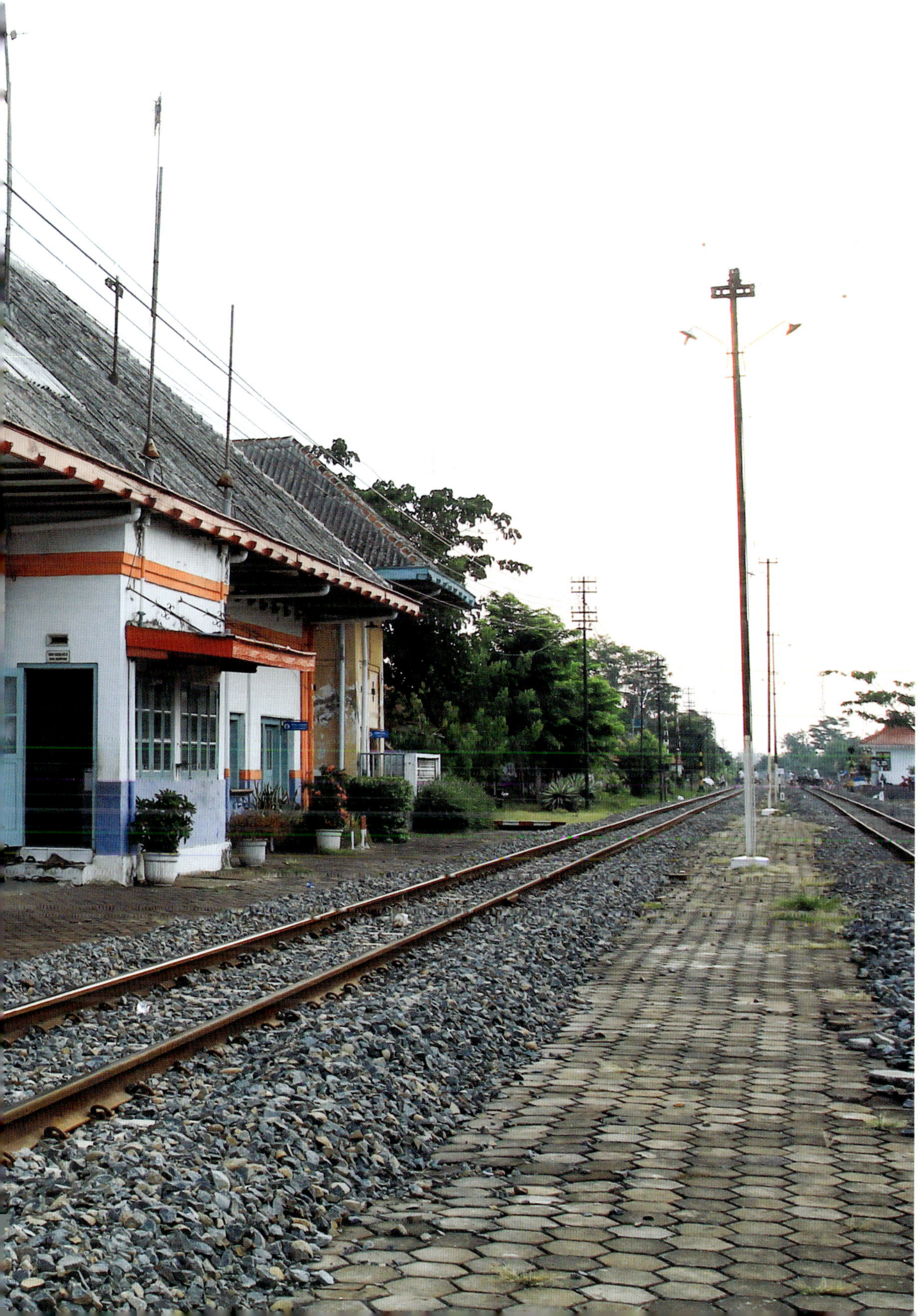

LEFT:

Batang, Java, Indonesia

With its flower tubs and intact appearance, this station on Java's north coast line does not look abandoned, but though intercity expresses and freight trains pass by, no trains stop here. By 1899 a railway had been established between Cheribon and Semarang, through Batang. The original Dutch-style buildings remain, owned but unused by Kereta Api Indonesia, the state railway system.

OVERLEAF:

Helensburgh, New South Wales, Australia

Helensburgh's old station was built in 1889 between two tunnels on a single-track section of the Illawara Railway. This was coal-mining country, with steep grades and poorly-ventilated tunnels, and new lines were constructed to avoid these in 1914–20. By 1915, a new station had been built on the diversion and the original one was abandoned. Wild nature has largely taken over the site.

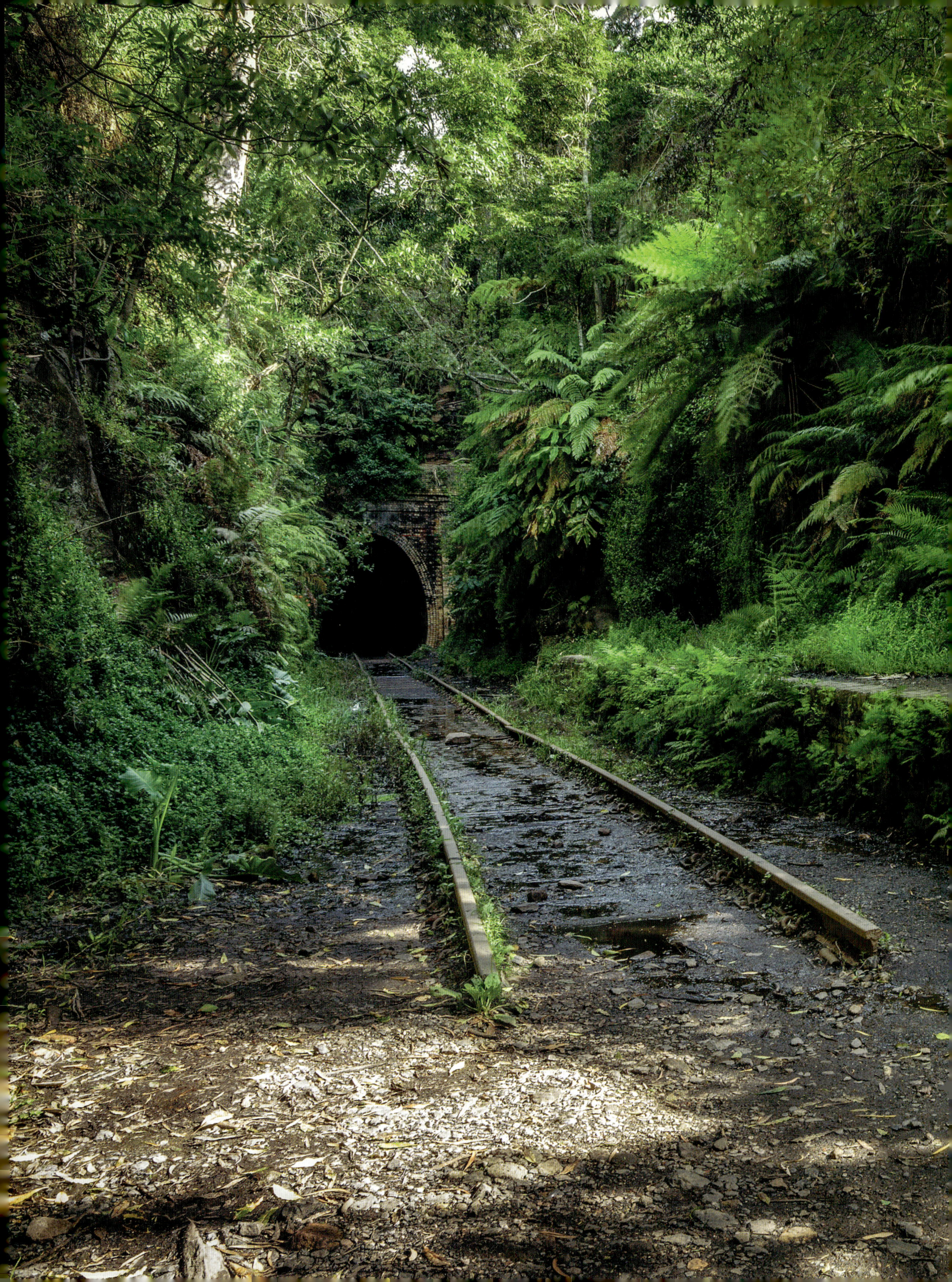

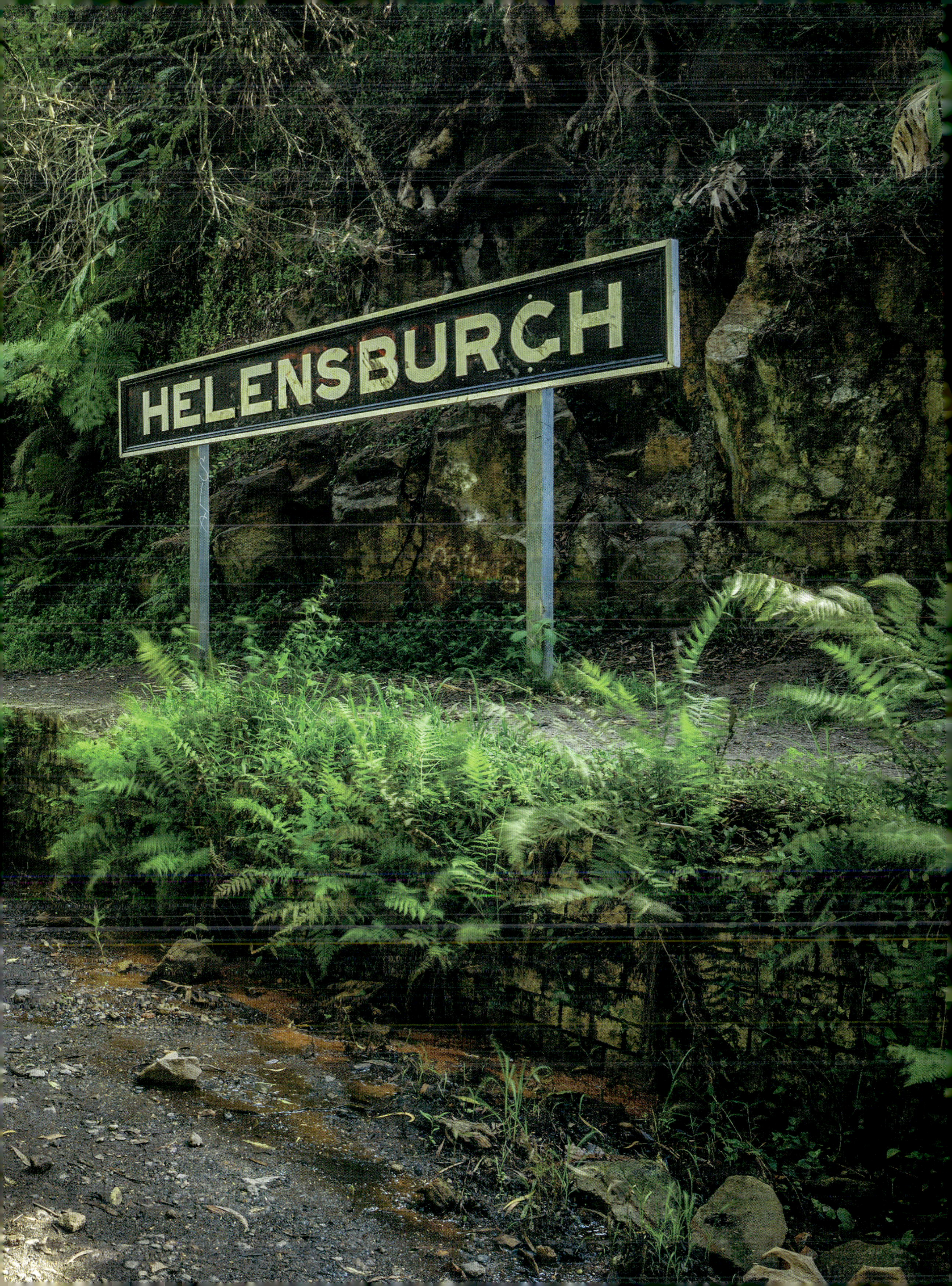
HELENSBURGH

RIGHT:

Jincumbilly, New South Wales, Australia

The Bombala branch line in southern NSW was begun in 1884 but completed only in 1921 when the last section, including Jincumbilly, was opened. About as minimal as a station can be, it had a waiting room, a short platform, and a siding to load or unload livestock. Passenger trains ceased in 1988, freight in 1989.

BELOW:

Leongatha, Victoria, Australia

A new era of prosperity for what had been a remote district of farmers and loggers was marked by the arrival of the railway here in 1891. Since the 1980s, though, the line has had a chequered time: closed and reopened on two occasions before definitive closure in 1993. Since then, as the grass grows over the tracks, there have been a number of schemes for reopening and even extension of the line.

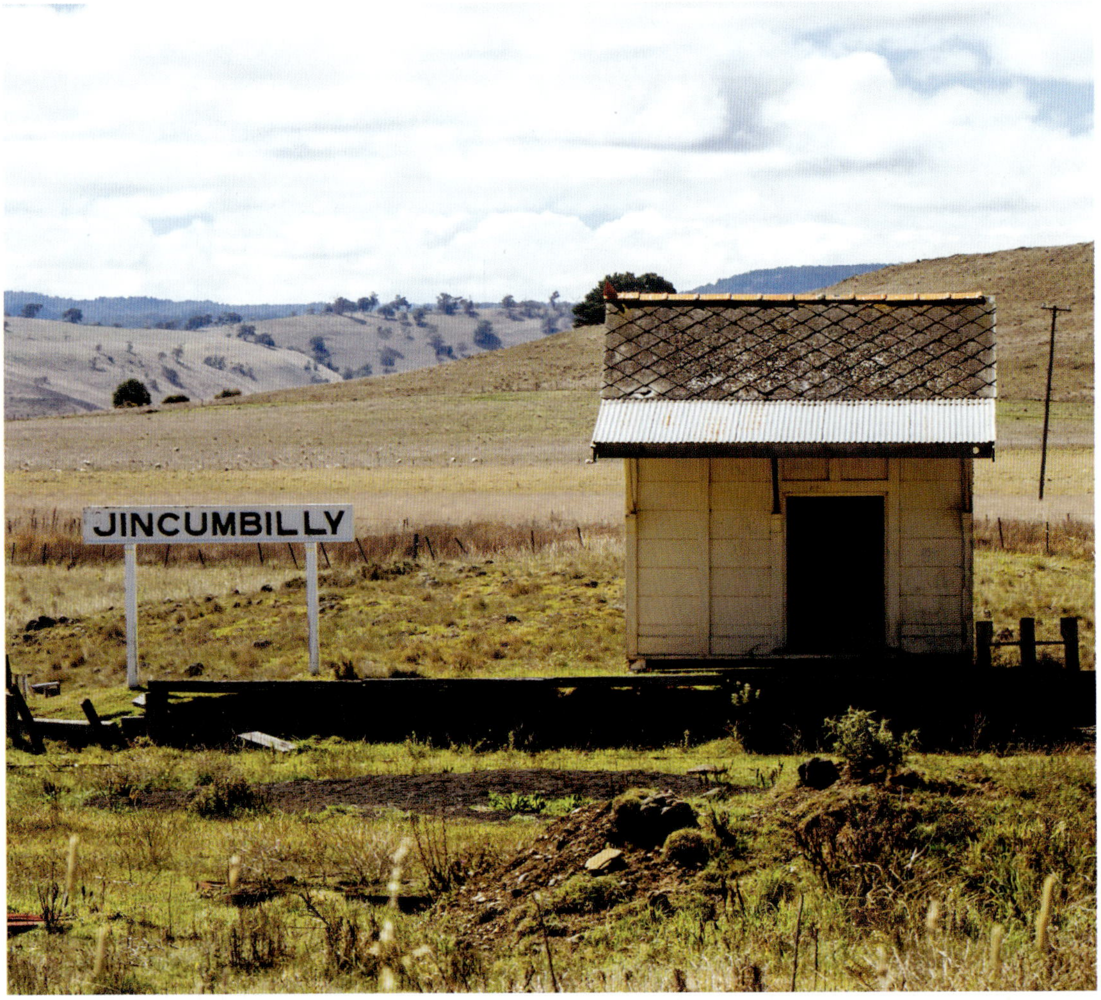

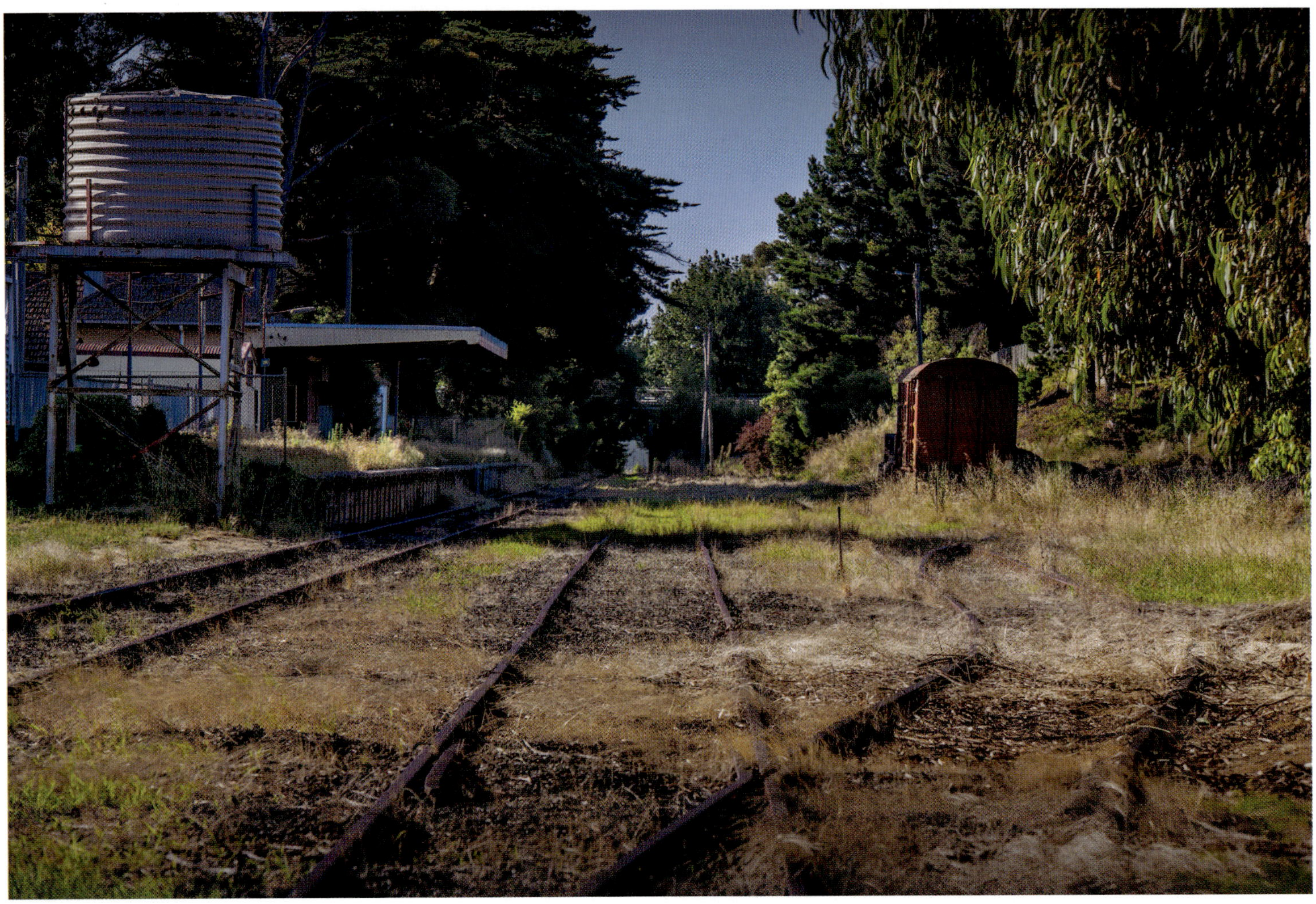

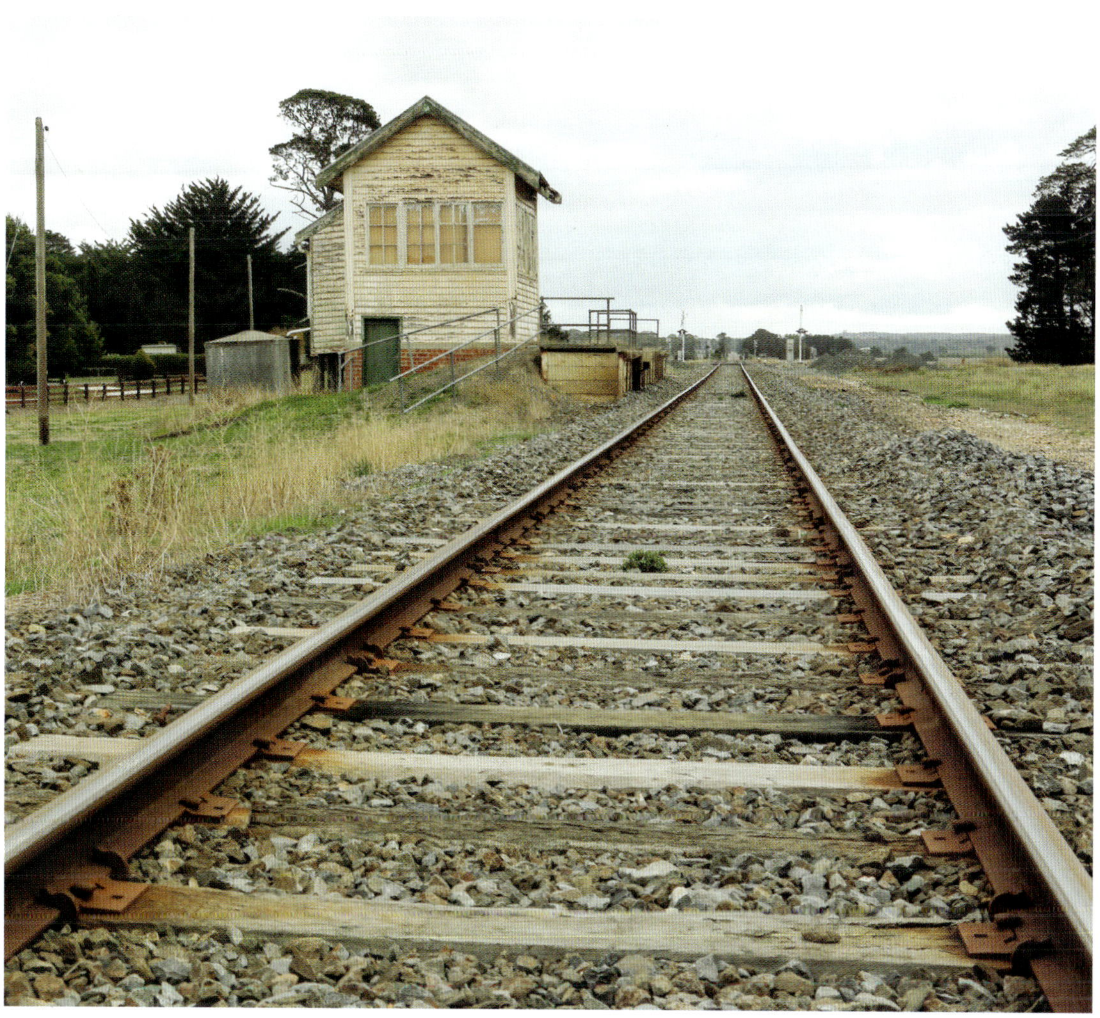

LEFT:

Trawalla, Victoria, Australia

The line from Ballarat to Ararat was completed in 1875. Passing through thinly-populated pastoral country, its local stations were built of wood at minimal cost. With other stations, Trawalla was closed in 1981; its section of line was closed in 1995 but reopened in 2004 as part of a reorganization of services in the Murray Basin region. Diesel-electric units now speed by where steam trains stood in the now lifted crossing loop to effect crew switches.

BELOW:

Raglan, New South Wales, Australia

Raglan's first station (1873) was dangerously located at the top of a steep grade and was replaced by this island-platform station in 1890. Like the great majority of rural stations in Australia, it was built to a standard pattern (Type 9 in this case), with a corrugated iron roof. Some maintenance work has been done on the heritage-listed buildings, and there are plans to restore it for community use and as a tourist attraction.

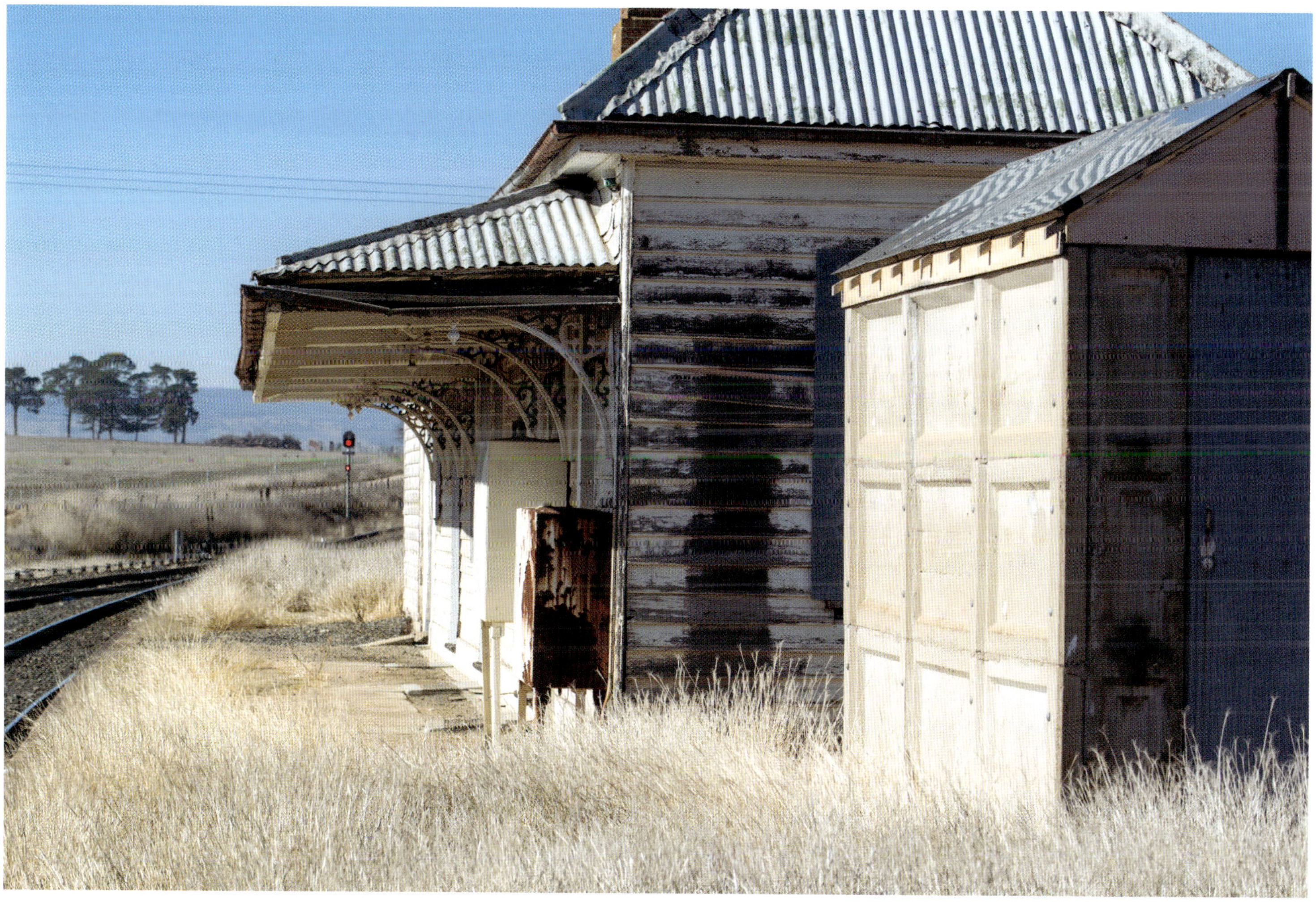

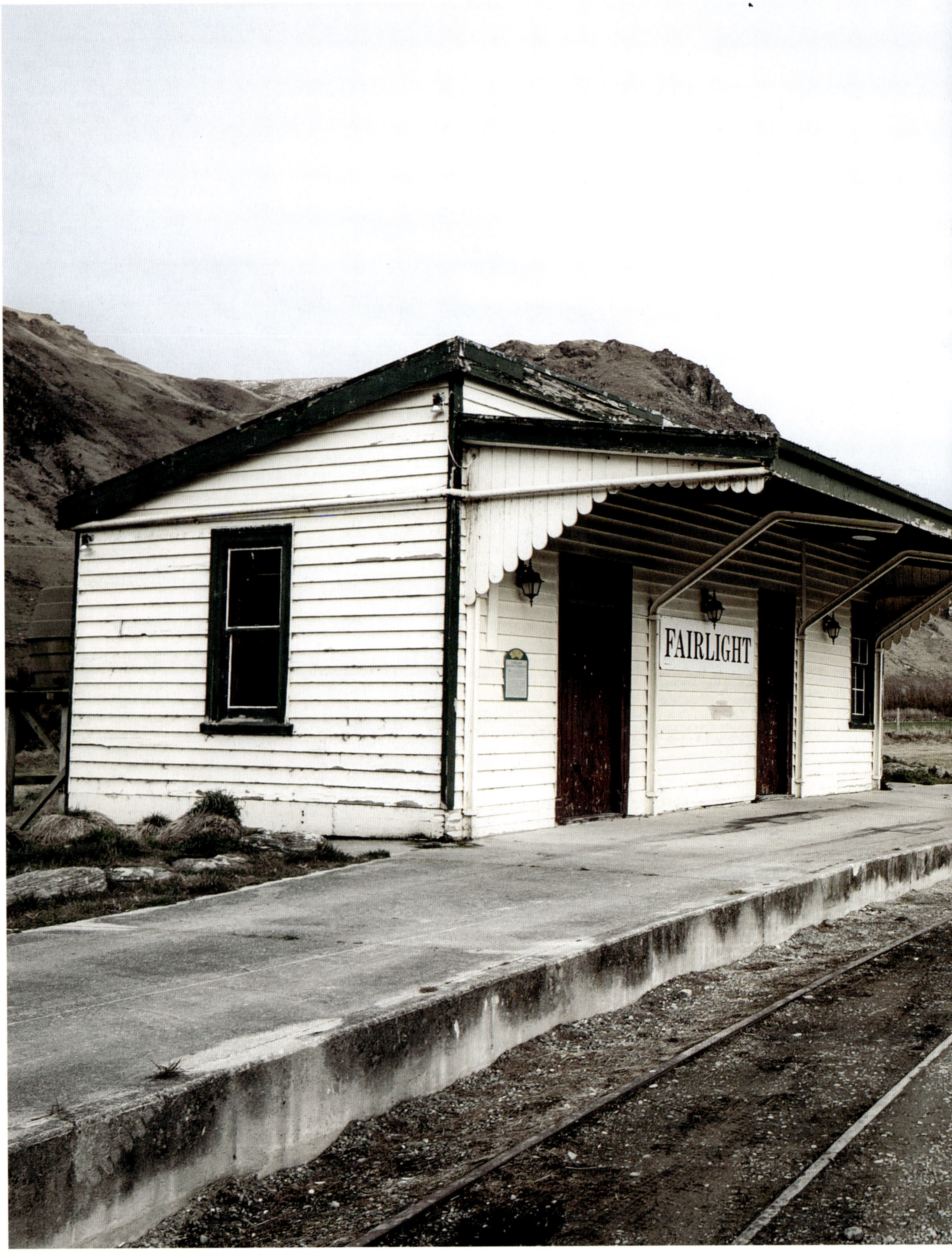
FAIRLIGHT

Fairlight Station, Otago Province, New Zealand
The penultimate station on the line from Invercargill to Kingston, Fairlight opened in 1878. Regular services ended in 1937. An occasional steam-hauled tourist train, the 'Kingston Flyer', ran until 1957. In 1982 the branch track was lifted, except for the Fairlight–Kingston section, on which the 'Kingston Flyer' still makes occasional runs. The present station building, dating from around 1920, was moved here from Otautau, on the Ohai branch, in 1996.

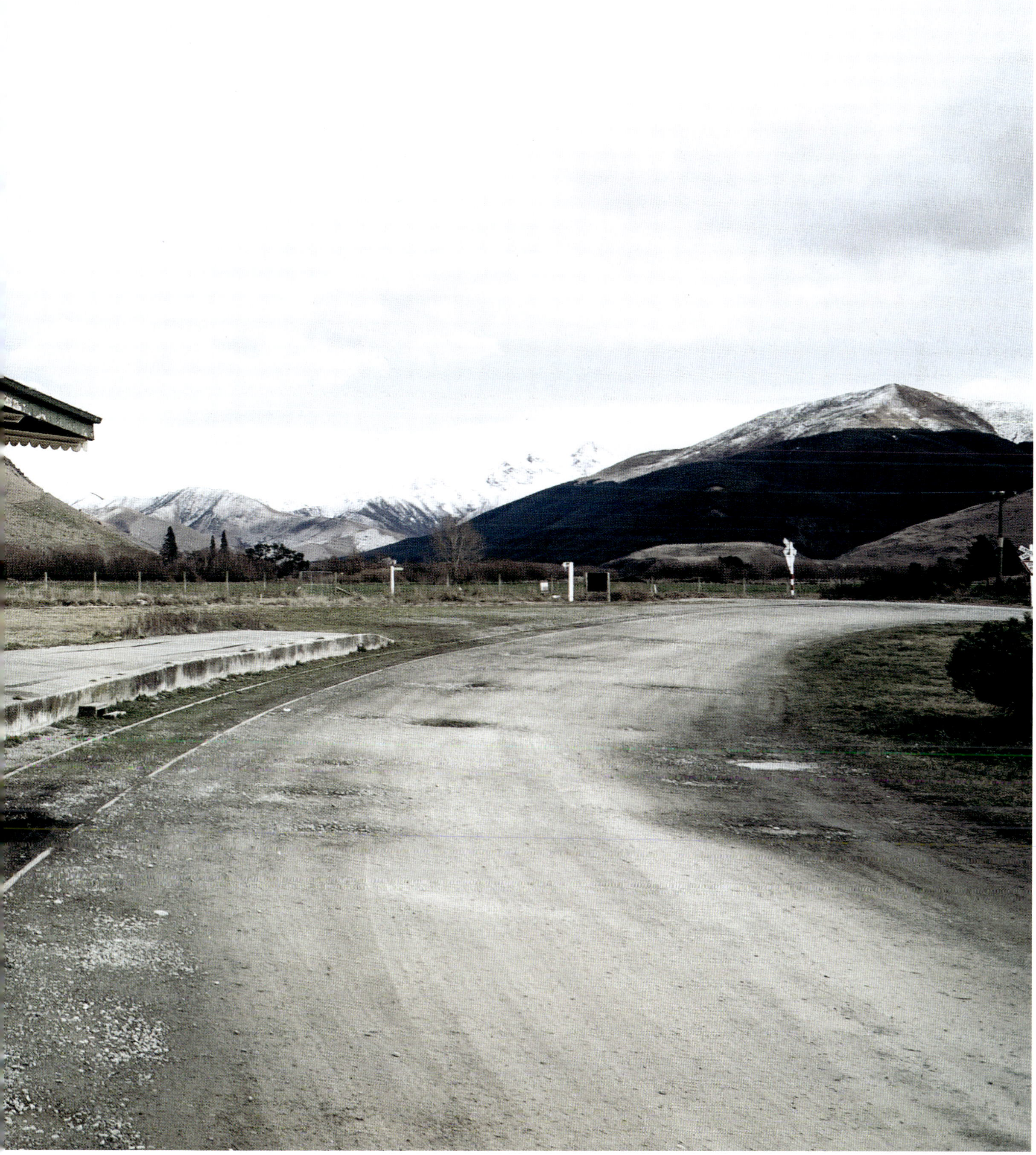

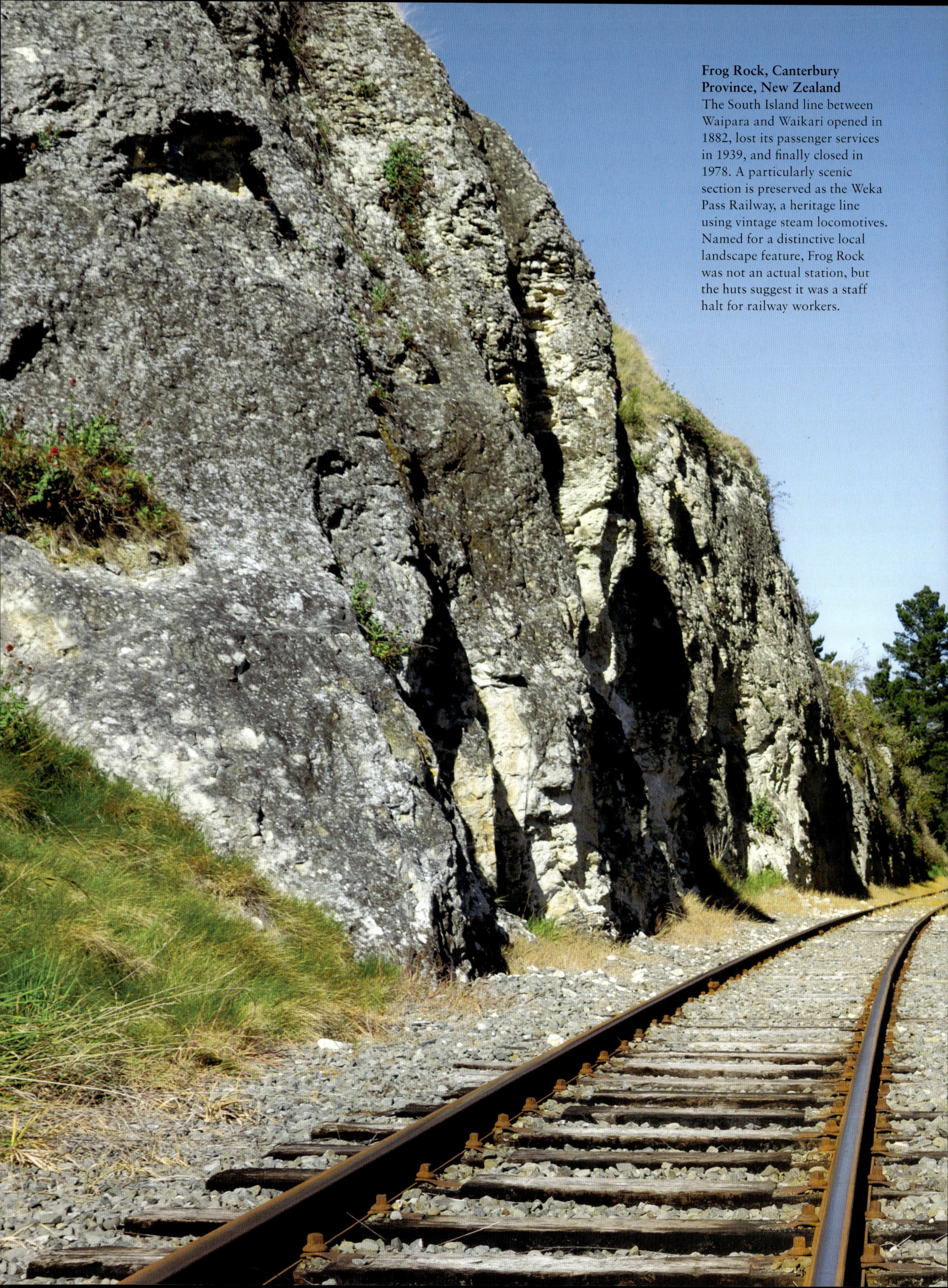

Frog Rock, Canterbury Province, New Zealand
The South Island line between Waipara and Waikari opened in 1882, lost its passenger services in 1939, and finally closed in 1978. A particularly scenic section is preserved as the Weka Pass Railway, a heritage line using vintage steam locomotives. Named for a distinctive local landscape feature, Frog Rock was not an actual station, but the huts suggest it was a staff halt for railway workers.

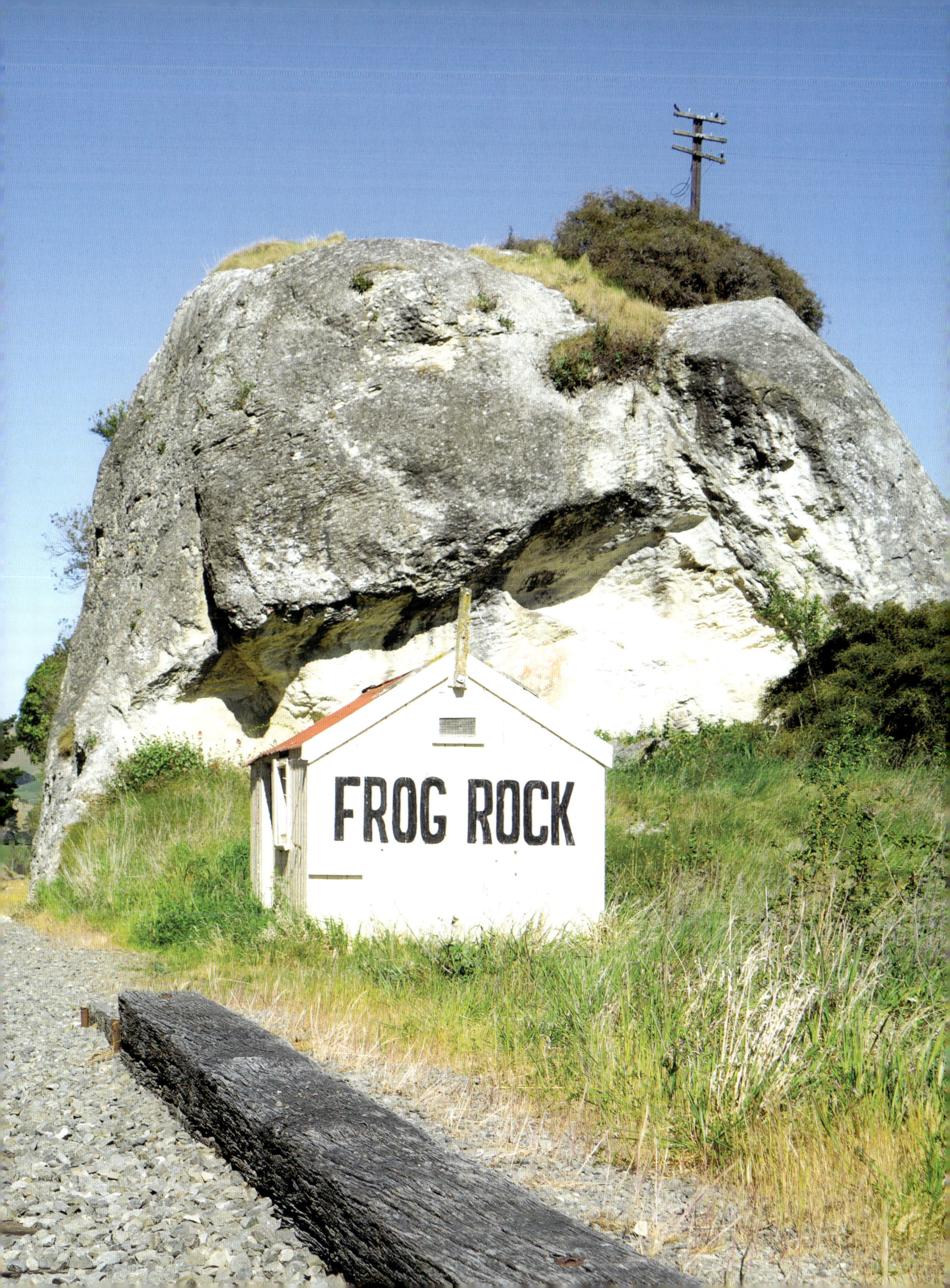
FROG ROCK

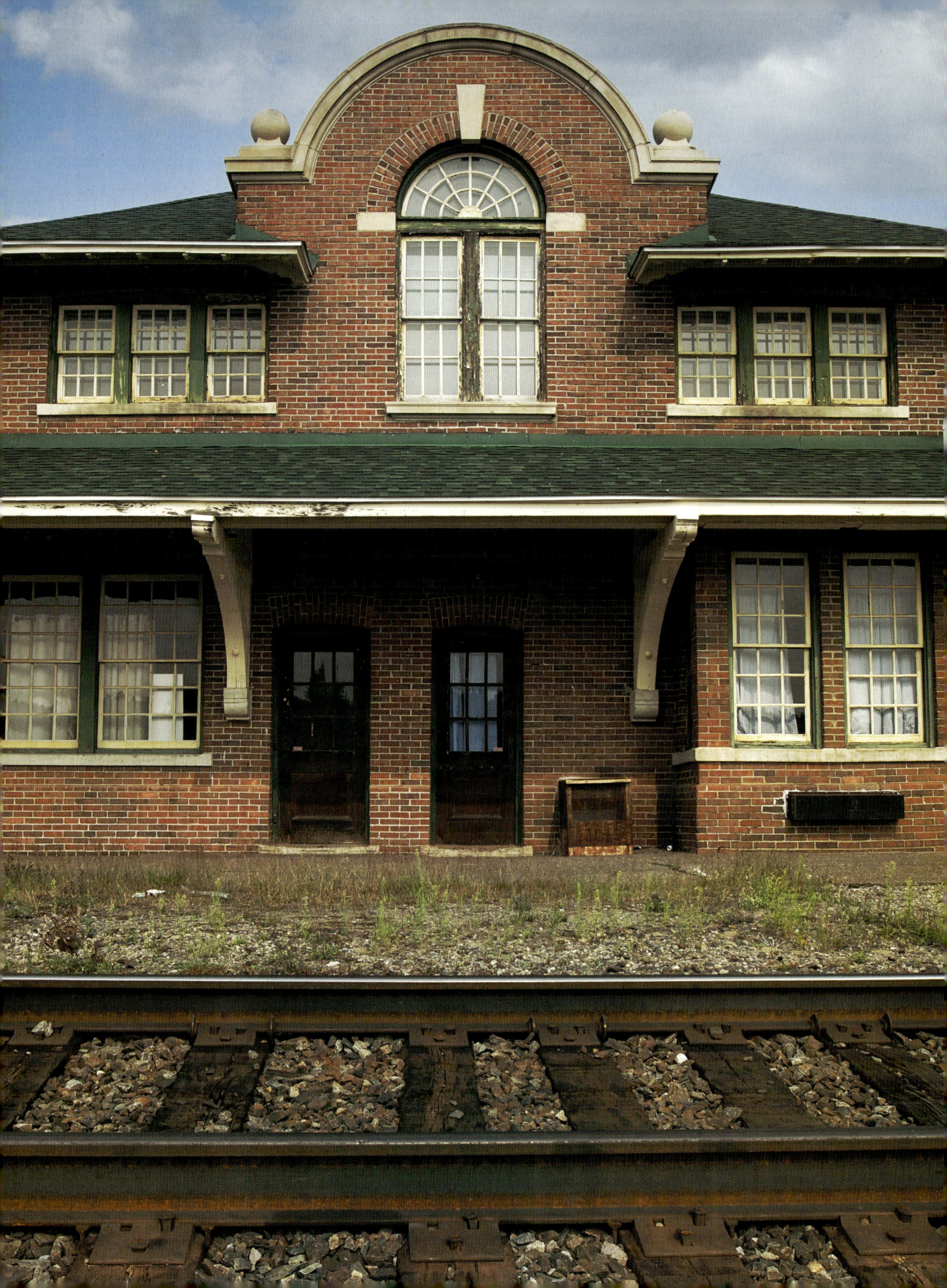

North America

North America shows the same wide variety of abandoned stations as the other regions of the world. What is most different here is the number of big city terminals that are now out of use. In the heyday of the American railroad as a long-distance transport service, companies vied with one other to build imposing city-centre stations. These were necessary to cope with large numbers of travellers but also displayed the prestige and solid reliability of the company.

With the huge growth of internal air travel in the 1950s and 60s, causing railroad company amalgamations and large-scale closures, many of these grand termini became increasingly redundant, and cities now struggle to find a use for them. At the other end of the population scale, past times and failed ambitions are strongly evoked by the numerous 'ghost town' depots, mostly in western states, where a simple clapboard building stands alongside a rail-less, grass-grown trackbed.

Currently the scene has been changing as far as abandoned stations are concerned. Most communities have a strong sense of local history and the local railroad depot, though now closed or not in active use, is seen as a key marker in the growth and development of the place, and so worthy of preservation. Buildings long neglected have been renovated or earmarked for preservation. Some of the stations illustrated here have already been spruced up as shops, restaurants, museums or community centres.

OPPOSITE:

New Liskeard, Ontario, Canada

The Ontario Northland line between North Bay and Moosonee passes through Liskeard, and the brick building dates from 1905, when the line opened. Originally proposed as the Lake Nipissing and James Bay Railway, in 1902 it became the Temiskaming and Northern Ontario Railway, and Ontario Northland in 1946. The station was closed in 2012 and only freight trains now pass by.

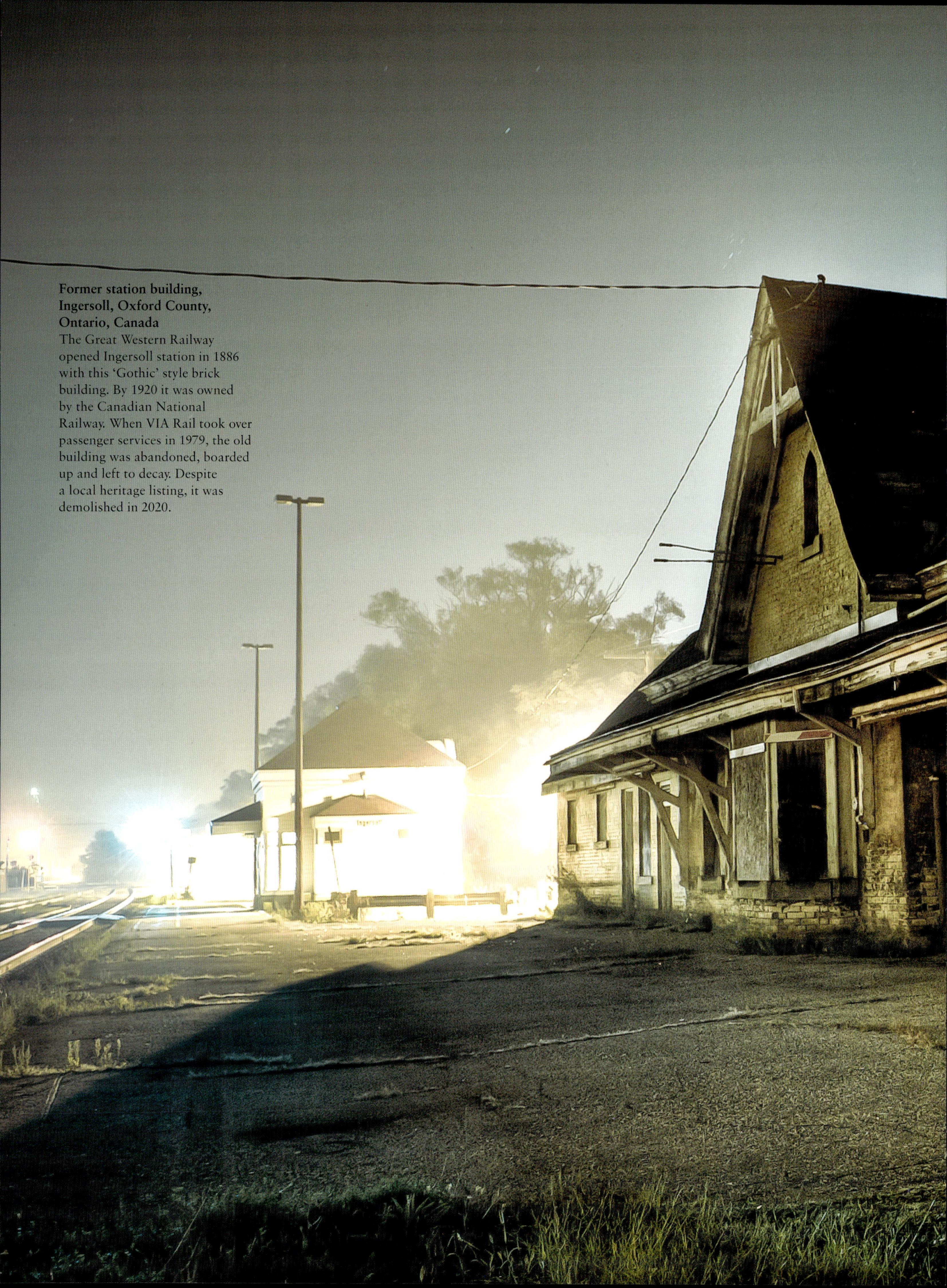

Former station building, Ingersoll, Oxford County, Ontario, Canada
The Great Western Railway opened Ingersoll station in 1886 with this 'Gothic' style brick building. By 1920 it was owned by the Canadian National Railway. When VIA Rail took over passenger services in 1979, the old building was abandoned, boarded up and left to decay. Despite a local heritage listing, it was demolished in 2020.

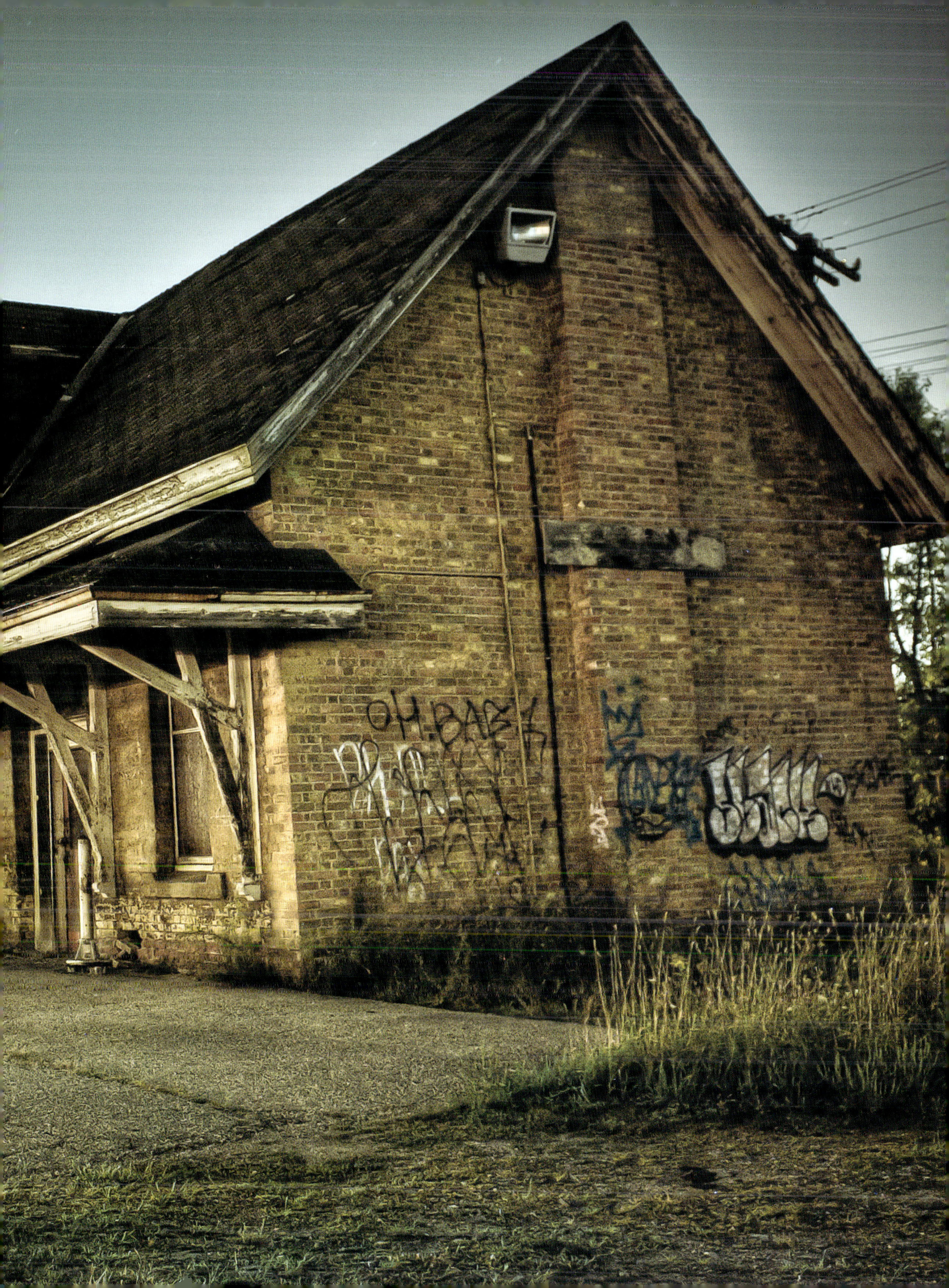
OH.BABY

BELOW TOP:

Ernestown, Ontario, Canada
Built in 1855–56 to a standard design used by the Grand Trunk Railway of Canada, Ernestown is one of the country's oldest surviving station buildings and was designated a heritage site in 1992. This does not commit anyone to preservation and the building's interior has been vandalised, though the sturdy 'Italianate' stone structure remains sound.

BELOW LOWER:

Vegreville, Alberta, Canada
On the Canadian National transcontinental line, this station was opened in 1905, and the present building was completed in 1930. Like most country stations it was built to a standard plan, with a stucco exterior, and a hipped roof with brackets supporting the massive eaves. The upper floor had living accommodation while the ground level had a ticket office, telegraph office, washrooms, ladies' waiting room and smoking room. Trains ceased to call in 1978.

OPPOSITE ABOVE:

Courtenay, British Columbia, Canada
Intended as a wayside station when it was built in 1914 by the Esquimalt & Nanaimo Valley Railway (owned by the Canadian Pacific since 1905), the Comox Valley line was never extended, and Courtenay became the terminus of the line. The upper floor provided overnight accommodation for train crews. Passenger and freight services ended in 2011.

OPPOSITE BELOW:

Boston Bar, British Columbia, Canada
The Canadian National Railway opened the station in 1914. The station house was moved away from the railway line in 1994, and though it is in need of repair and restoration, an active programme for this purpose is in hand, with the aim of creating a rest-stop and museum. It is also rumoured to be haunted.

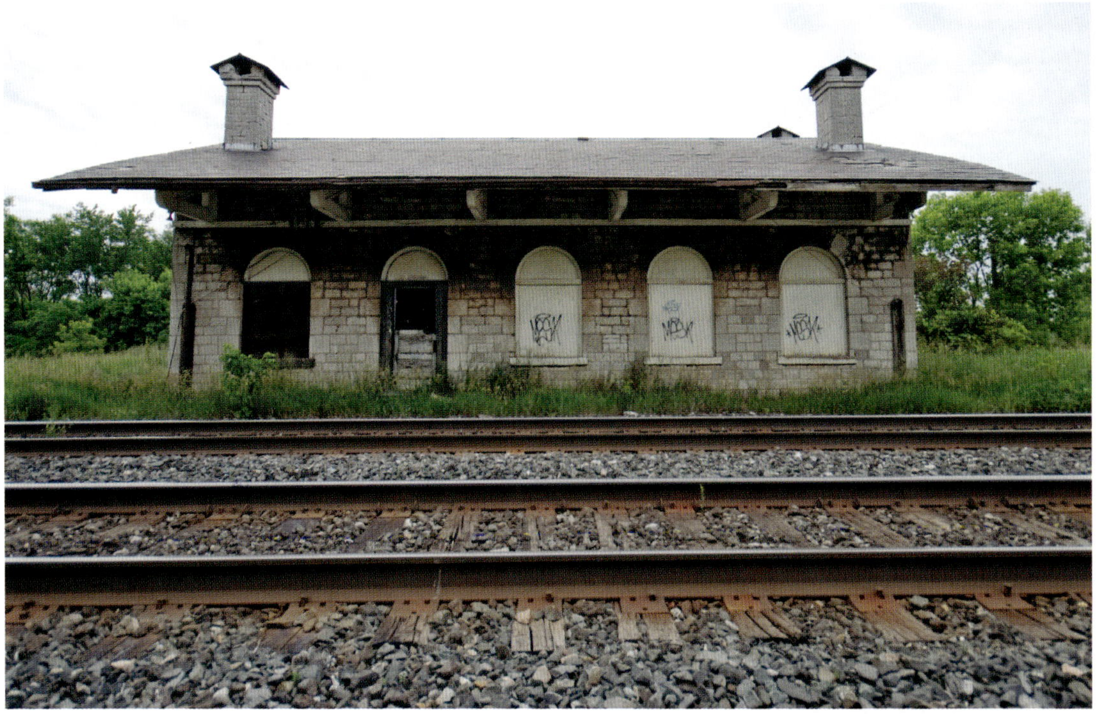

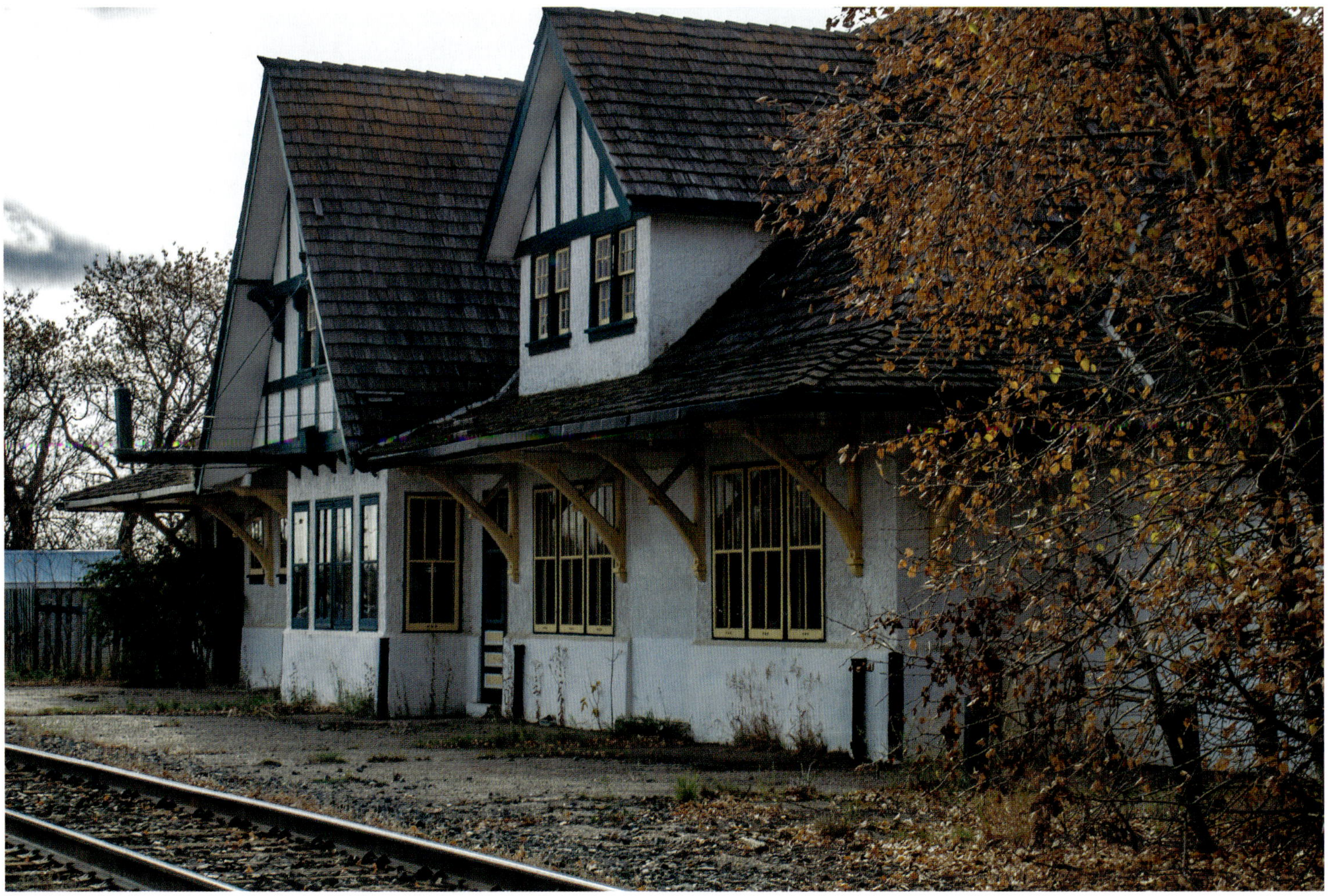

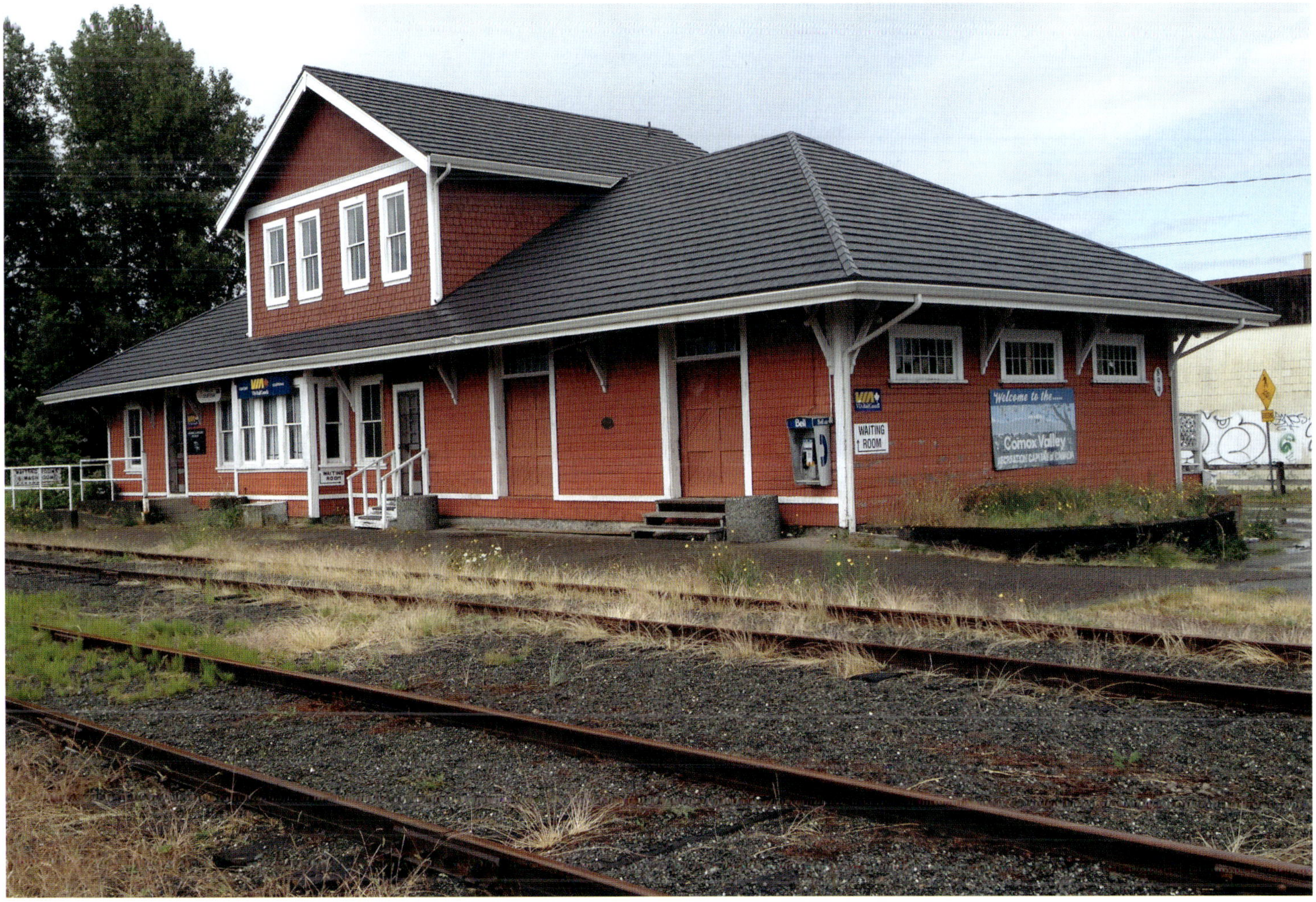
Welcome to the...
Comox Valley
WAITING
ROOM

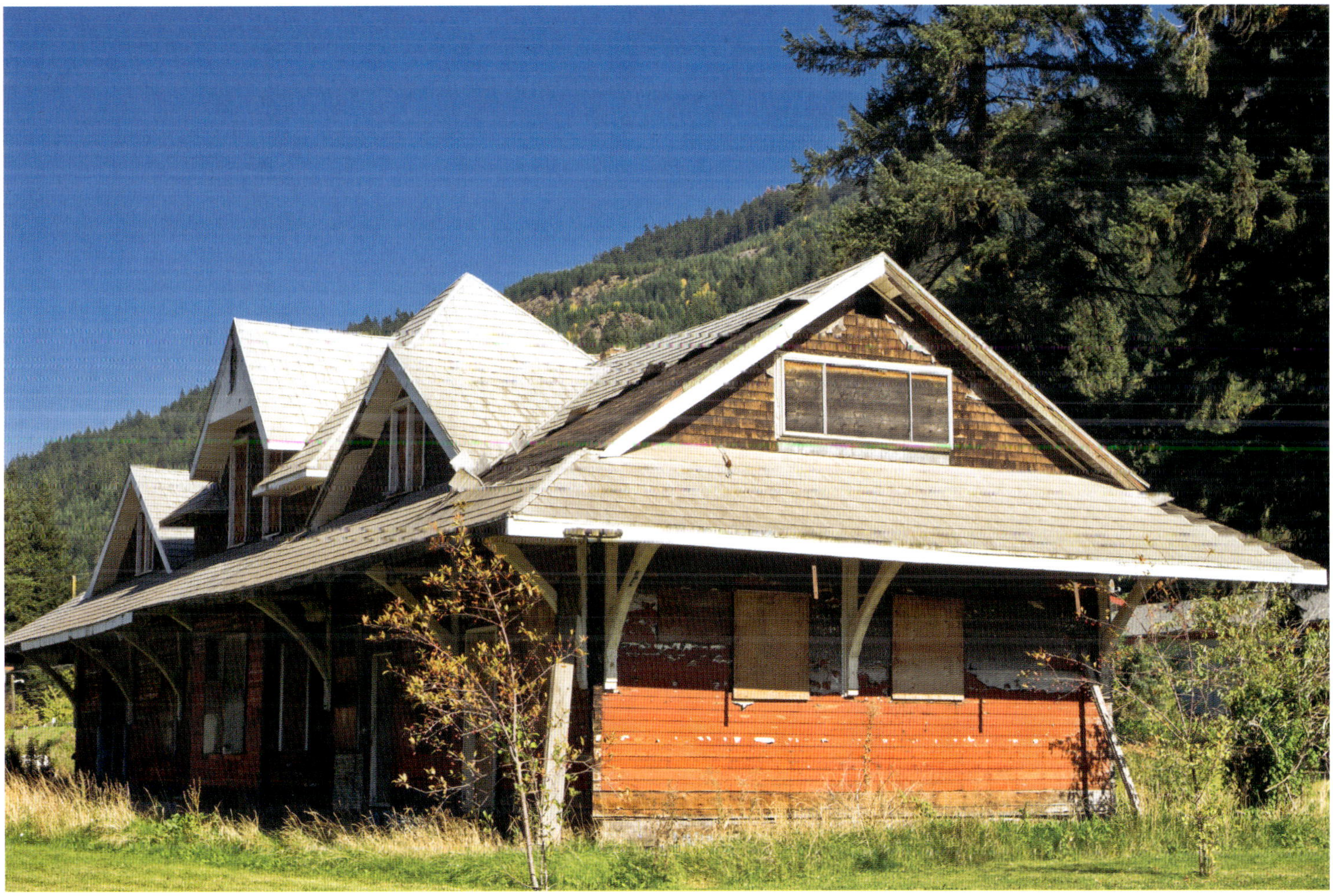

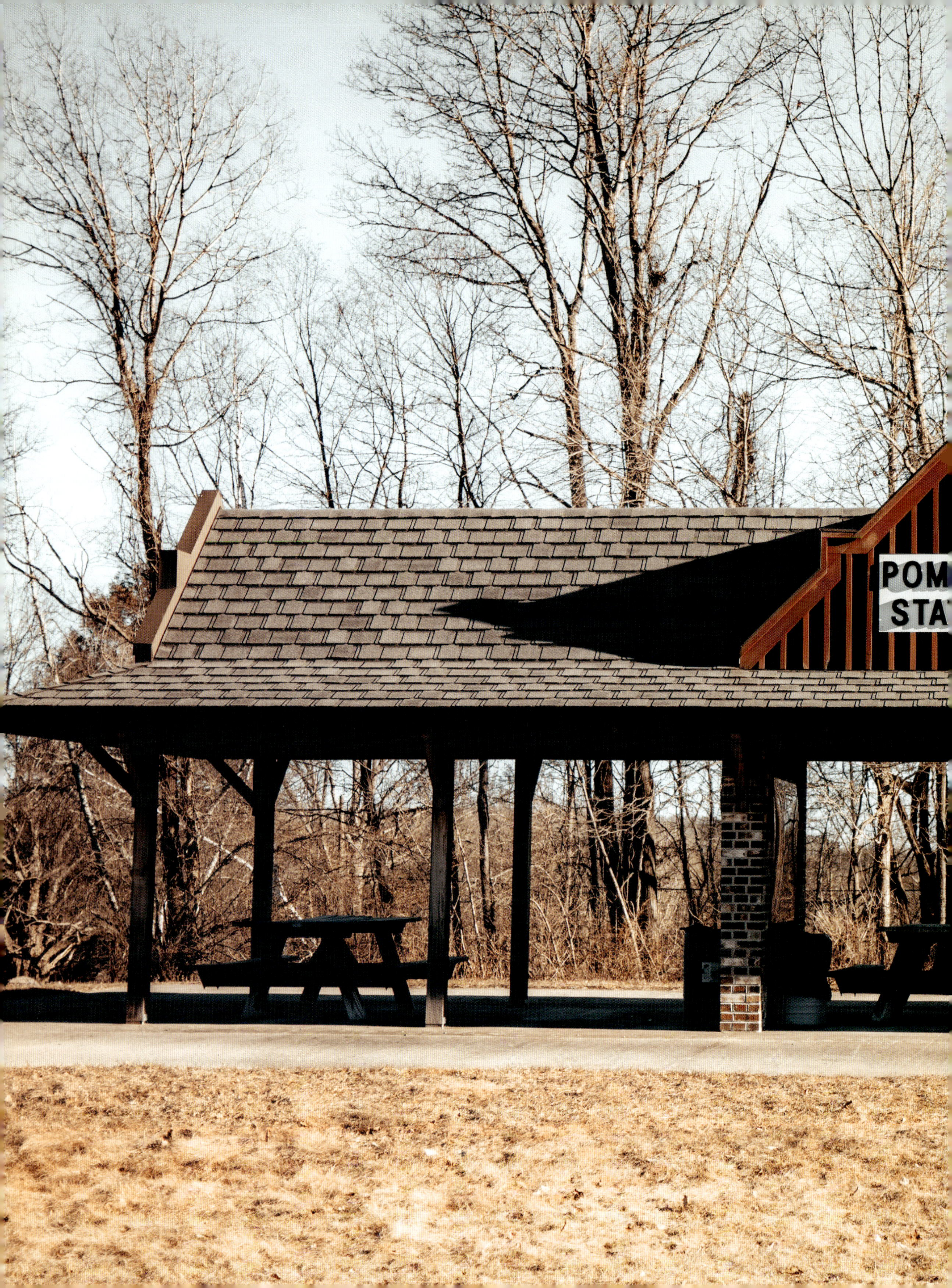

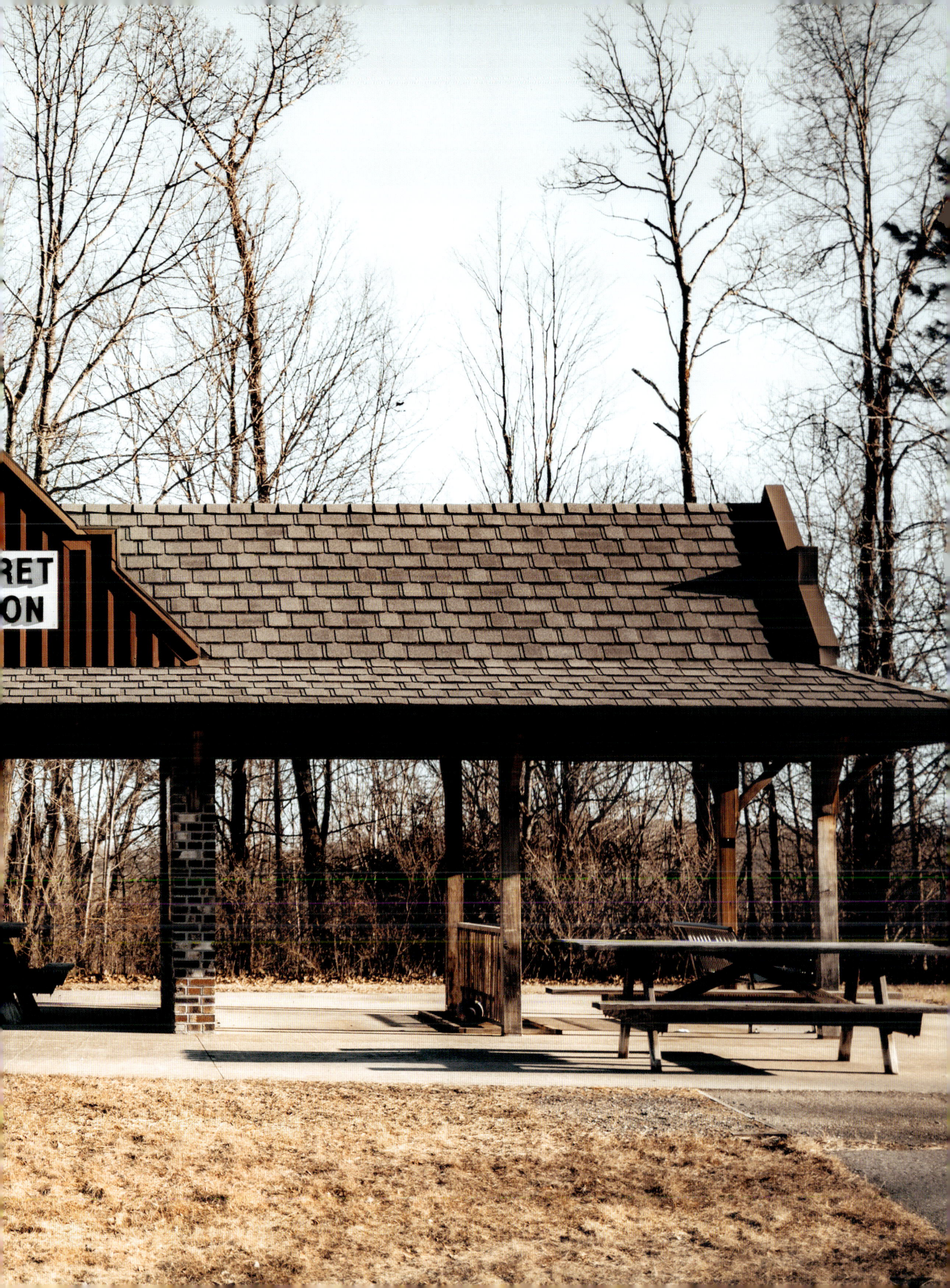

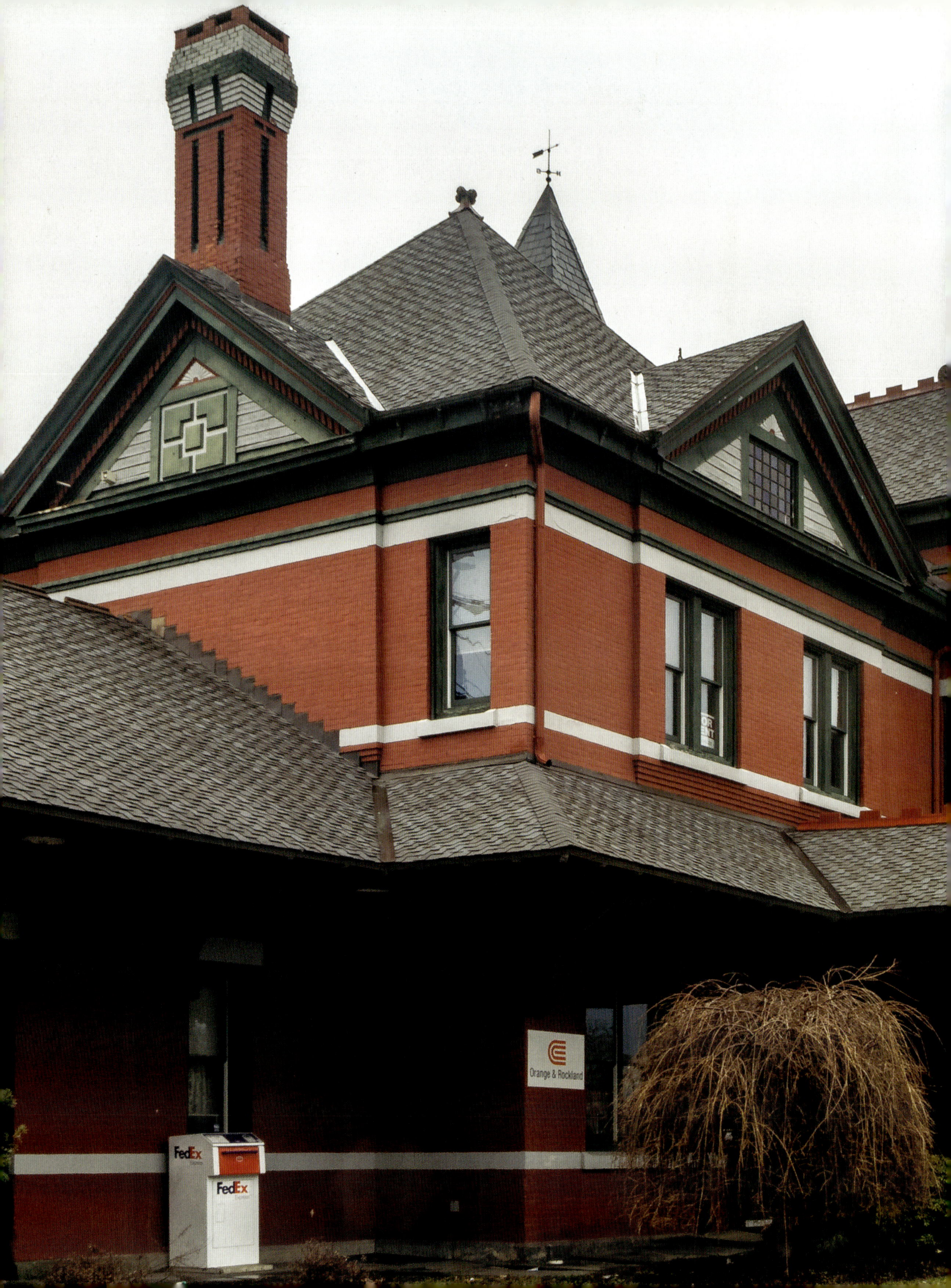
Orange & Rockland
FedEx
FedEx

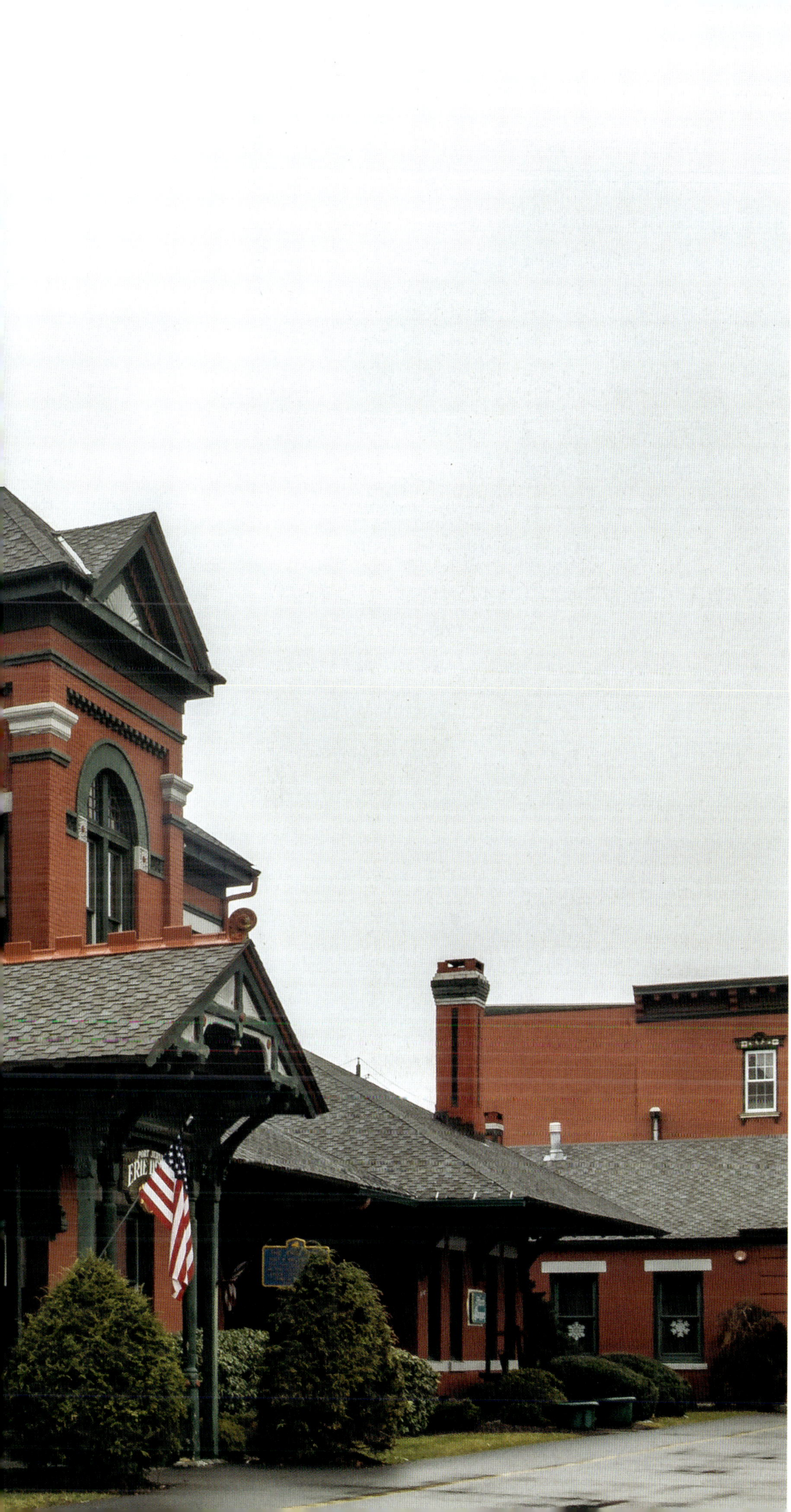

PREVIOUS PAGES:

Pomfret Station, Connecticut, USA

This airy picnic shelter stands on the site of the station opened on the Air Line Railroad in 1872. Successive buildings were put up at the depot, and the shelter's roofing style is modelled on the final one, which closed with the railway in the mid-1960s and was subsequently burned down.

LEFT:

Port Jervis, New York State, USA

The New York & Erie Railroad reached Port Jervis in 1847 but this substantial depot building dates from 1892 and was extended in 1912. The largest station on the Erie's Delaware Division, its size indicates the importance of railway communication to this small industrial town in the pre-motor era. Passenger services from here ended in 1970. After a few uncertain decades, the distinctive building has been restored as a commercial centre.

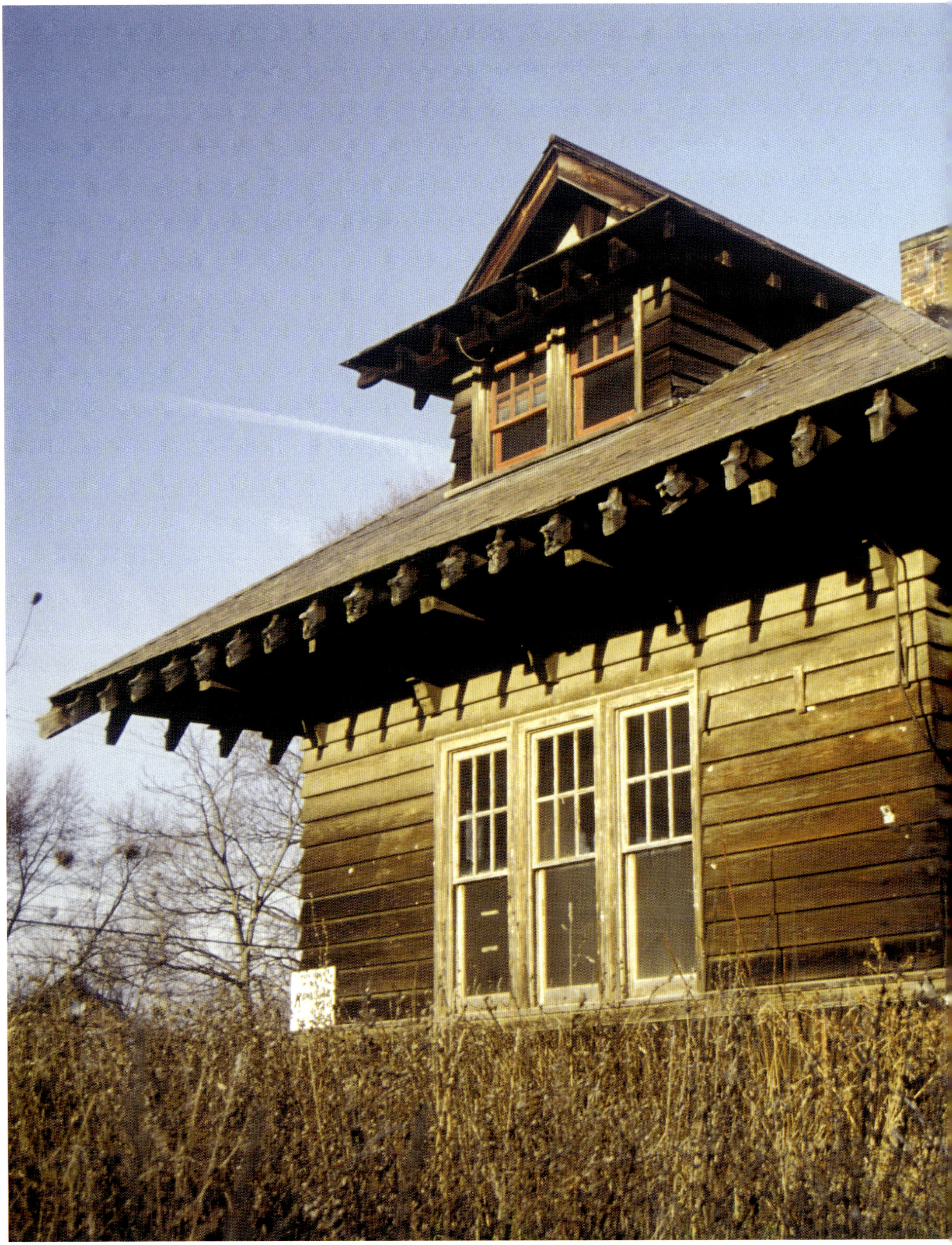

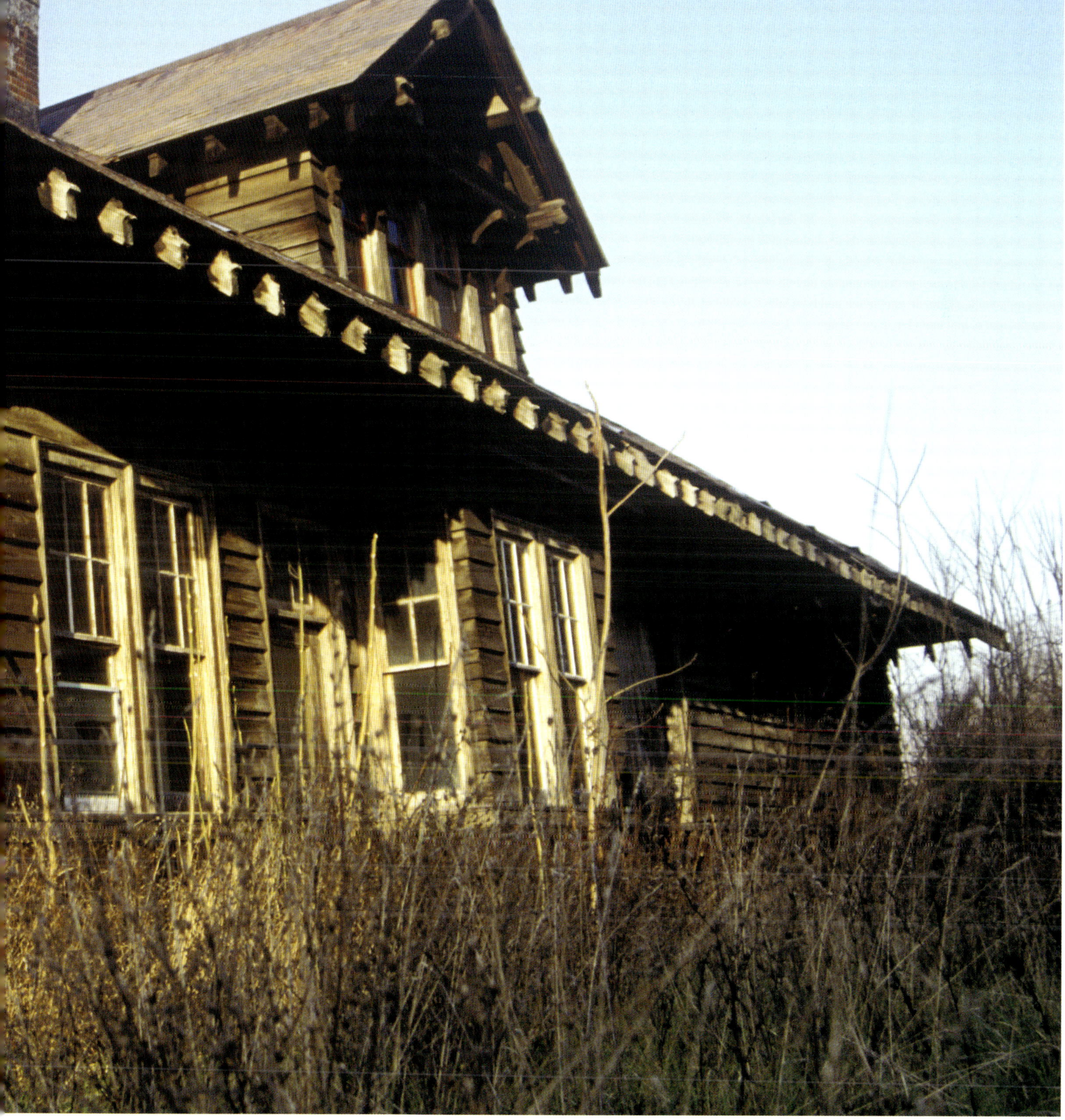

Accord Station, Ulster County, New York State, USA
Several railroad companies built lines in the Catskills from 1866, their stations almost invariably wood-built. Accord Station was opened in 1902, on the New York, Ontario & Western Railroad's Kingston branch. Rail services ended in 1958. The building, restored, is in use as a private residence.

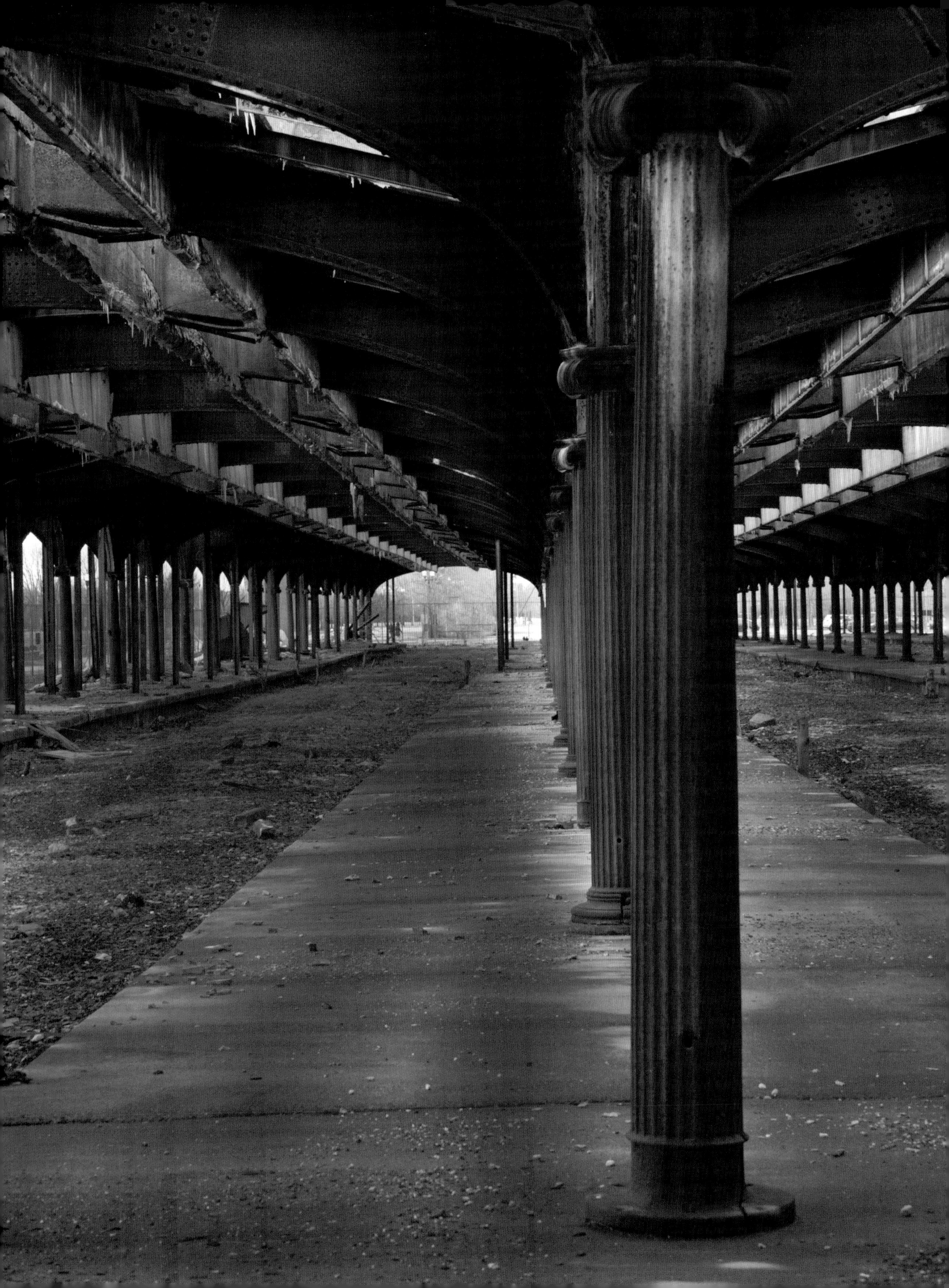

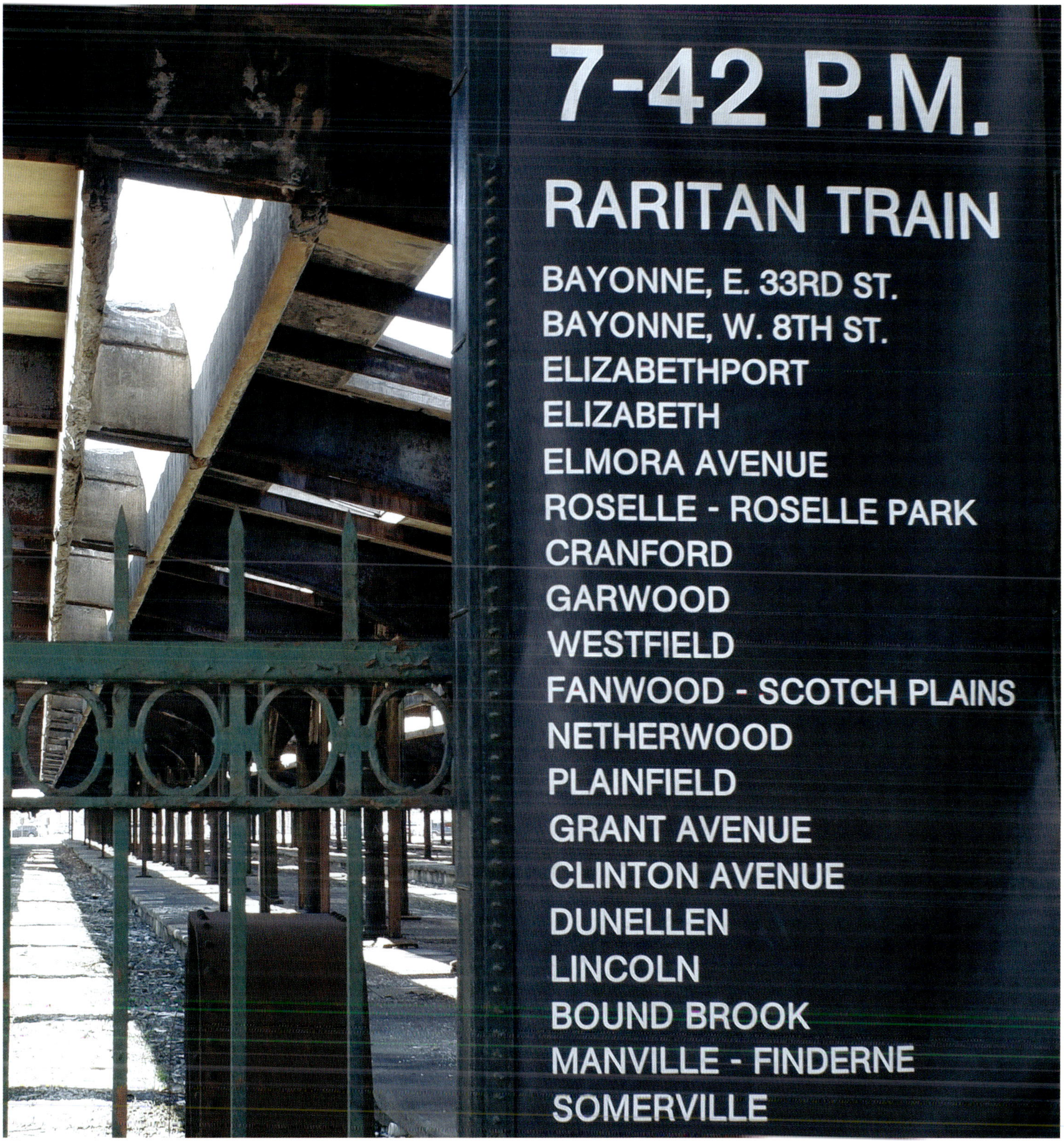

OPPOSITE, ABOVE & OVERLEAF:

Jersey City Terminal, Liberty State Park, New Jersey, USA

Uniquely among major termini, this vast station was built primarily for immigrant arrivals to the USA. Facing Ellis Island and the Statue of Liberty, it replaced a smaller terminal in 1889 and remained operational until 1967. While the main building has been renovated as a visitor centre, the entire platform area, with its awnings and ironwork, remains as it was more than 50 years ago, though the tracks have been lifted.

Restoration of the main hall includes refurbished destination boards for long-gone trains (pictured above).

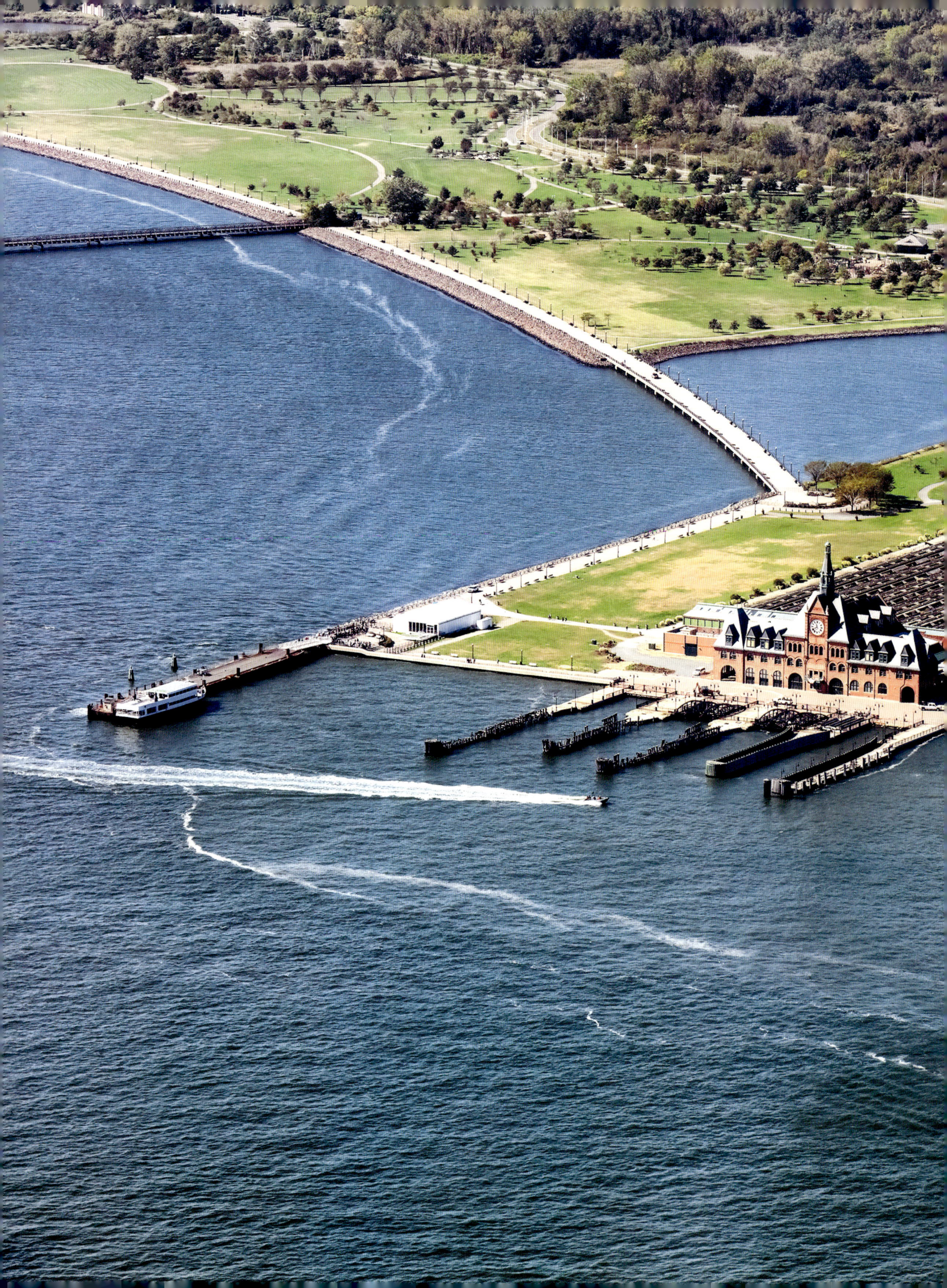

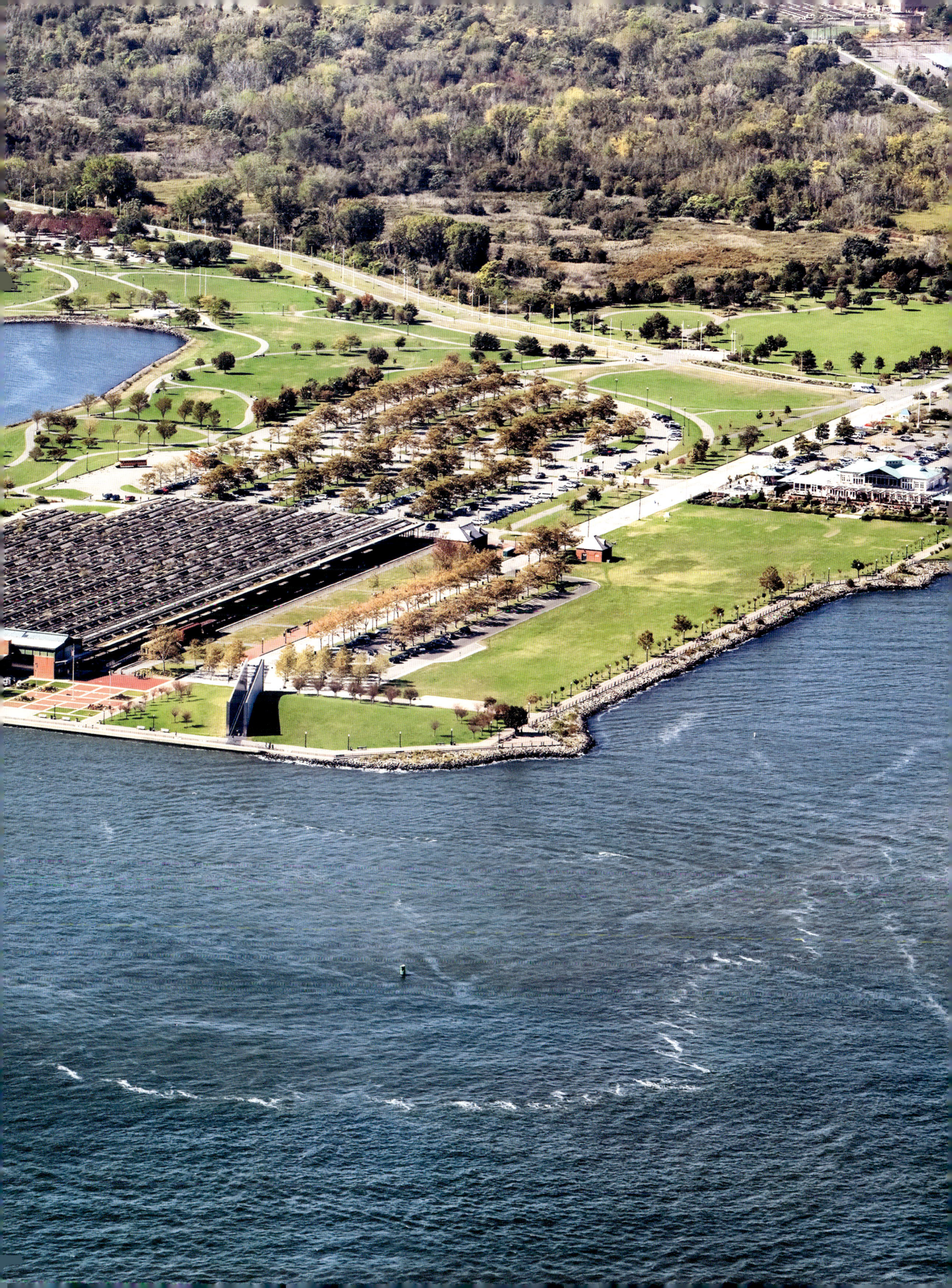

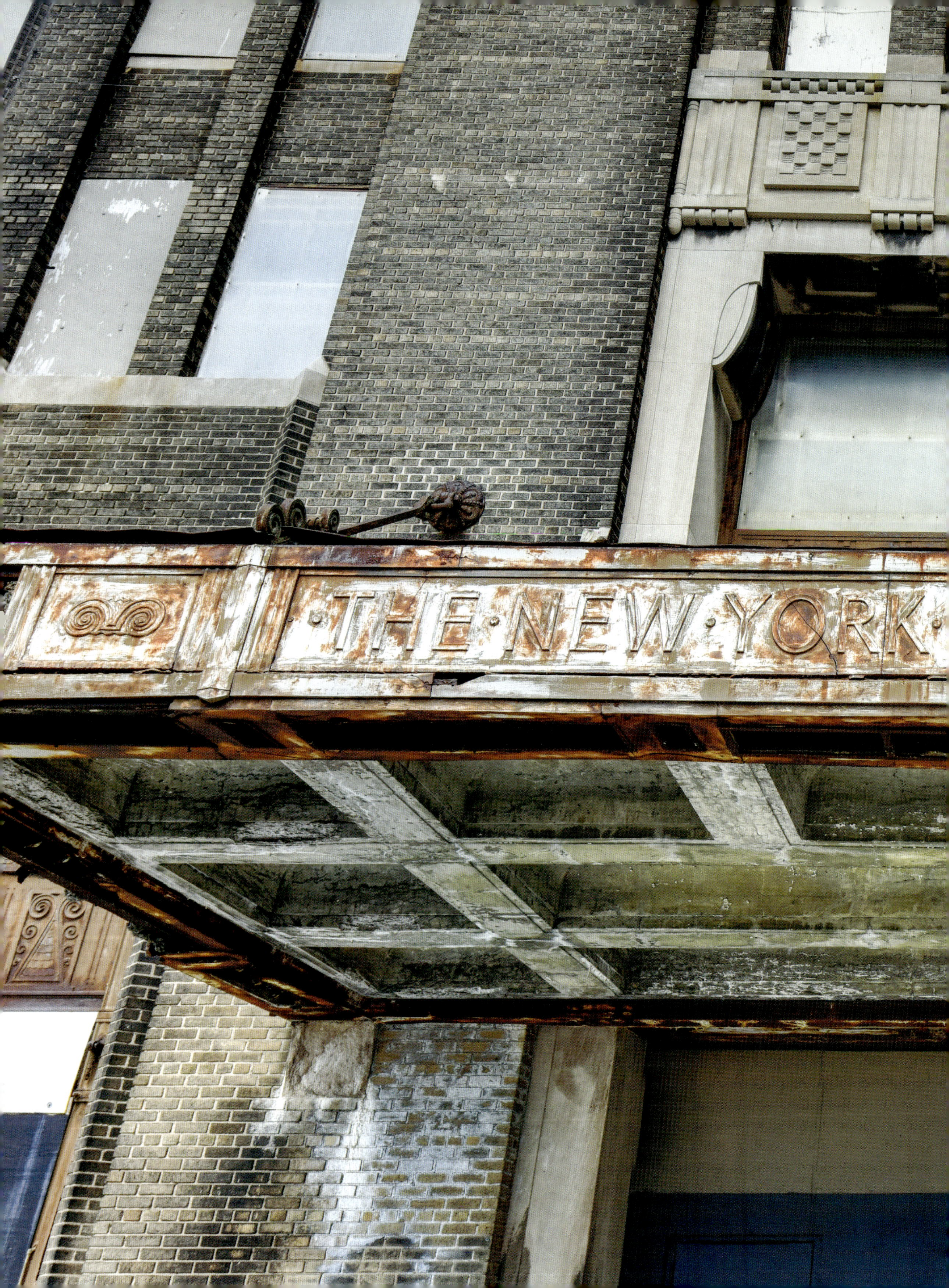
THE NEW YORK

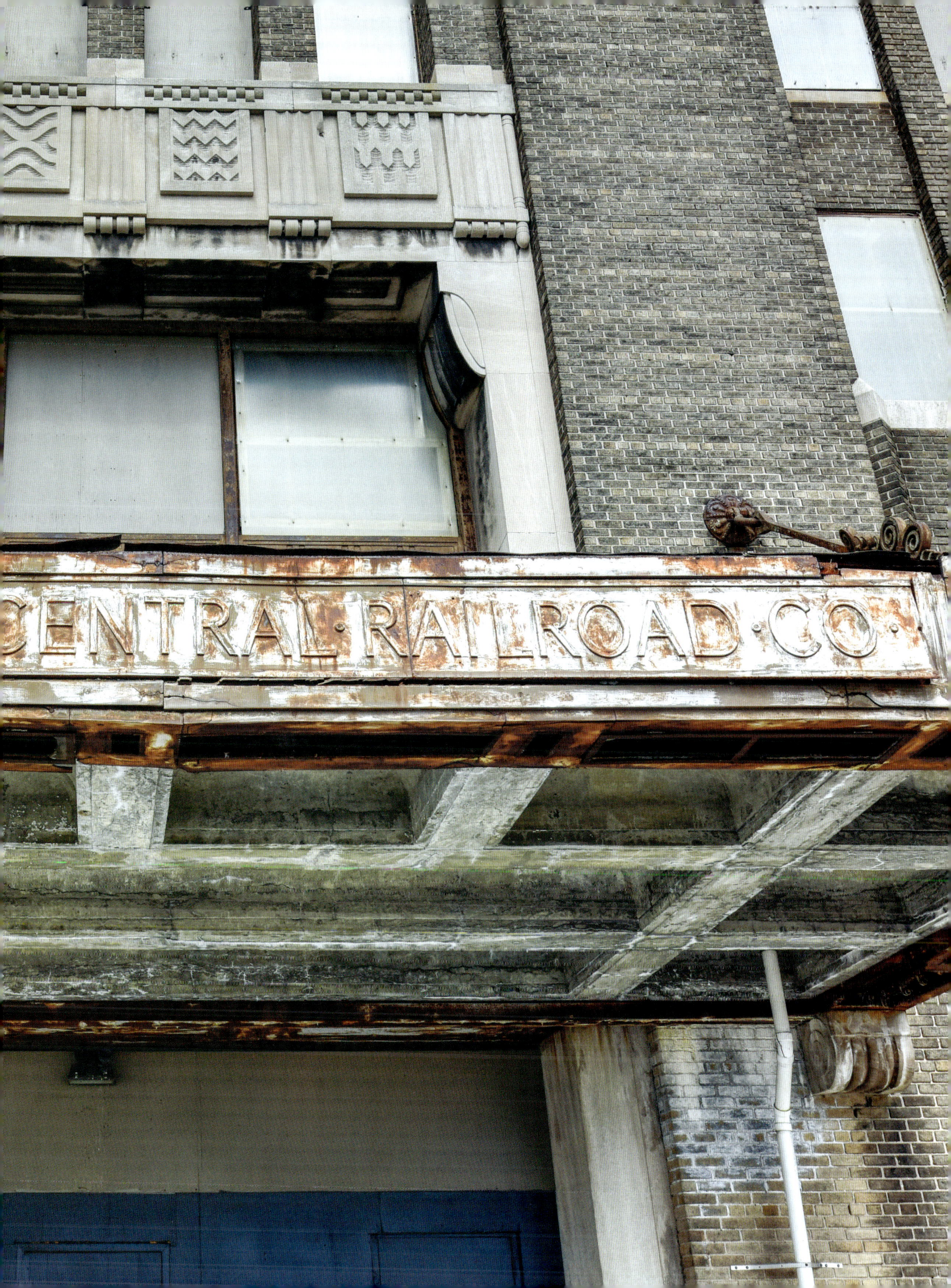
CENTRAL·RAILROAD·CO·

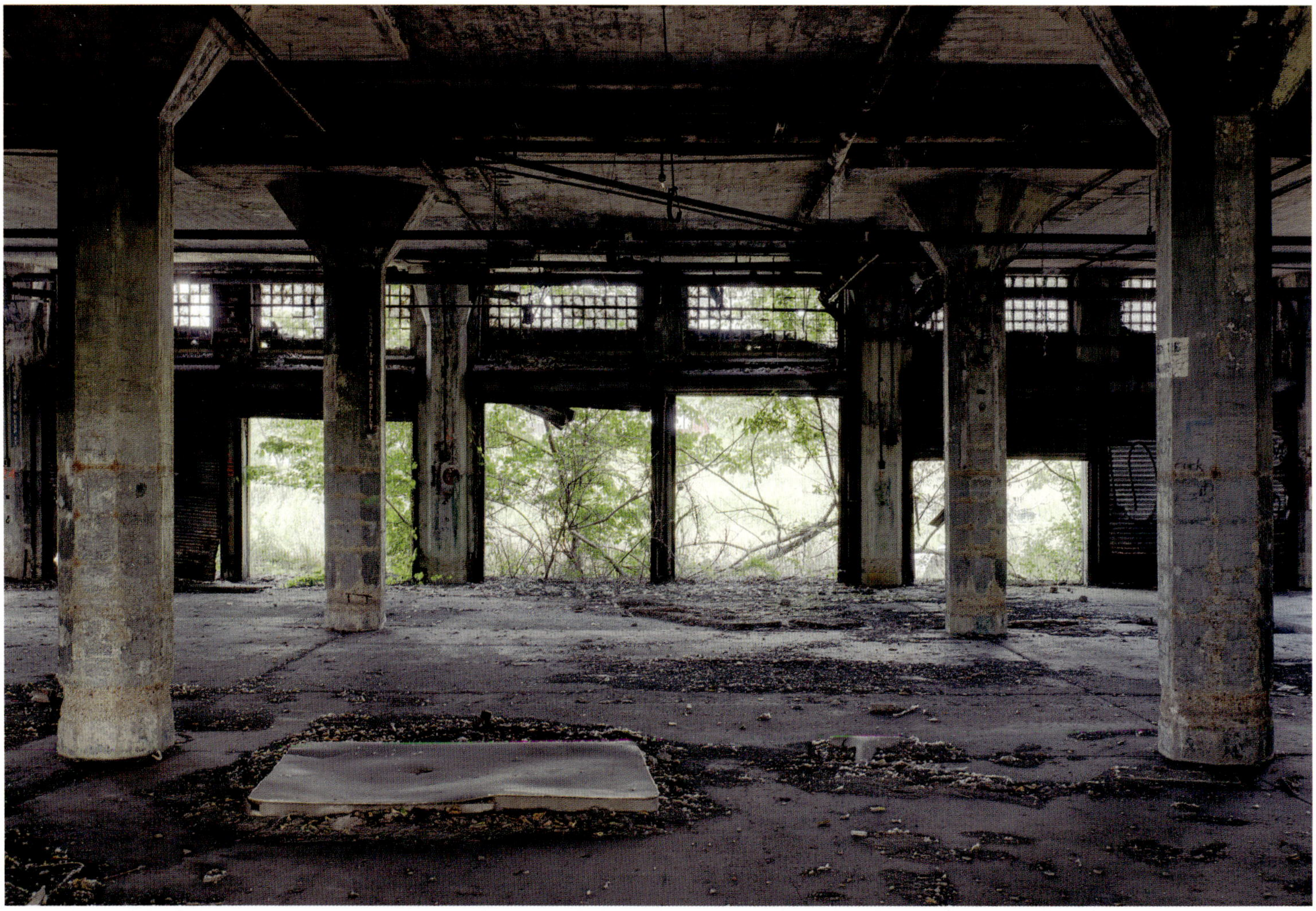

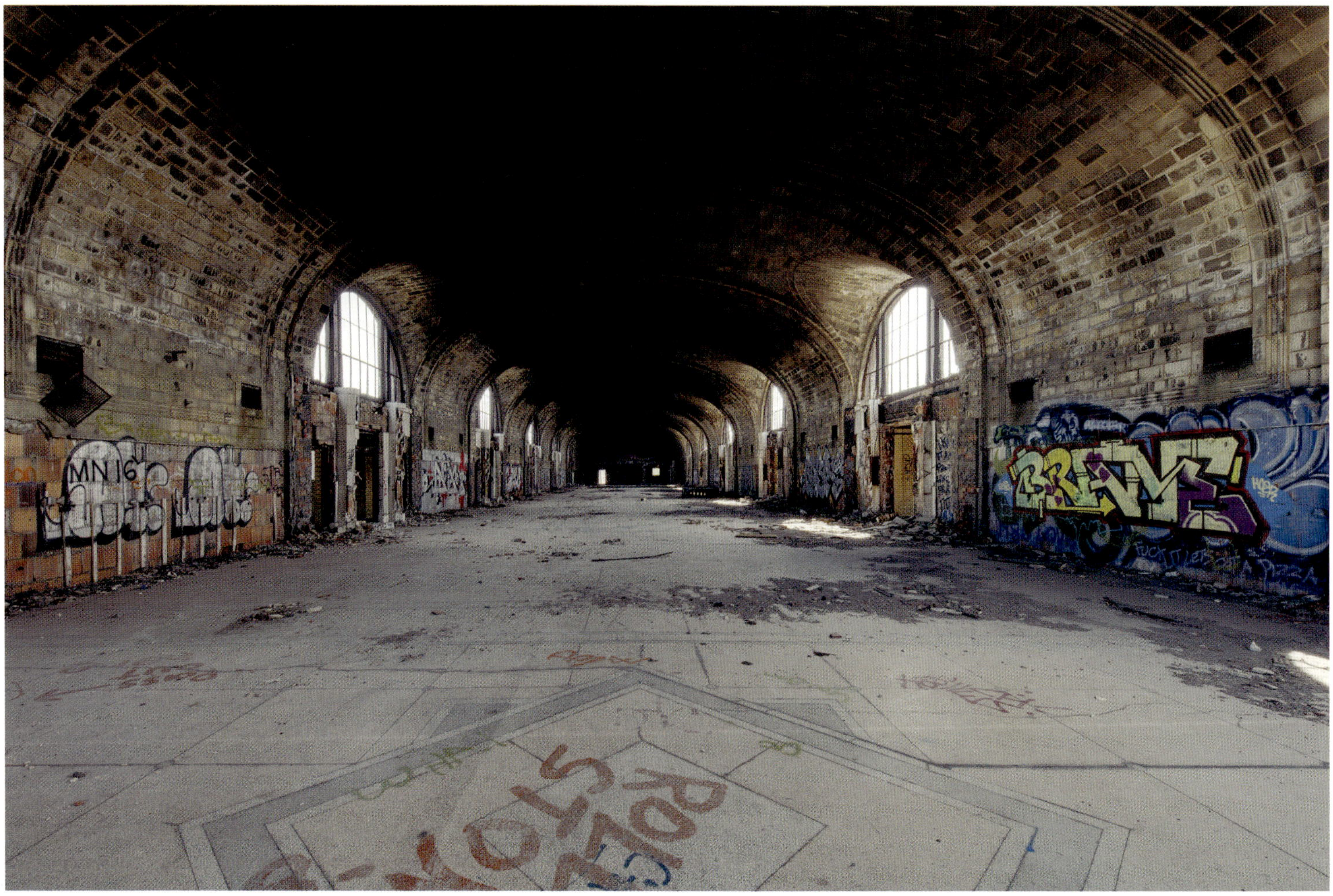

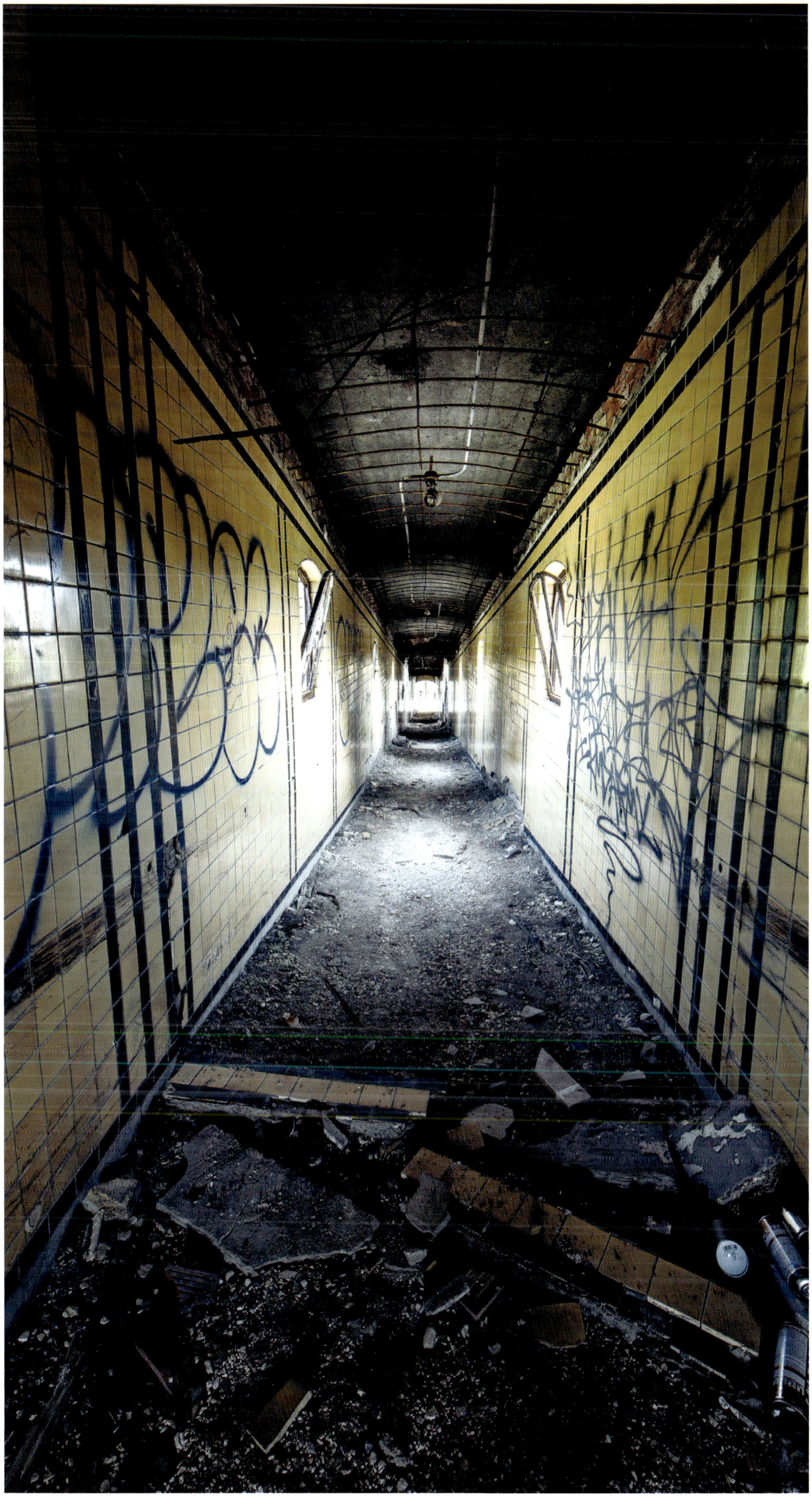

PREVIOUS PAGES AND ALL PHOTOGRAPHS ON THESE PAGES:

Central Station, Buffalo, New York State, USA

Nine railroad companies served Buffalo but the New York Central was the one that provided most services, including the 'Twentieth Century Limited' and 'Empire State Express'. Its main station was not a terminal but a through one allowing for New York–Chicago trains. Inspired by Eliel Saarinen's great Art Nouveau station in Helsinki, it was an opulent building.

Unfortunately, it was not particularly central to the city, and never achieved the passenger numbers it was built for. Opened in 1929, it closed to traffic 50 years later. Looted and vandalised, it was rescued in 1997 by a preservation group, but a use for it has yet to be found.

Vandalism in the years of neglect removed or destroyed all the equipment and many other features, but the graffiti at low level cannot obscure the fine detail of the brickwork in this long ground-floor arched passageway (opposite below).

Partial restoration is now in progress but funds are limited and the task is immense.

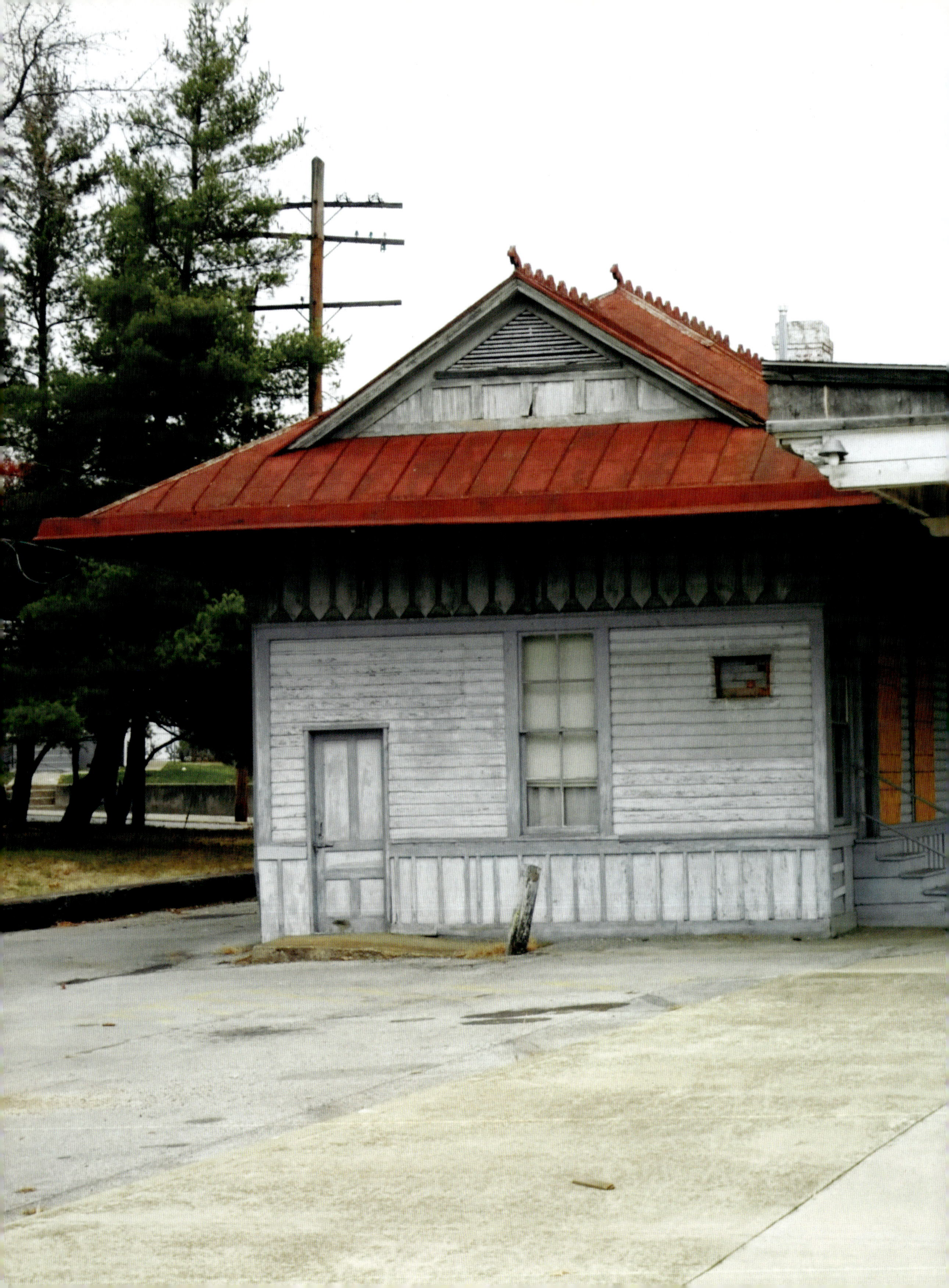

Paris, Bourbon County, Kentucky, USA
Opened in 1882 by the Louisville & Nashville Railroad, with a utilitarian awning tacked on later, Paris saw its last passenger train in 1968. The freight-only tracks of the Transkentucky Transportation Railroad adjoin the deserted passenger depot. Since this photograph was taken, the building has been restored as a restaurant and bourbon bar.

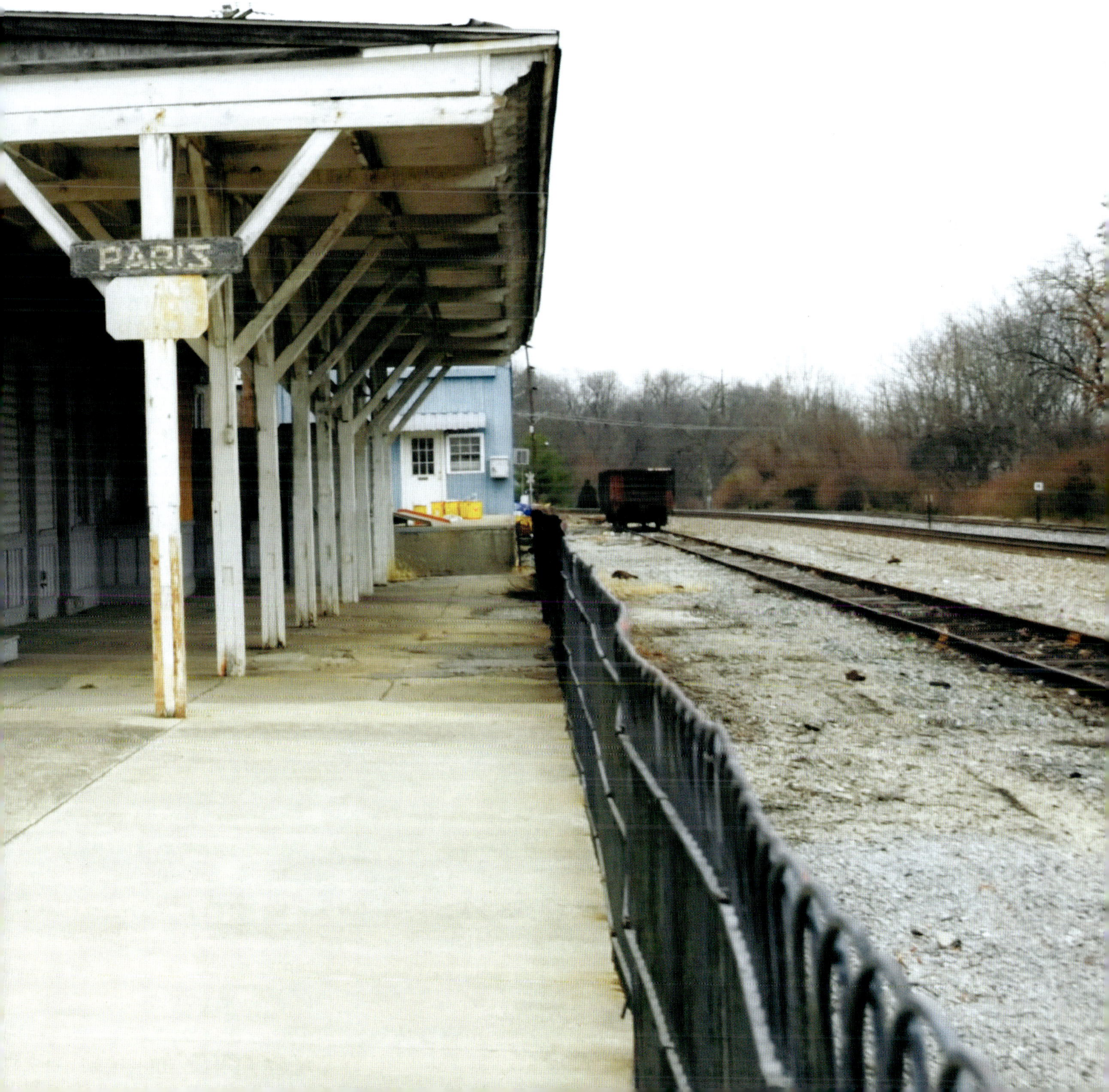

Bureau Junction, Bureau County, Illinois, USA

Here the Peoria & Rock Island Line's branch to Peoria diverged from the main Chicago–Rock Island line. After the branch opened in 1871 the village supported a community of railway workers, with a locomotive roundhouse and a depot house, which also served as a boarding house and restaurant. Only the clapboard station building remains.

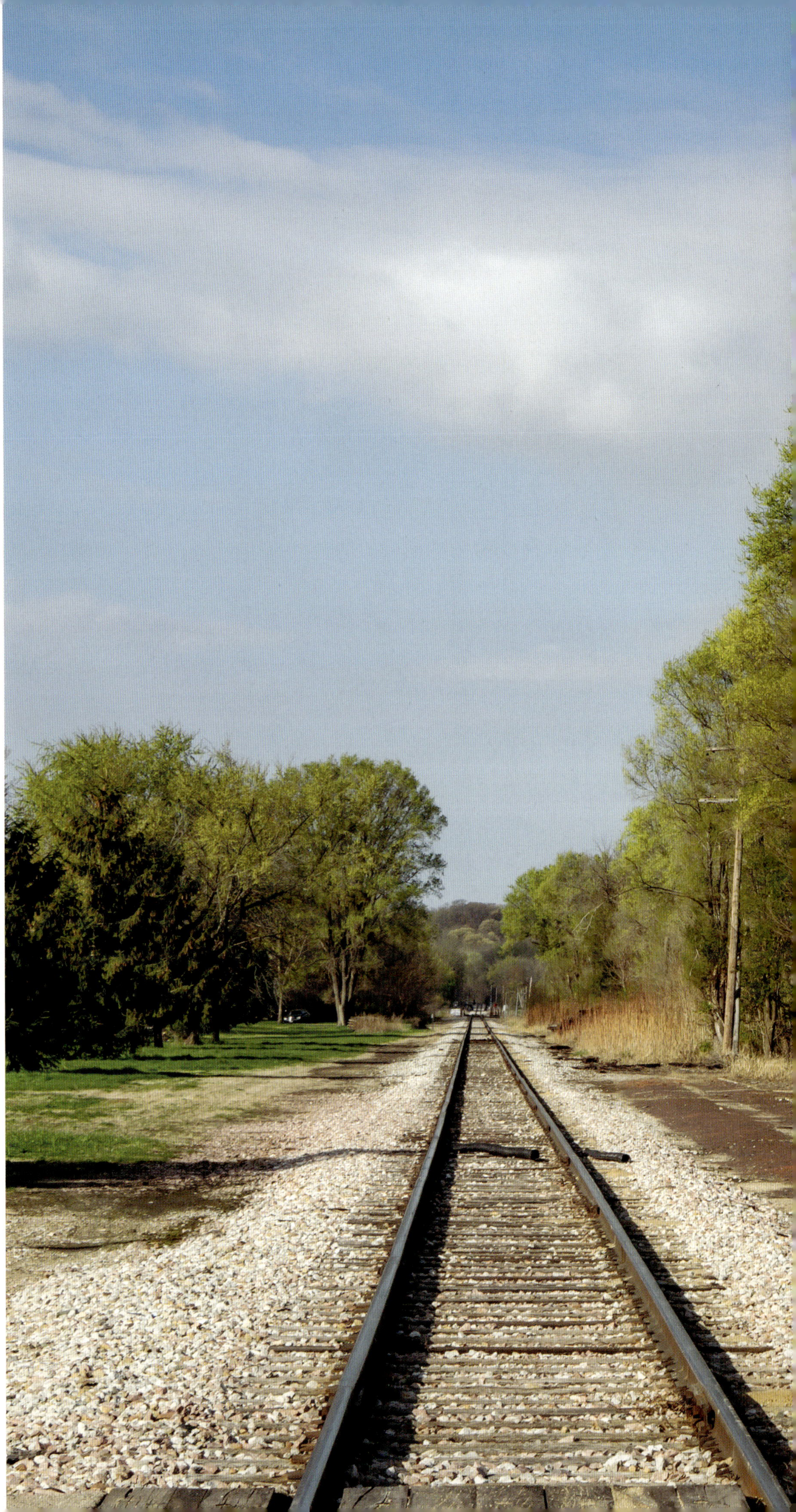

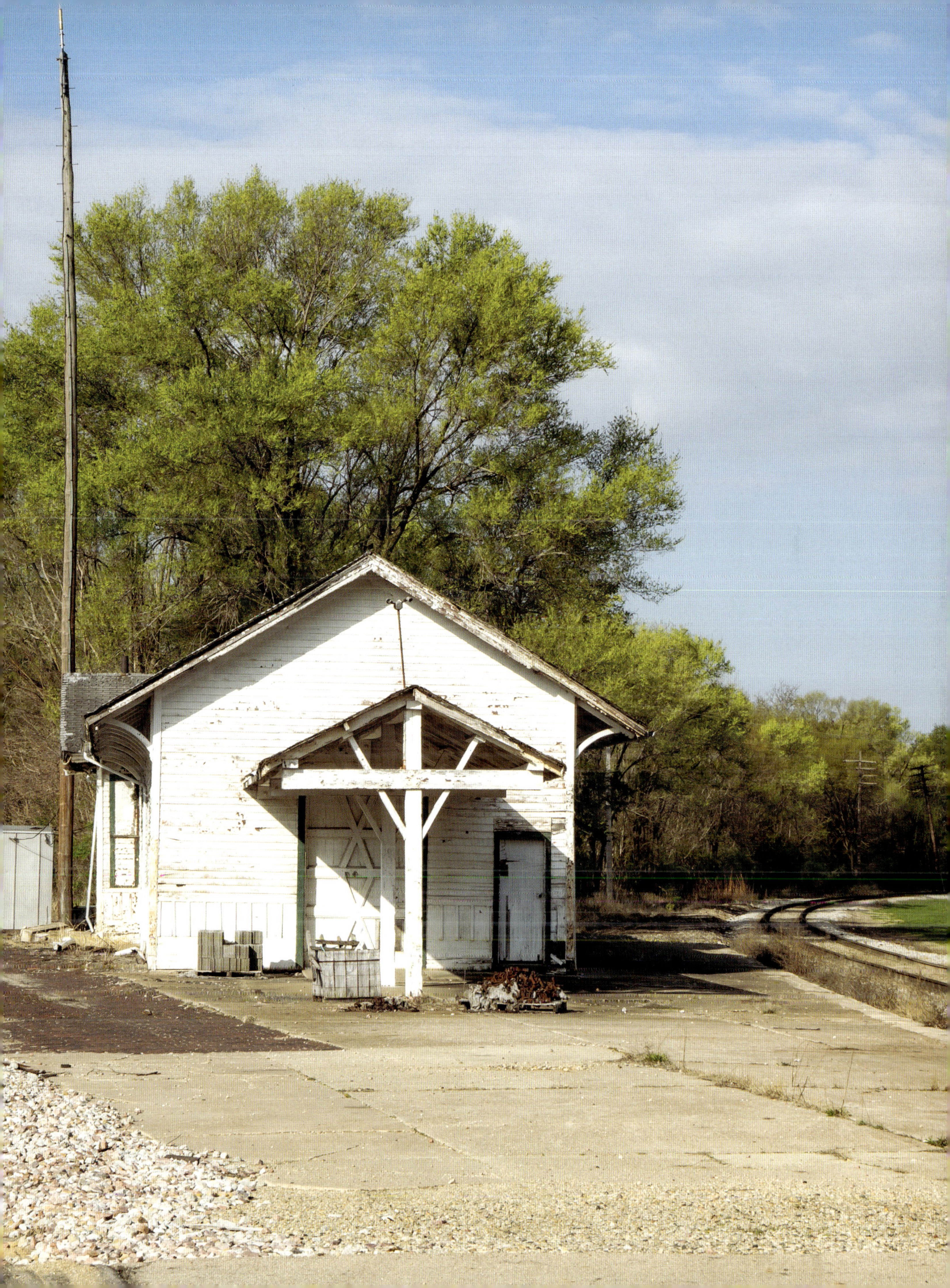

HERE & OVERLEAF:

Michigan Central Station, Detroit, Michigan, USA

From 1914 to its closure in 1988, this was Detroit's main station, built and owned by the Michigan Central Railway. Like many other city stations of the period, it was designed in the Beaux-Arts style. Its 13-storey office block made it the world's tallest station building at the time. After thirty years of neglect, the Ford Company bought the old station in 2018 to become the centrepiece of a new 'Michigan Centre' area in the run-down city.

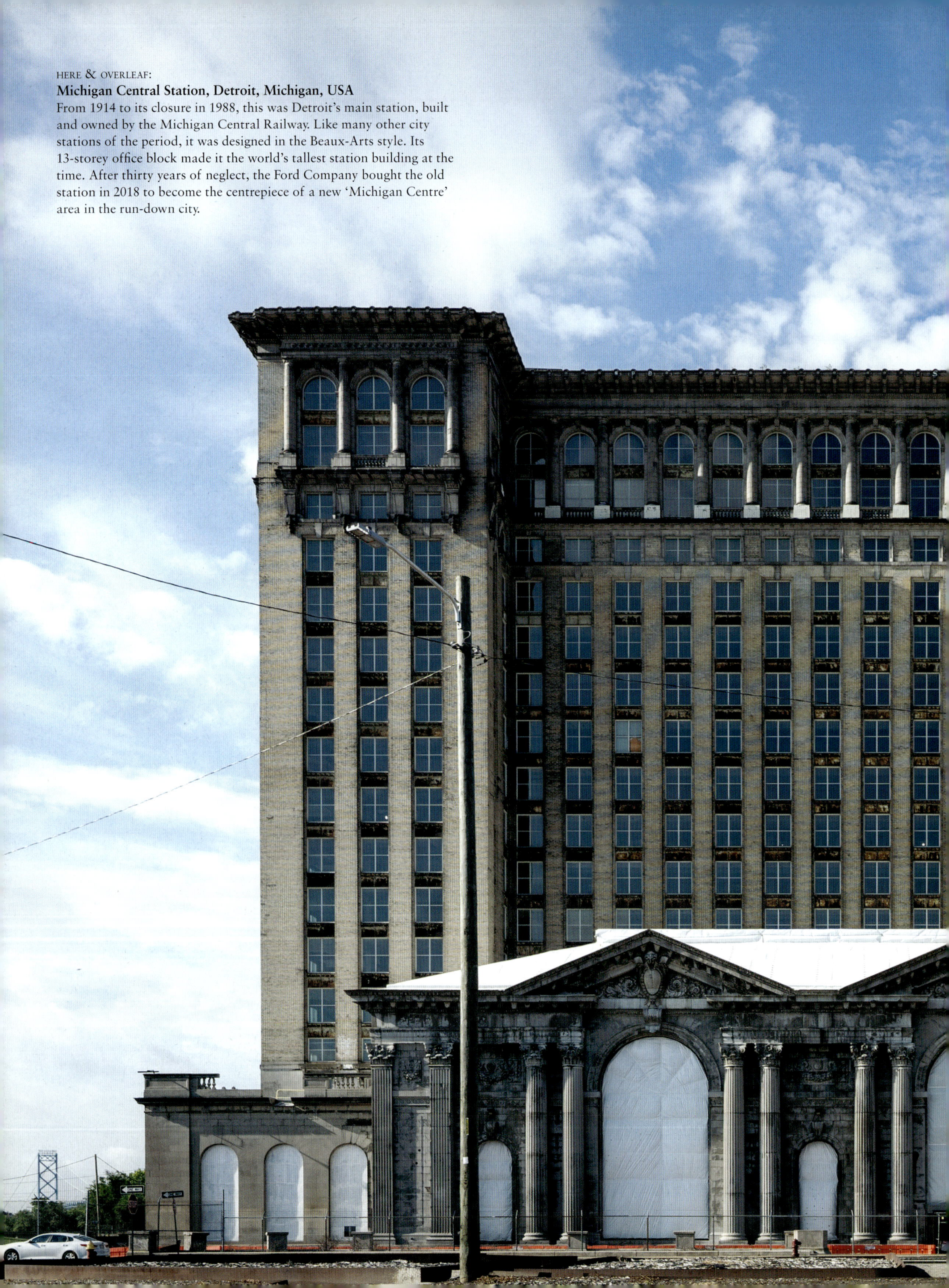

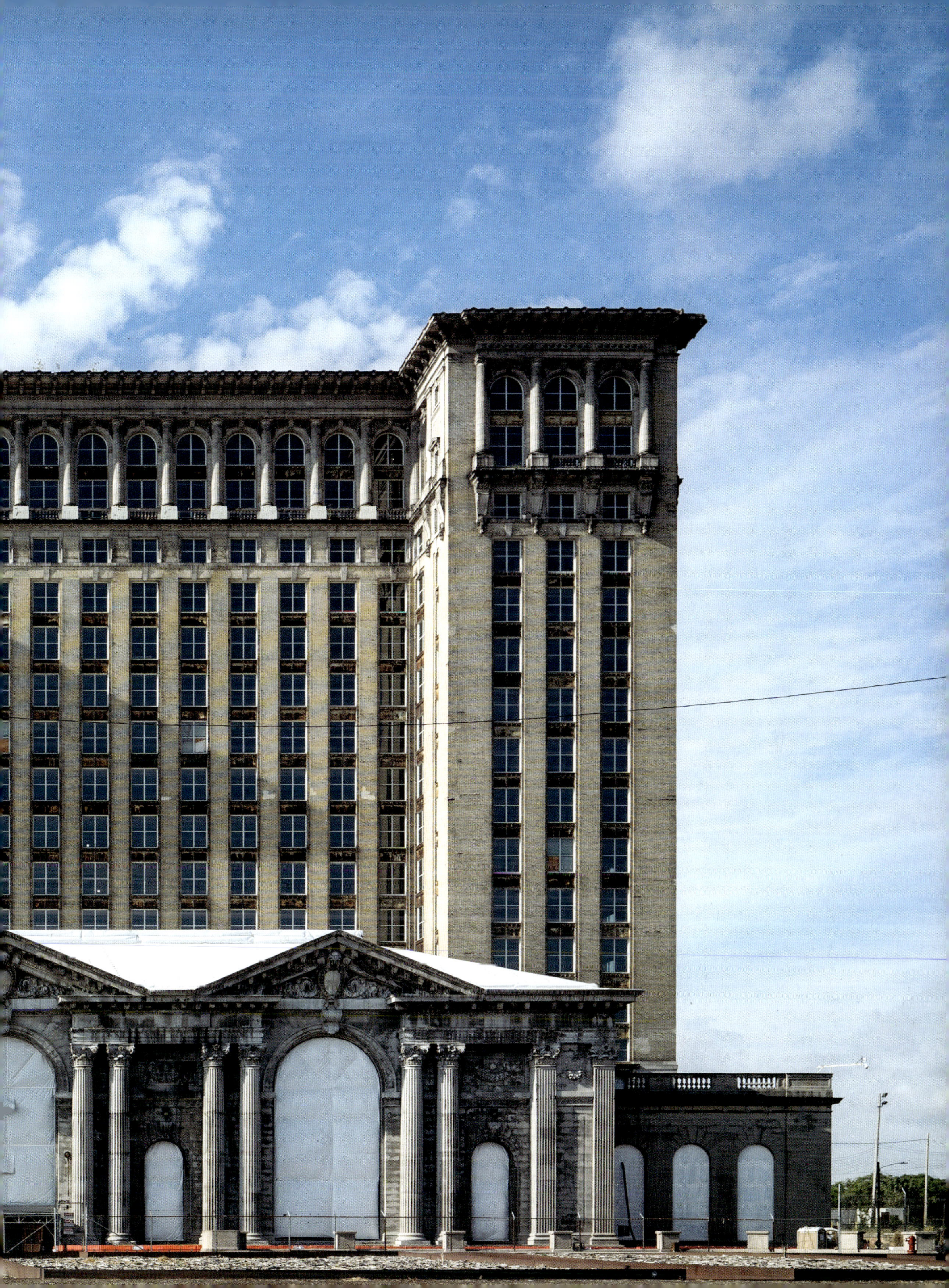

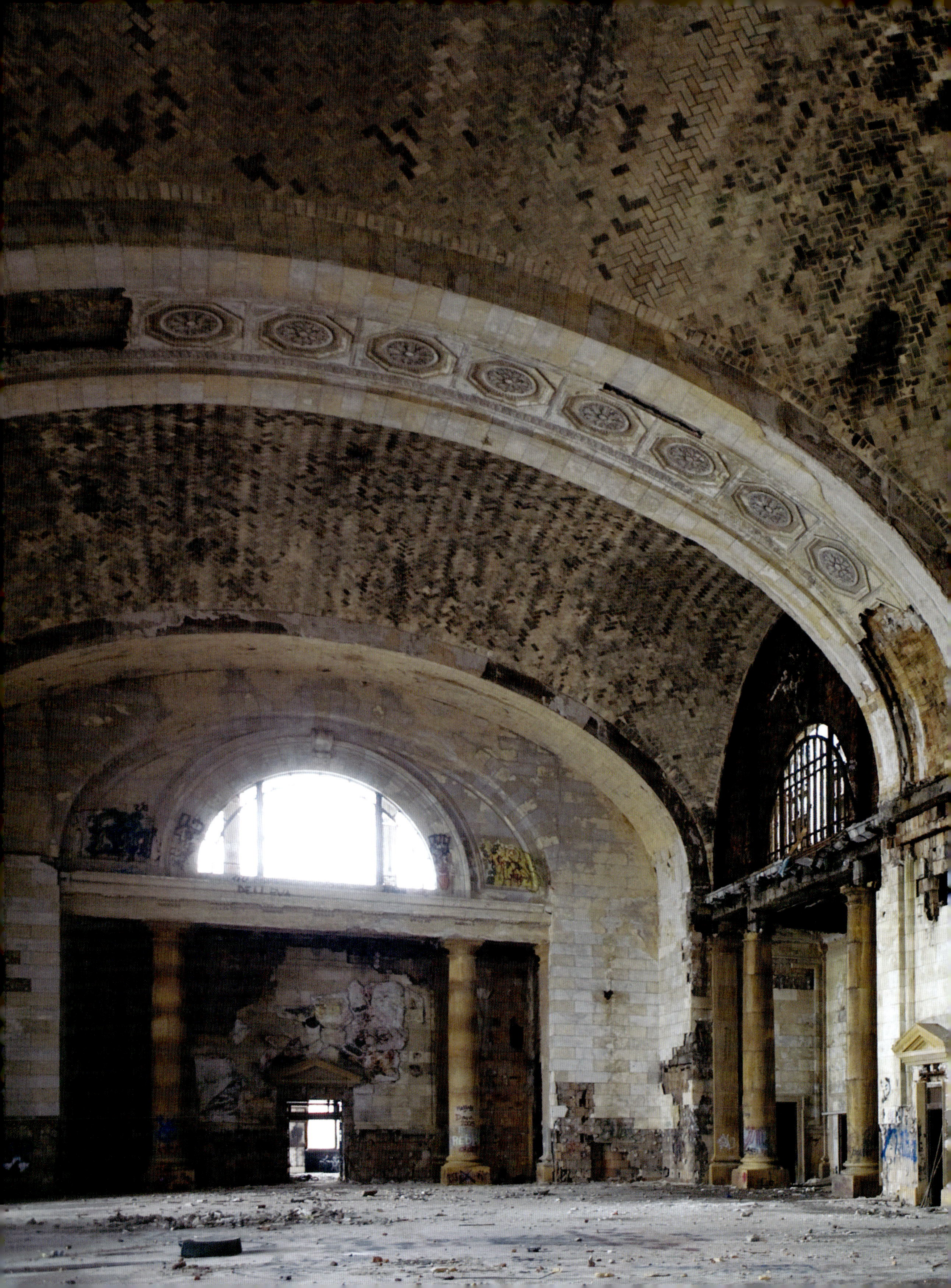

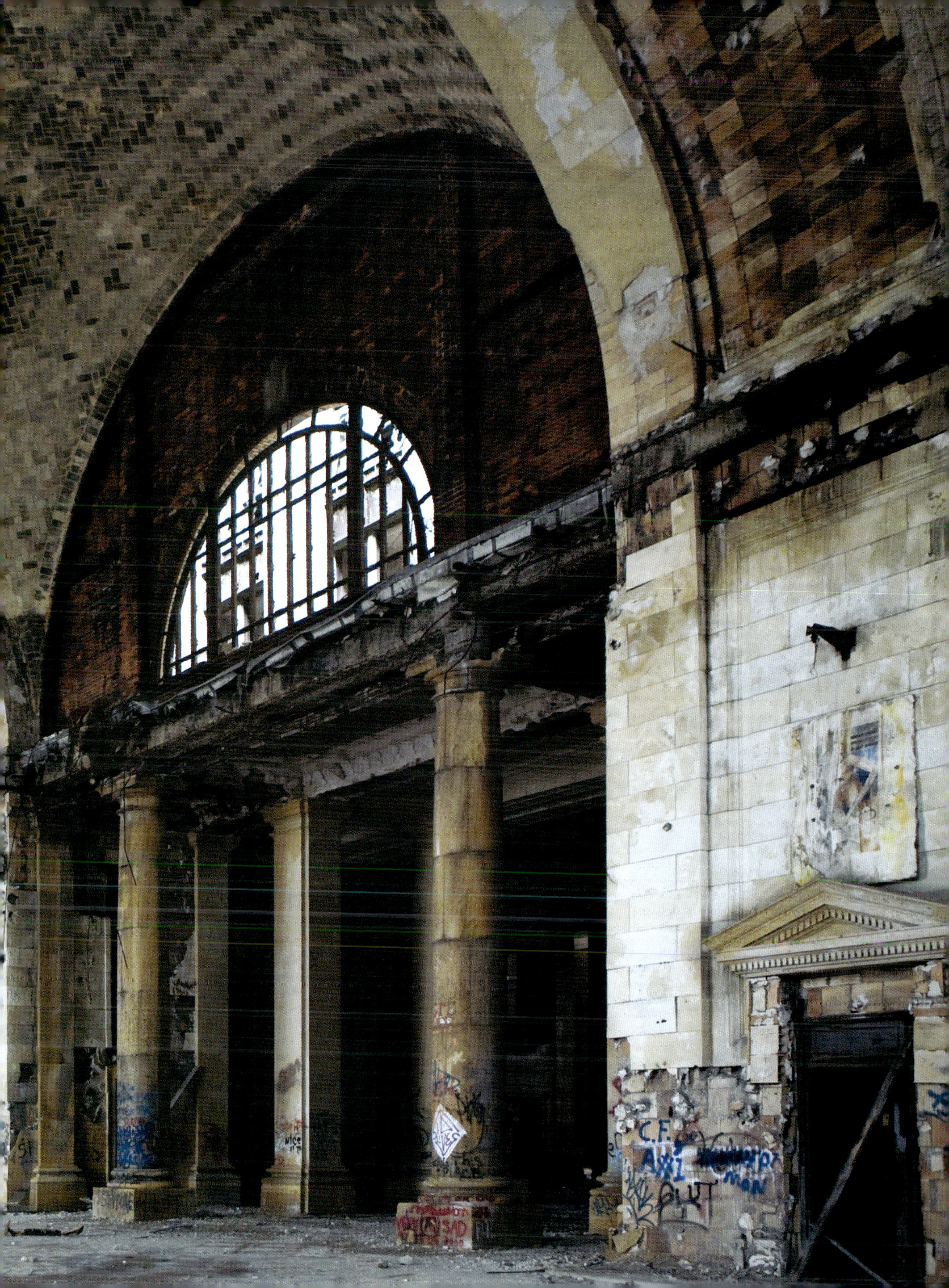

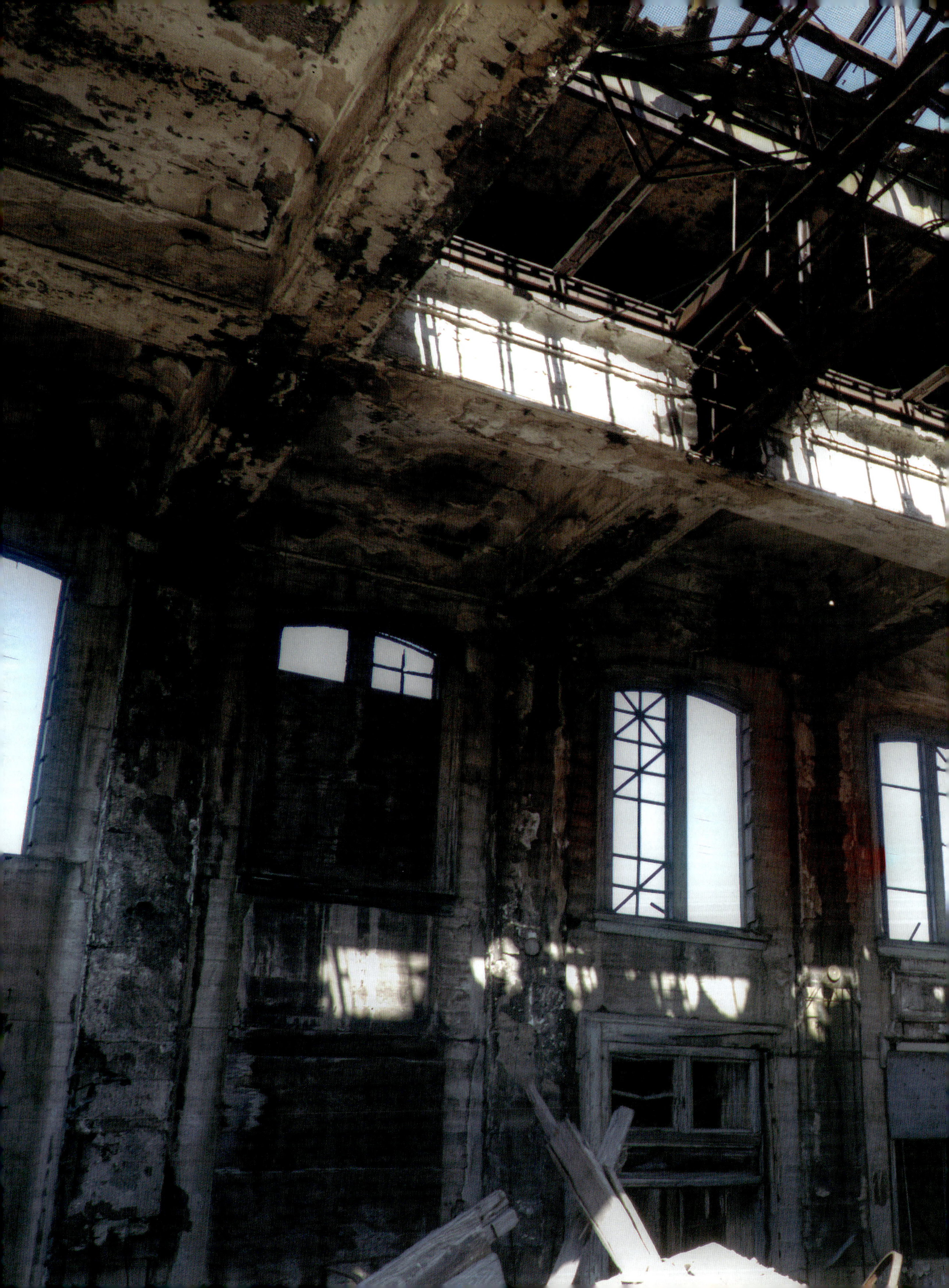

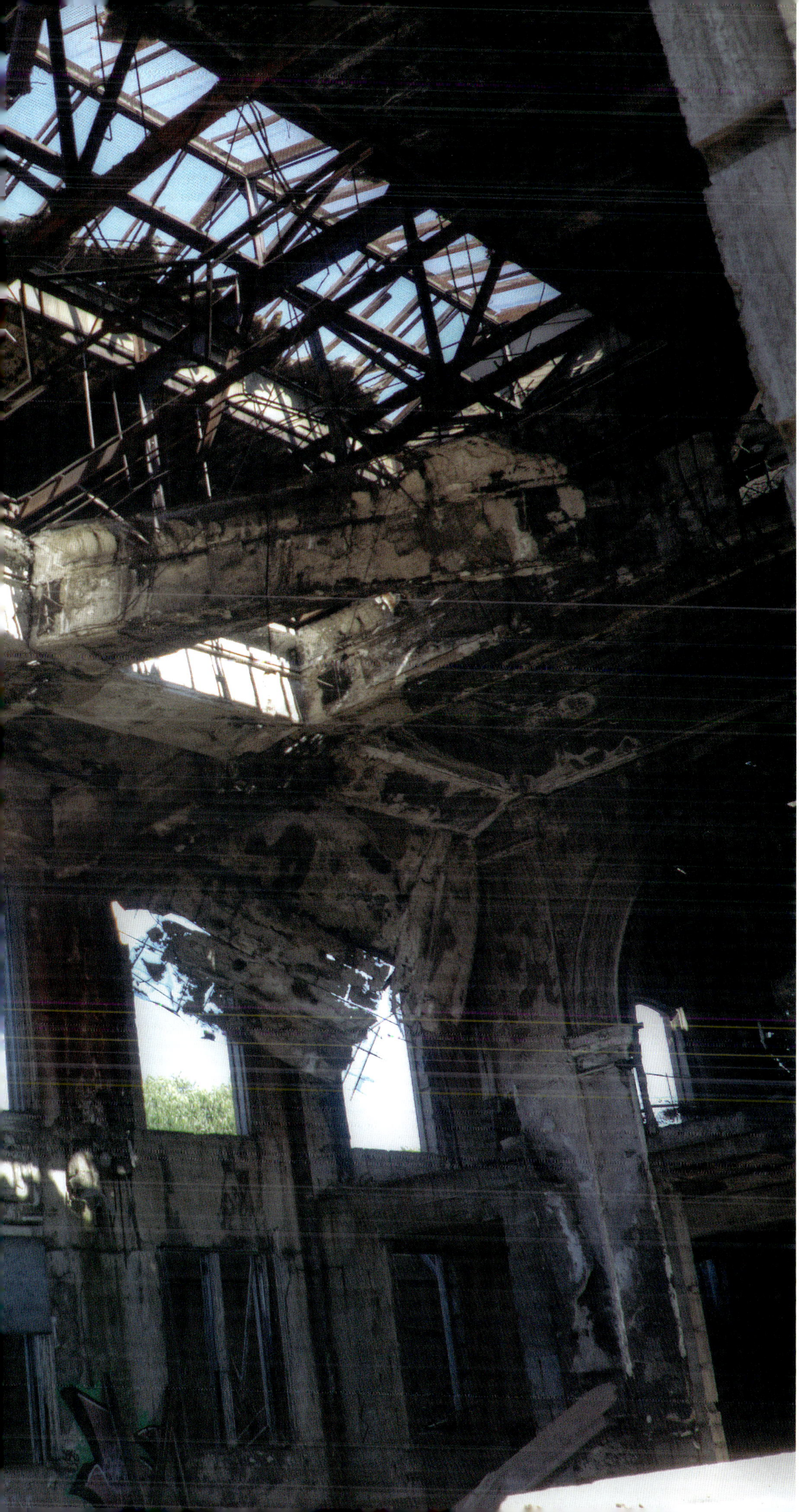

LEFT & OVERLEAF:
Union Station, Gary, Indiana, USA
Gary's classical-style station (concrete scored to look like stonework), built in 1910 by the Lake Shore & Michigan Southern and Baltimore & Ohio Railroads, remains one of the finest buildings in Gary. Internally, though, it is a wreck. Closed down and abandoned in the late 1950s, in 2018 it was sold to a renovation group for $10.

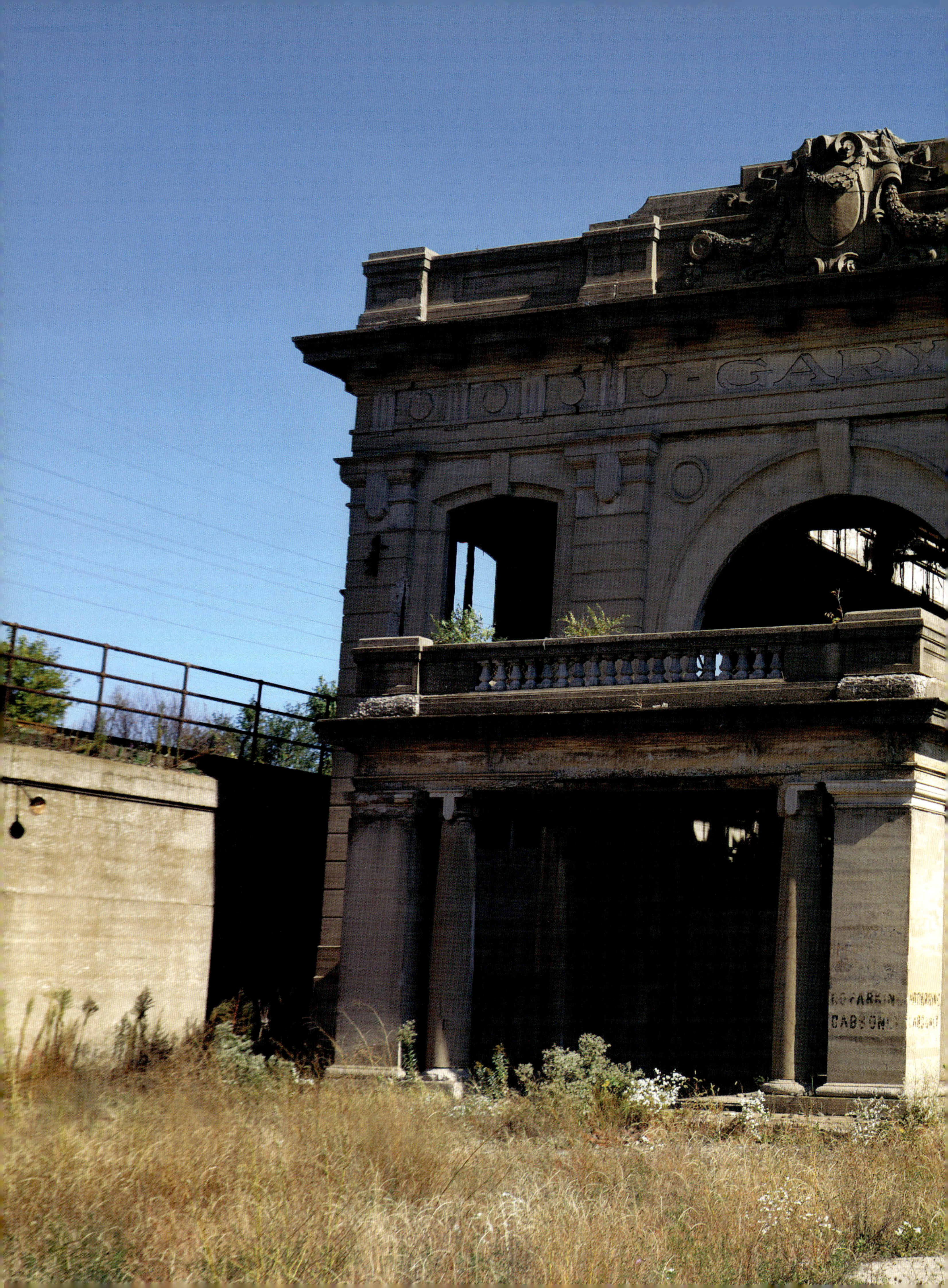
GARY

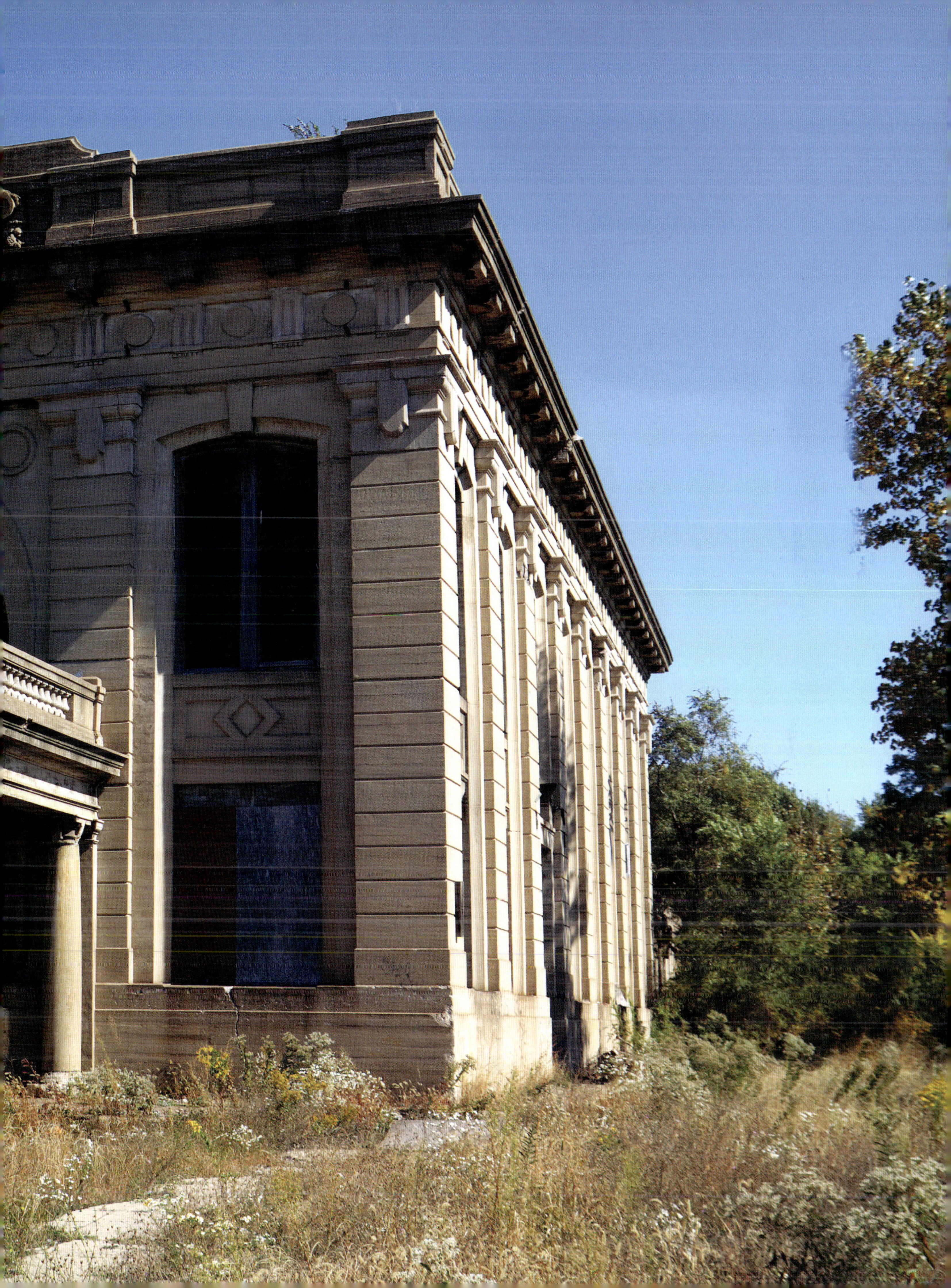

Union Station, Joplin, Missouri, USA
Opened in 1911, the disused station was once a busy place serving several railroads. Mining waste was incorporated in the concrete used to build it. The station was closed in 1969 and is in the National Register of Historic Places.

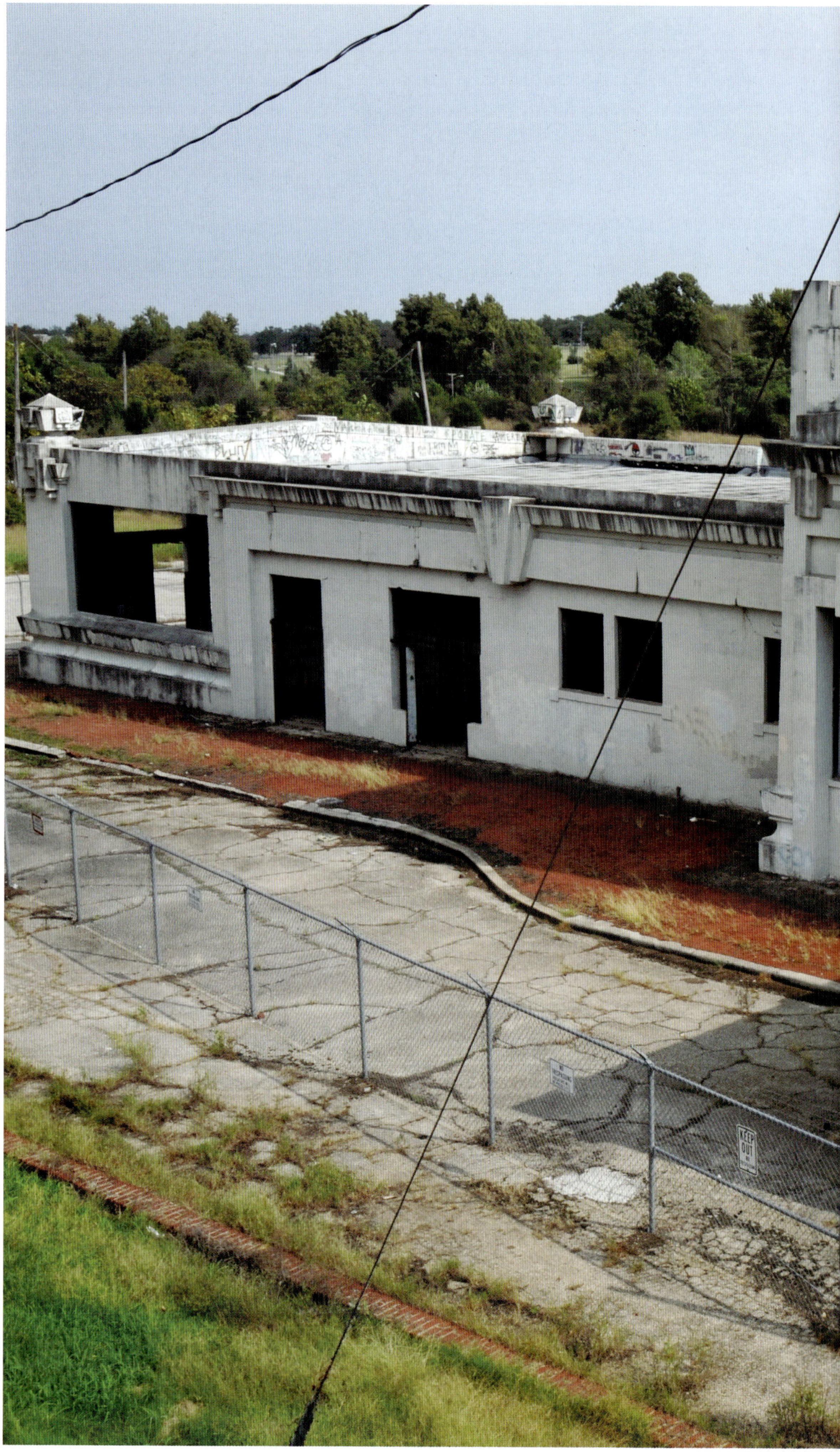

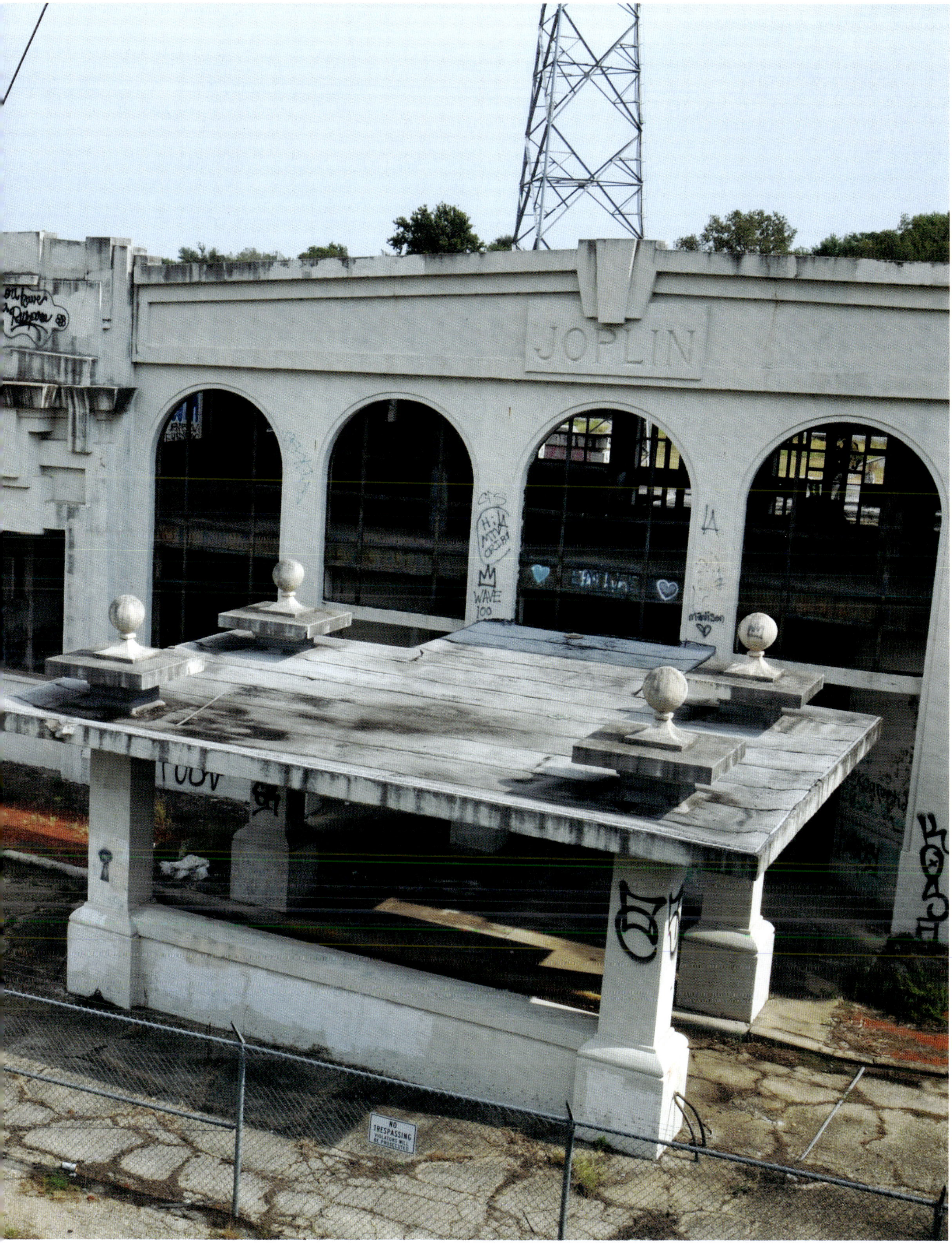
JOPLIN
NO TRESPASSING

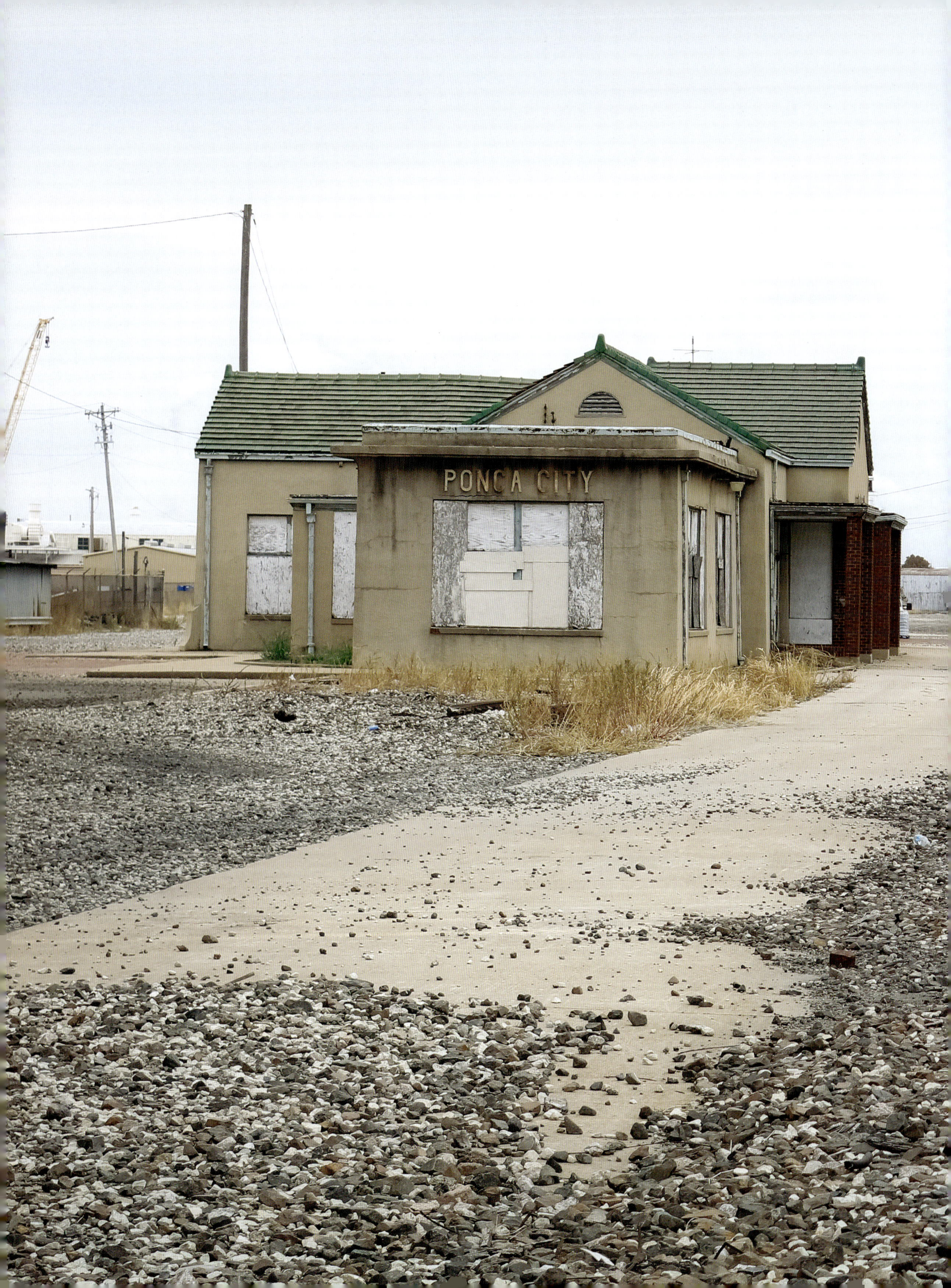
PONCA CITY

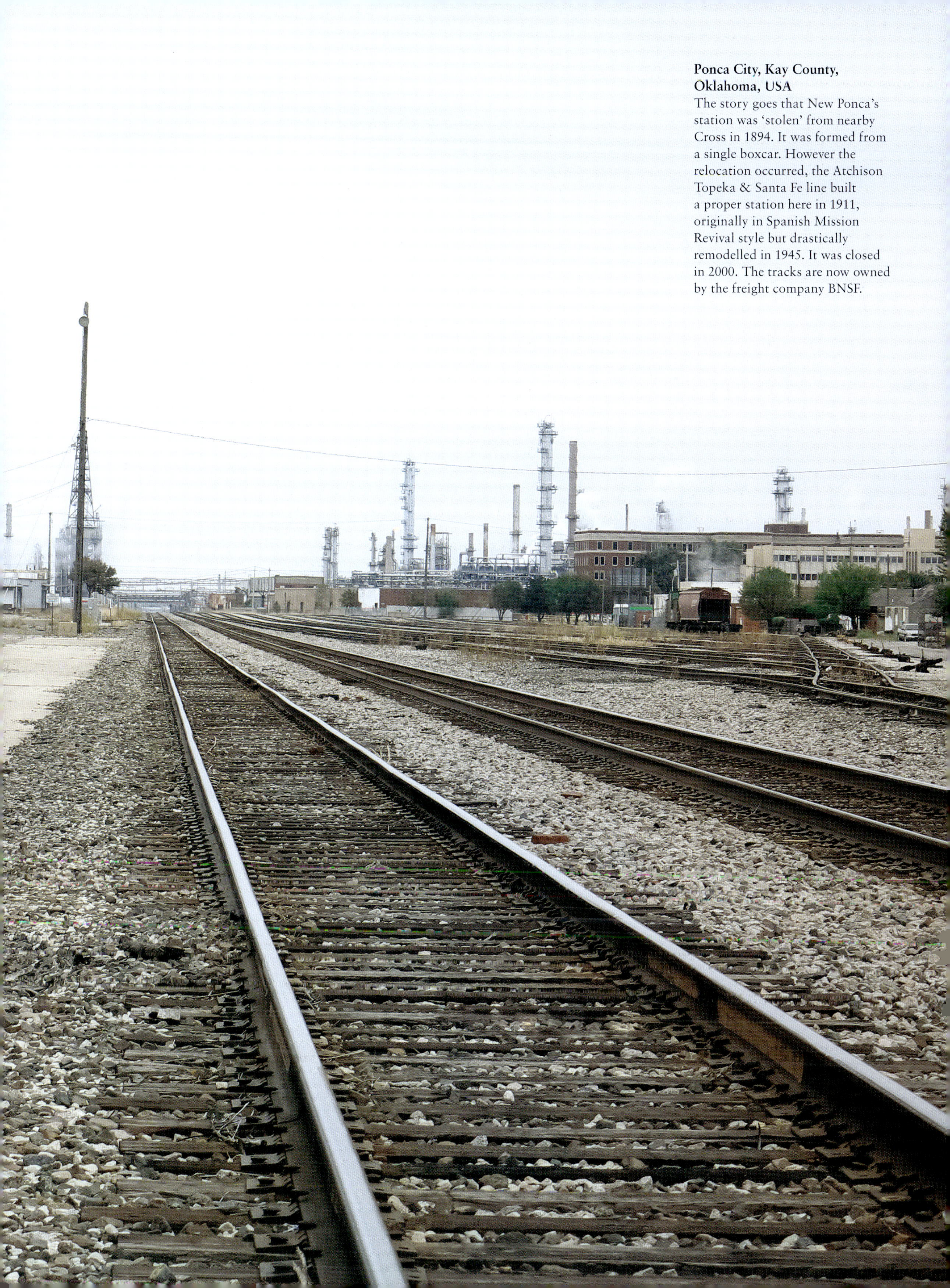

Ponca City, Kay County, Oklahoma, USA
The story goes that New Ponca's station was 'stolen' from nearby Cross in 1894. It was formed from a single boxcar. However the relocation occurred, the Atchison Topeka & Santa Fe line built a proper station here in 1911, originally in Spanish Mission Revival style but drastically remodelled in 1945. It was closed in 2000. The tracks are now owned by the freight company BNSF.

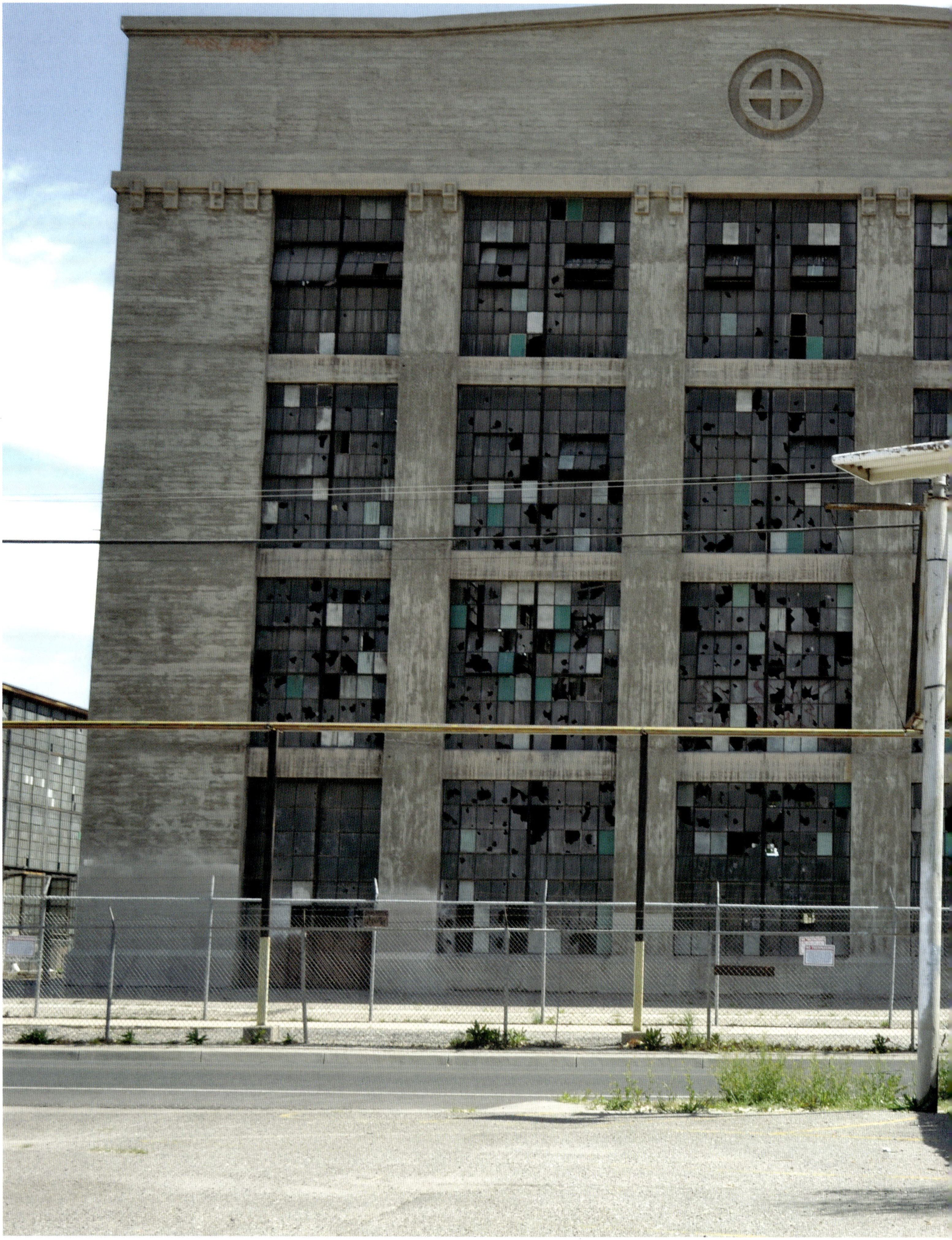

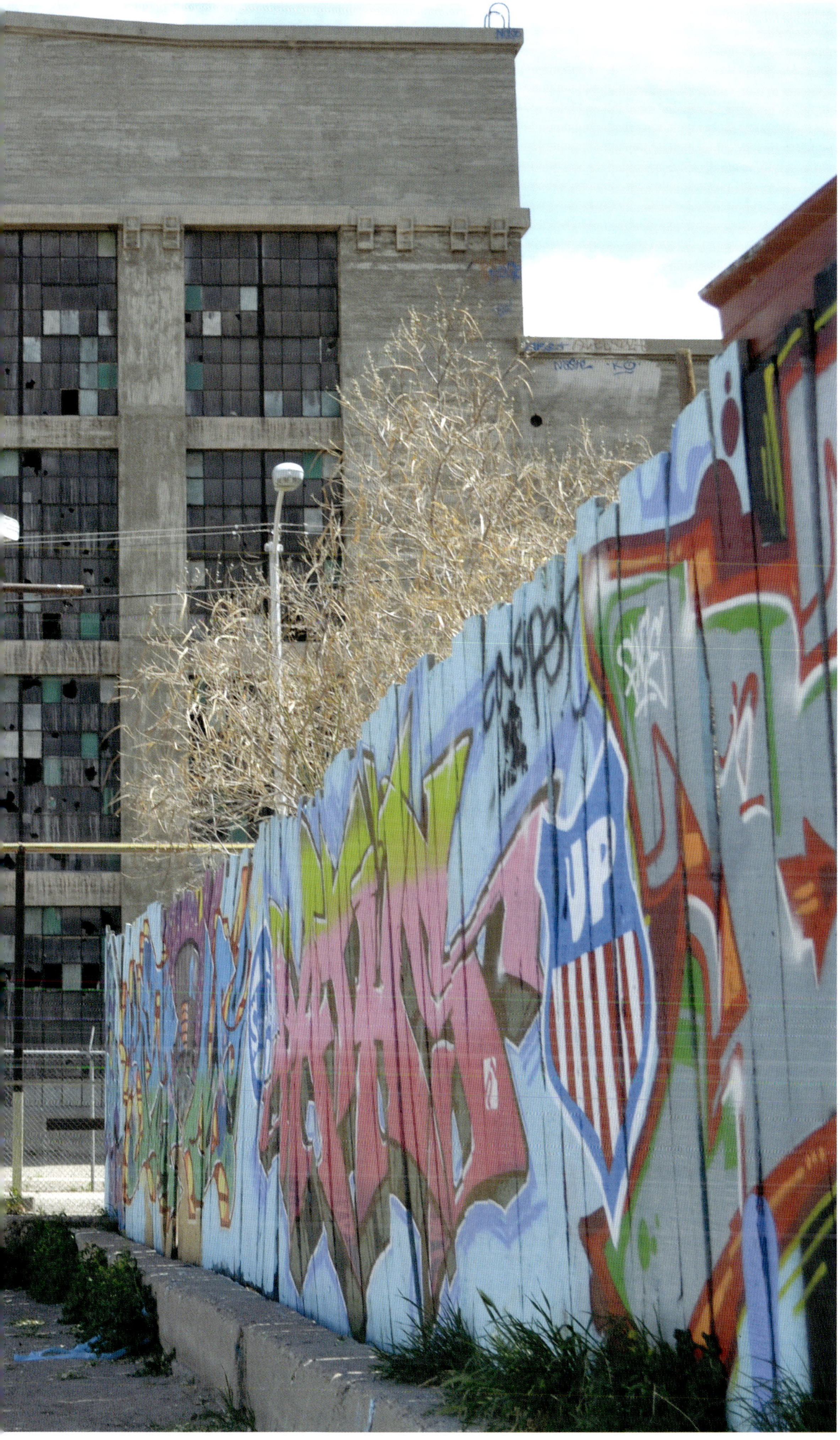

Santa Fe Workshops, Albuquerque, New Mexico, USA
Albuquerque was a principal base for the Santa Fe Railroad, and it had extensive workshops here to service its large fleet of steam, and later diesel-electric, locomotives, until the early 1980s. The buildings are mostly of 1915-25 vintage, including this modernistic façade of the machine shop.

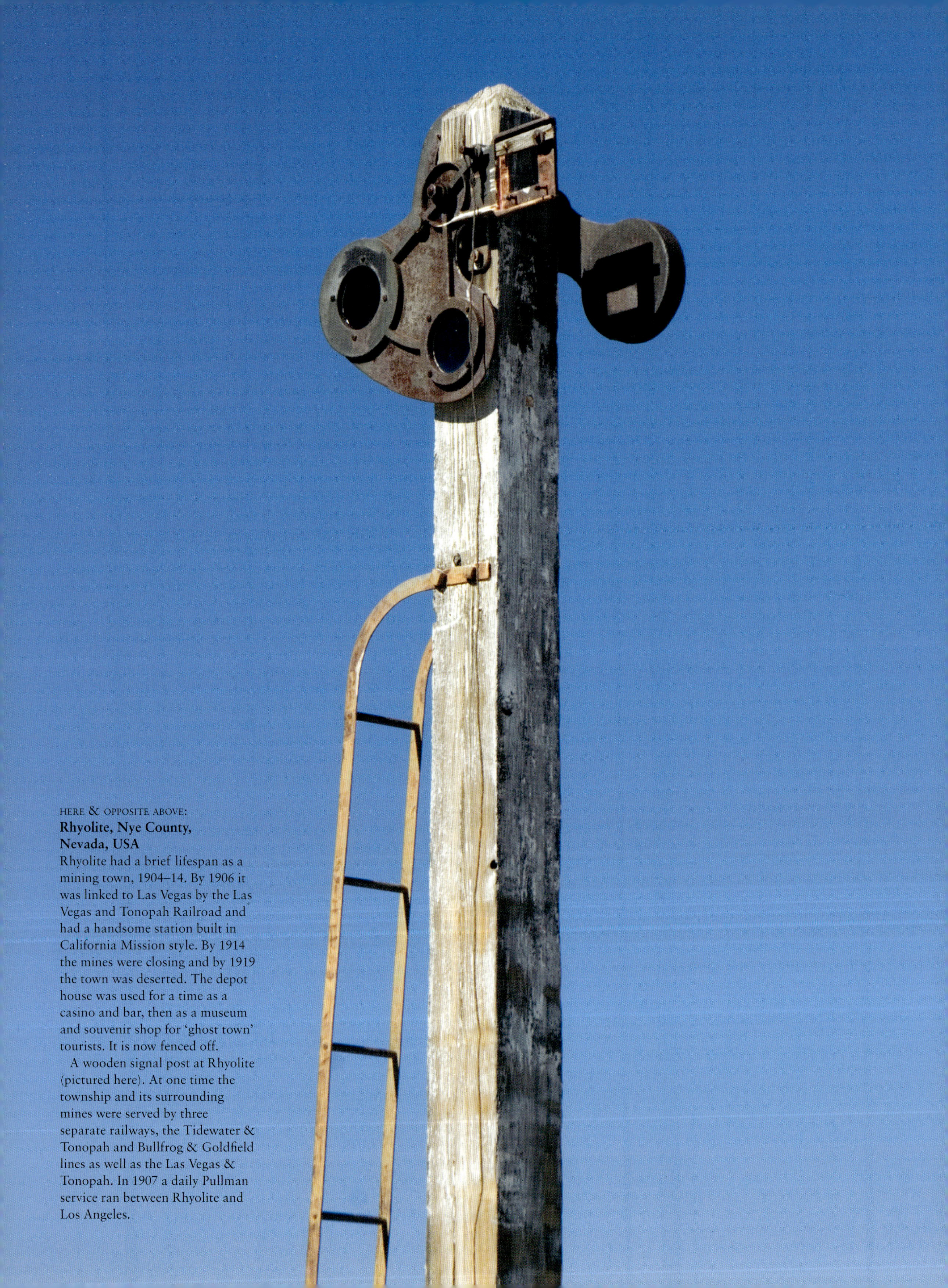

HERE & OPPOSITE ABOVE:
Rhyolite, Nye County, Nevada, USA
Rhyolite had a brief lifespan as a mining town, 1904–14. By 1906 it was linked to Las Vegas by the Las Vegas and Tonopah Railroad and had a handsome station built in California Mission style. By 1914 the mines were closing and by 1919 the town was deserted. The depot house was used for a time as a casino and bar, then as a museum and souvenir shop for 'ghost town' tourists. It is now fenced off.

A wooden signal post at Rhyolite (pictured here). At one time the township and its surrounding mines were served by three separate railways, the Tidewater & Tonopah and Bullfrog & Goldfield lines as well as the Las Vegas & Tonopah. In 1907 a daily Pullman service ran between Rhyolite and Los Angeles.

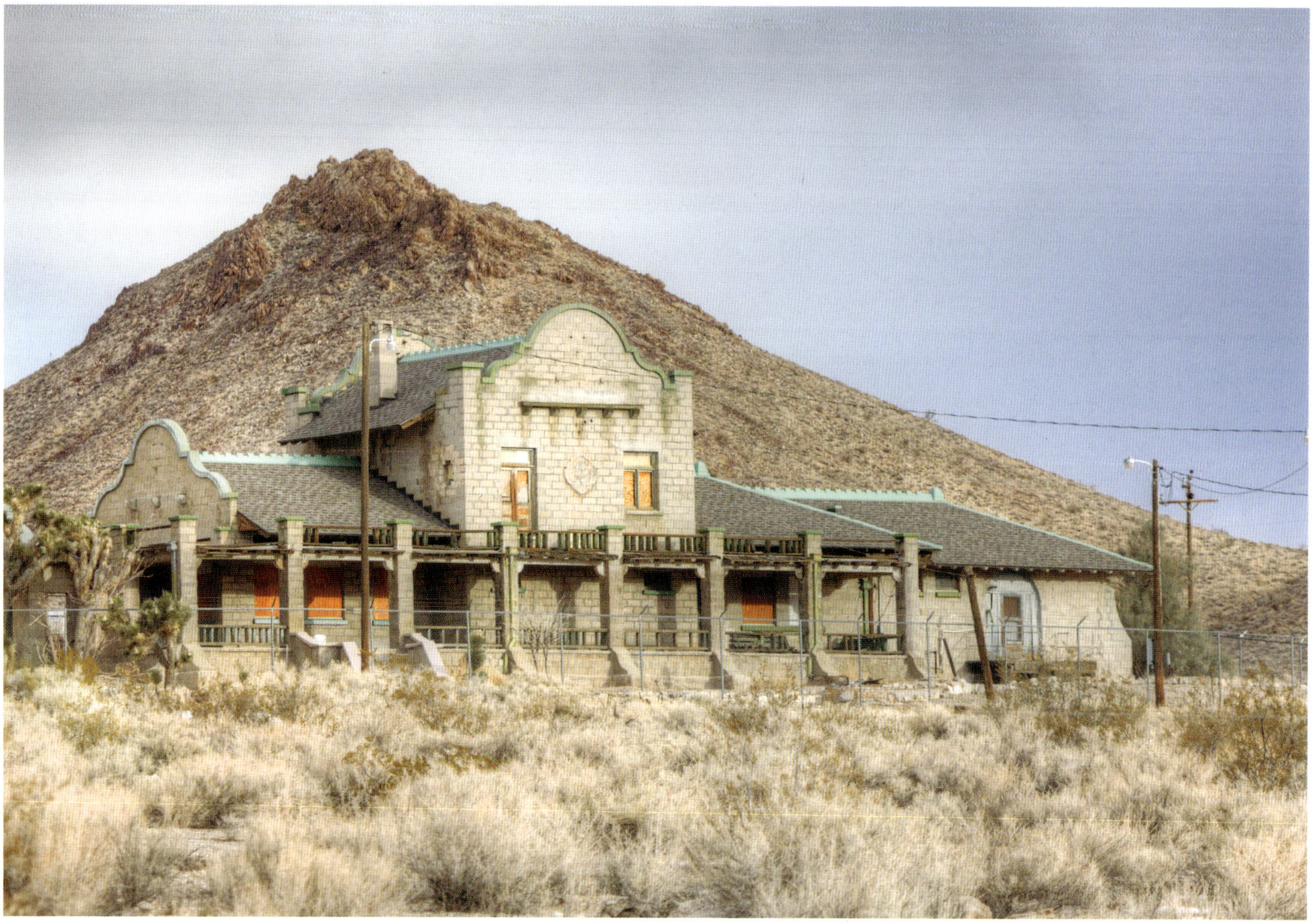

RIGHT:

Goldfield, Esmeralda County, Nevada, USA

The Tonopah & Goldfield Railroad reached Goldfield in 1905, and both line and station had longer careers than most others in this briefly flourishing mining region, but final closure came in 1947. These subway entrances are something of a mystery, looking like items from a city station. The Goldfield station site shows no evidence of a subway. It is possible, though unlikely, that they are relics of a film set, as a number of films were shot here.

OVERLEAF (BOTH PHOTOGRAPHS):

Ancho Station, Lincoln County, New Mexico, USA

A station was built here in 1901 by the El Paso & Northeastern Railway, and with the working of a nearby gypsum deposit and establishment of a brickworks, the community grew, but in the 1950s decline set in. The station was shut down in 1959, and for a time in the 1960s was the home of a museum. No longer in use, it still stands, though the rail tracks are long gone.

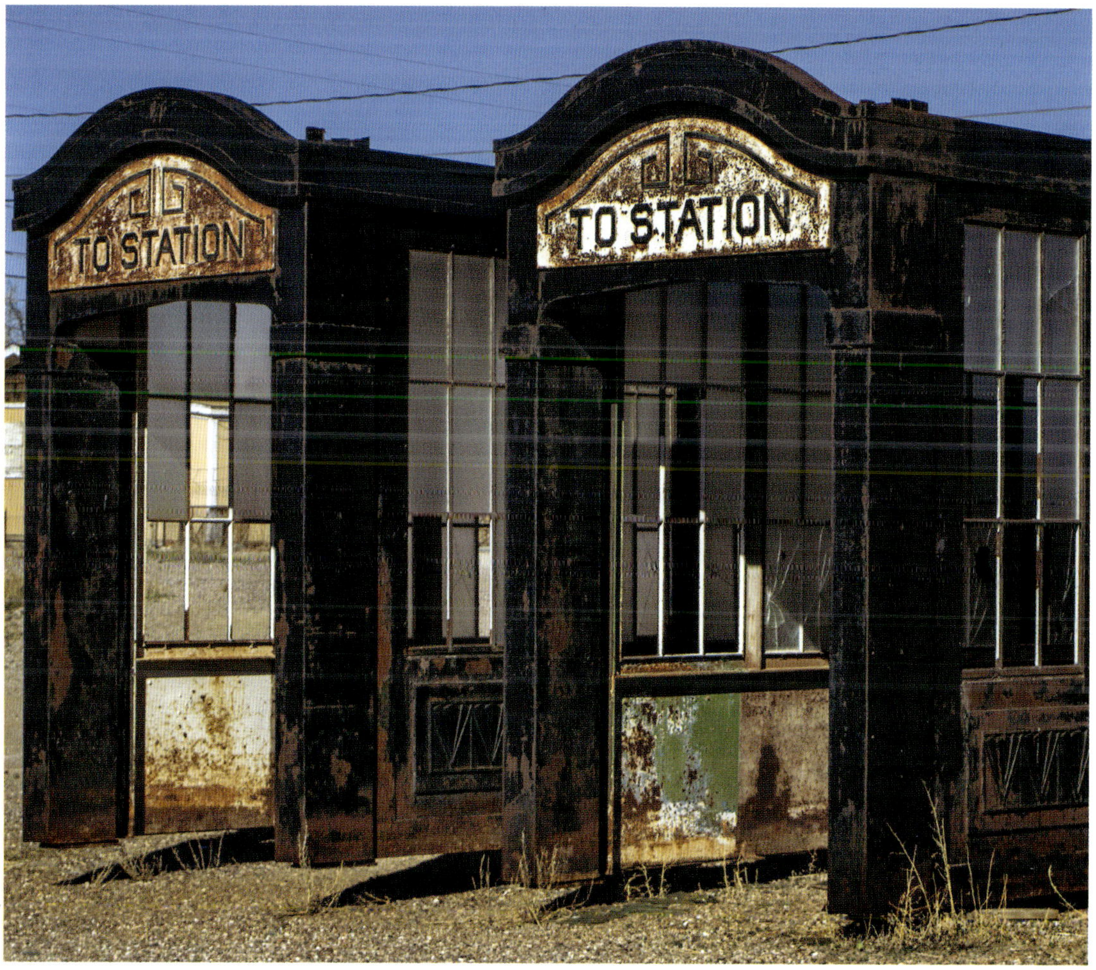

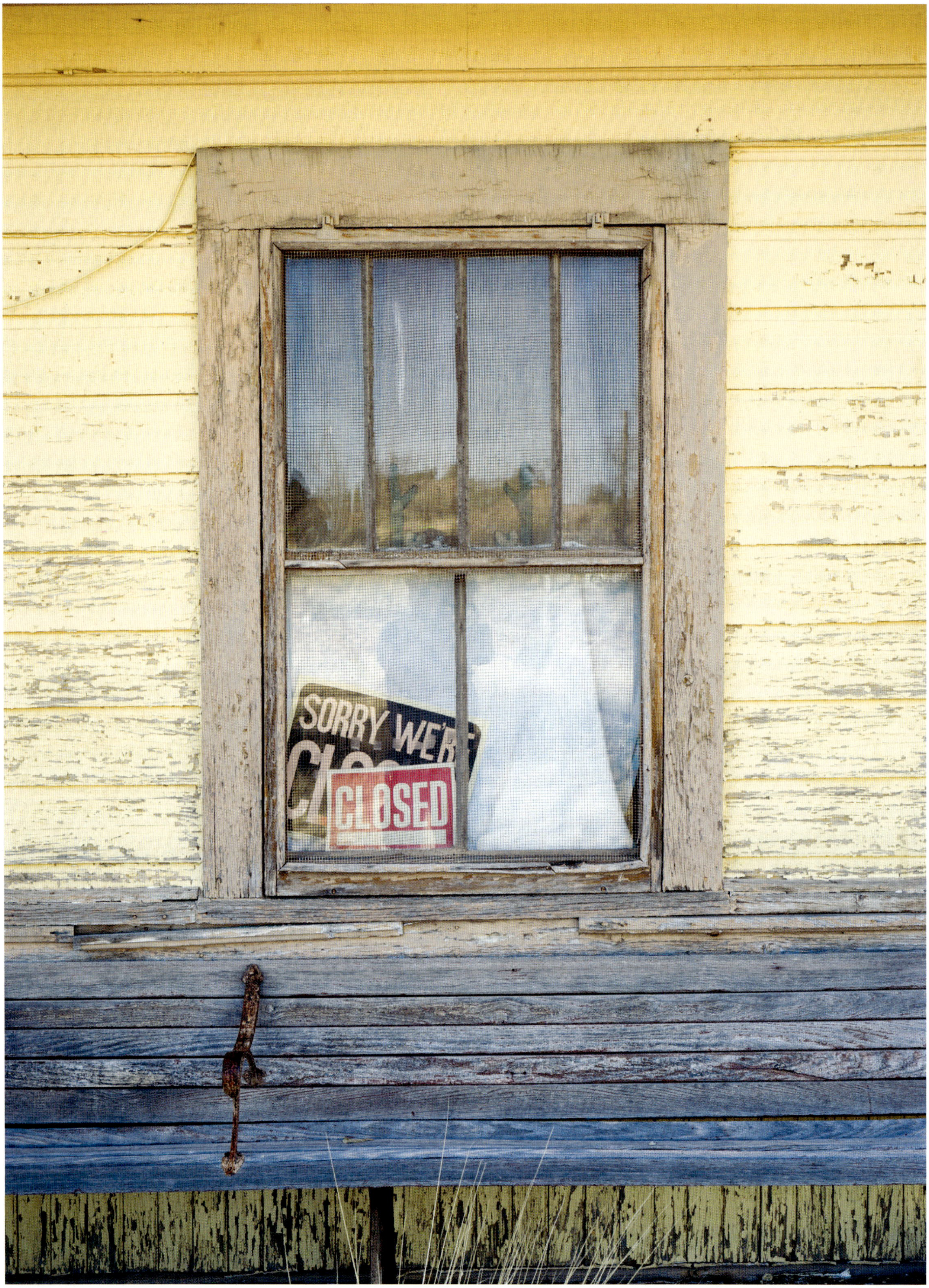
SORRY WE'RE
CLOSED

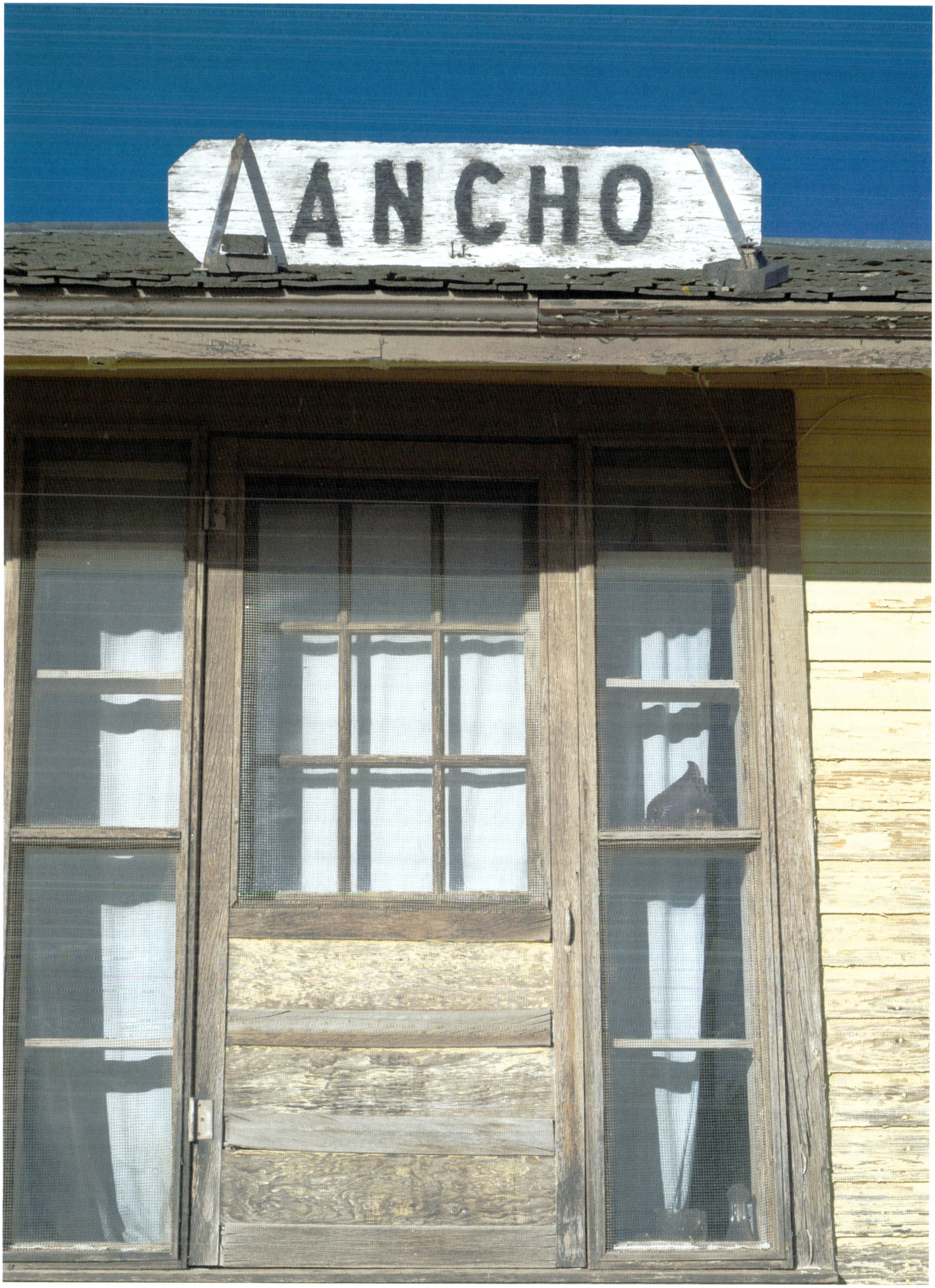
ANCHO

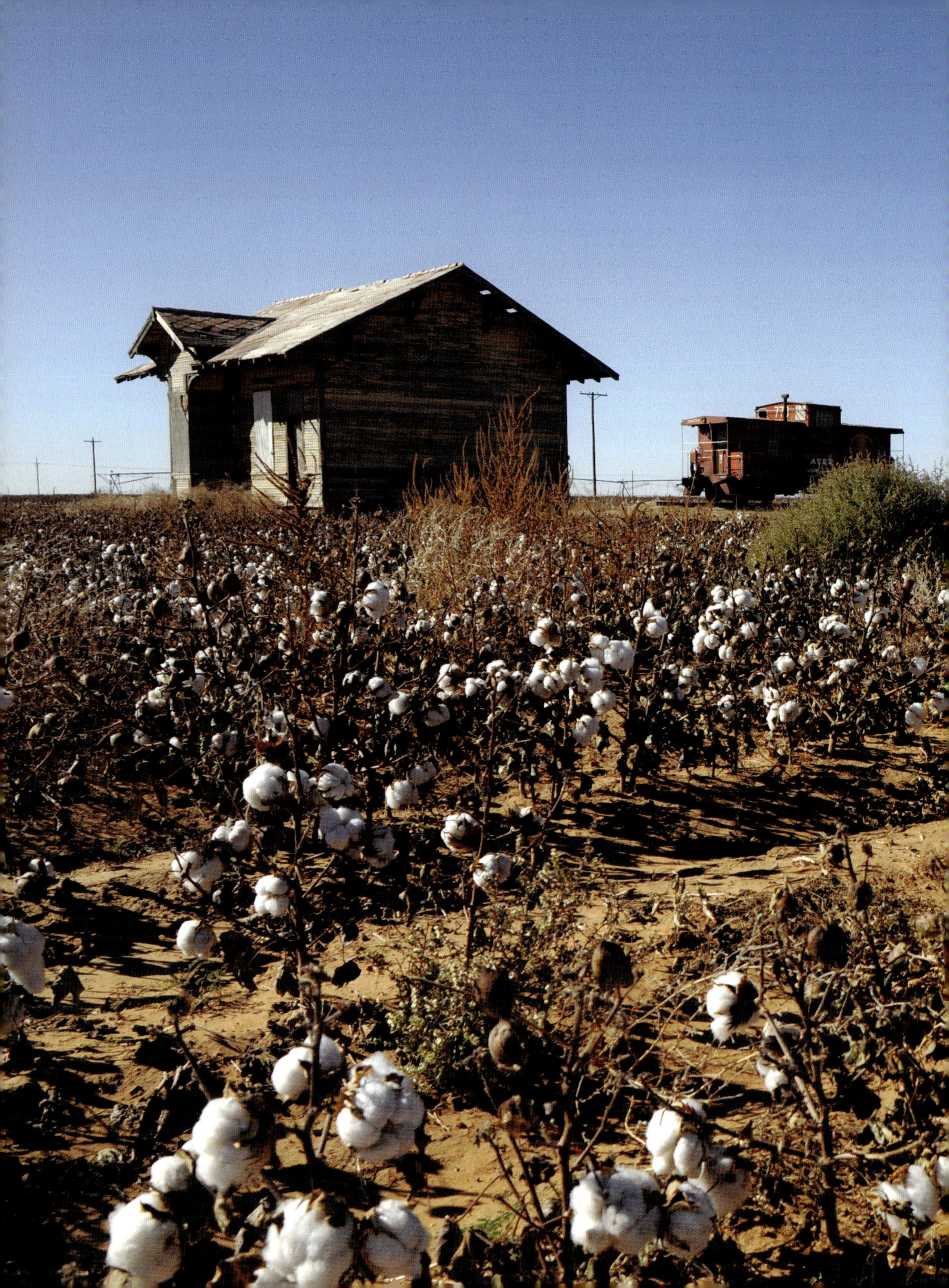

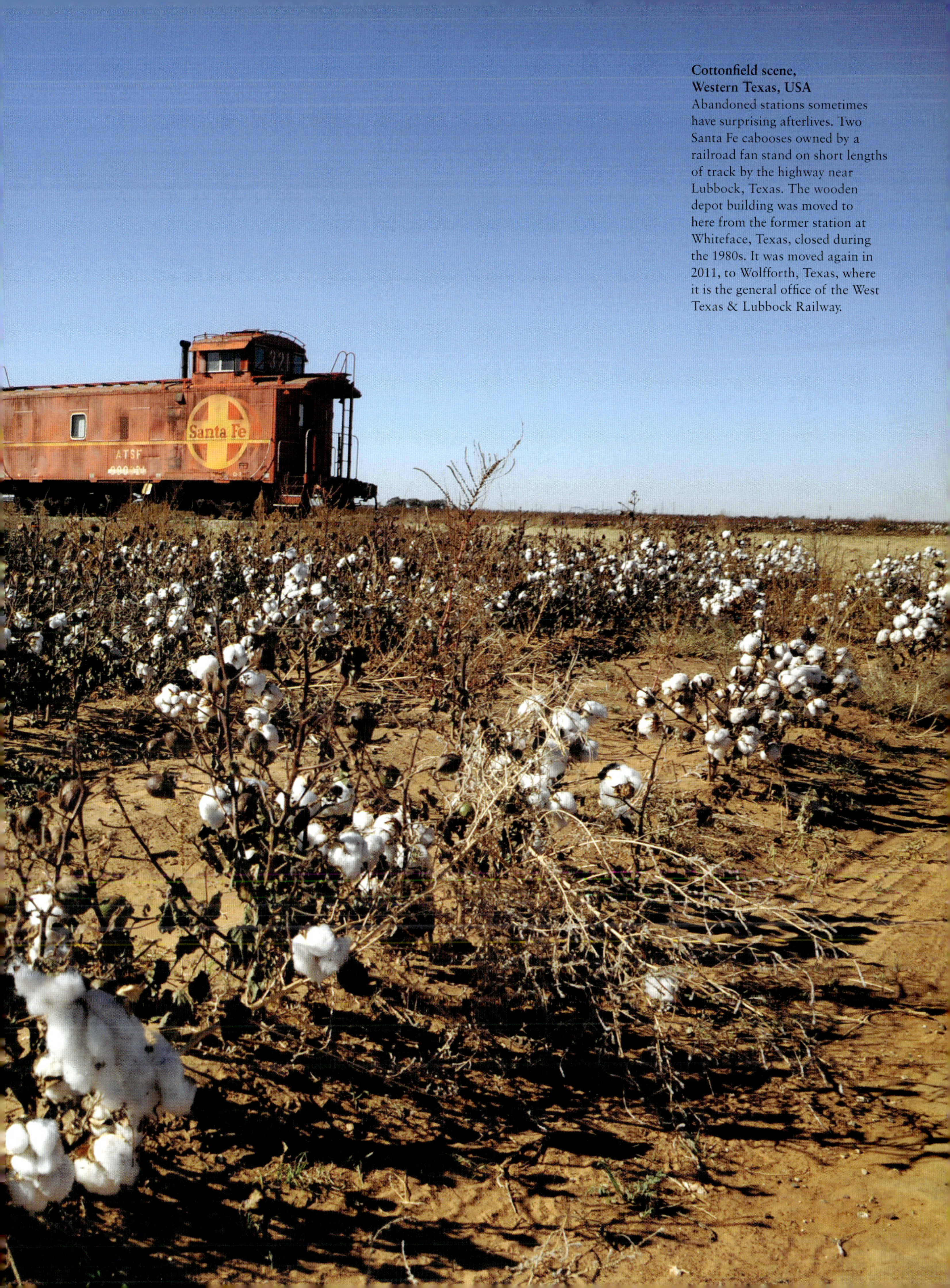

Cottonfield scene, Western Texas, USA

Abandoned stations sometimes have surprising afterlives. Two Santa Fe cabooses owned by a railroad fan stand on short lengths of track by the highway near Lubbock, Texas. The wooden depot building was moved to here from the former station at Whiteface, Texas, closed during the 1980s. It was moved again in 2011, to Wolfforth, Texas, where it is the general office of the West Texas & Lubbock Railway.

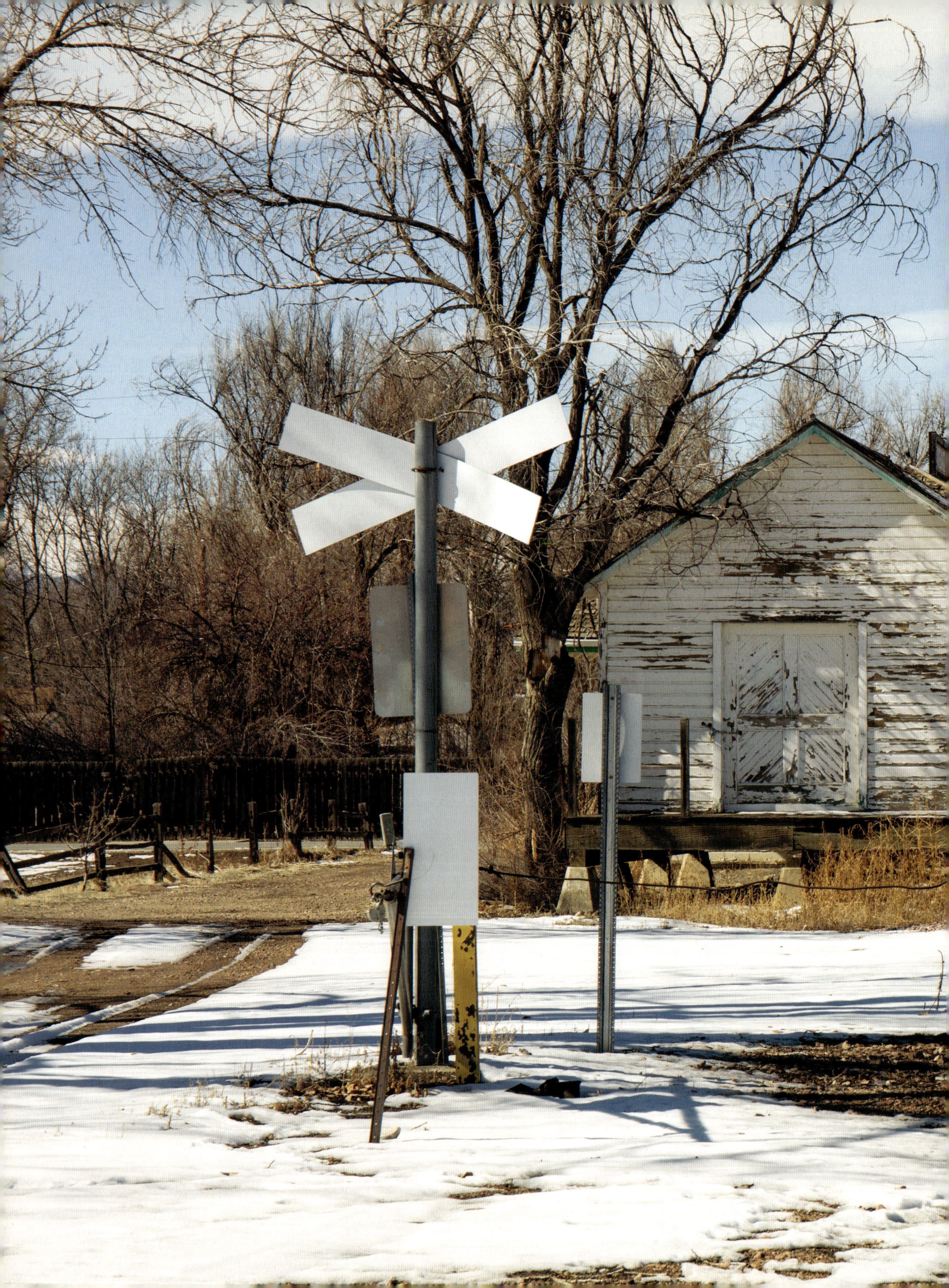

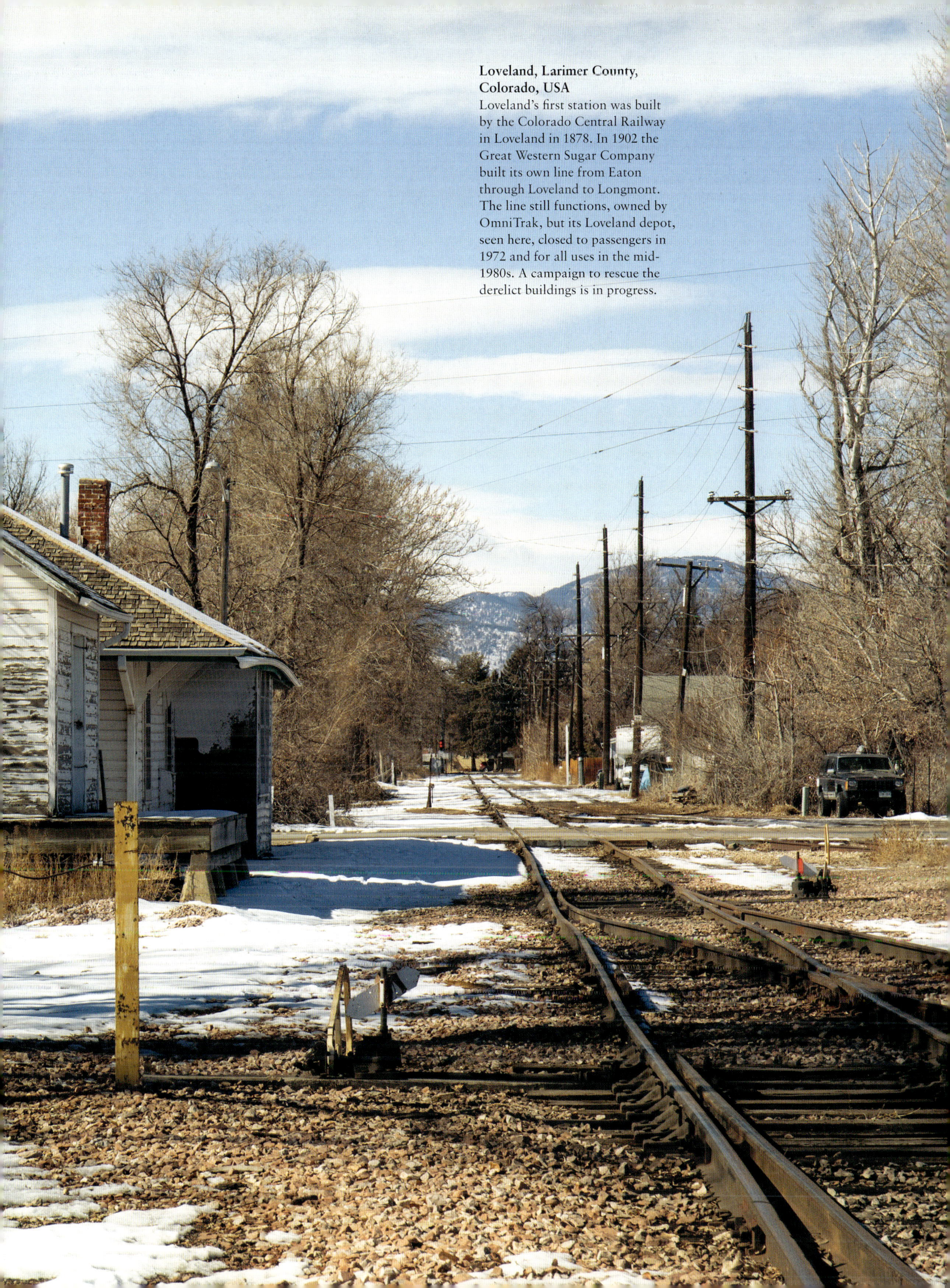

Loveland, Larimer County, Colorado, USA
Loveland's first station was built by the Colorado Central Railway in Loveland in 1878. In 1902 the Great Western Sugar Company built its own line from Eaton through Loveland to Longmont. The line still functions, owned by OmniTrak, but its Loveland depot, seen here, closed to passengers in 1972 and for all uses in the mid-1980s. A campaign to rescue the derelict buildings is in progress.

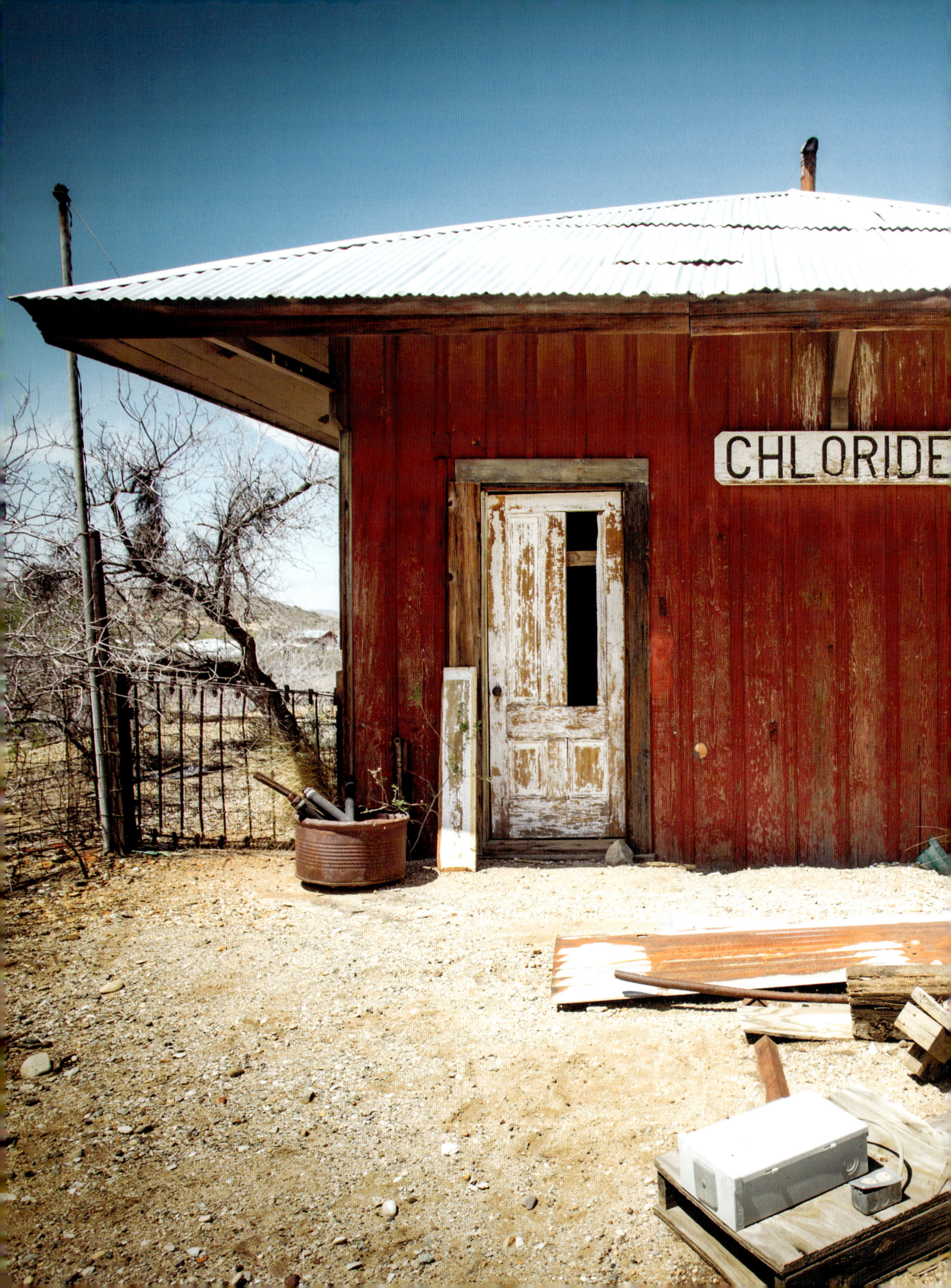
CHLORIDE

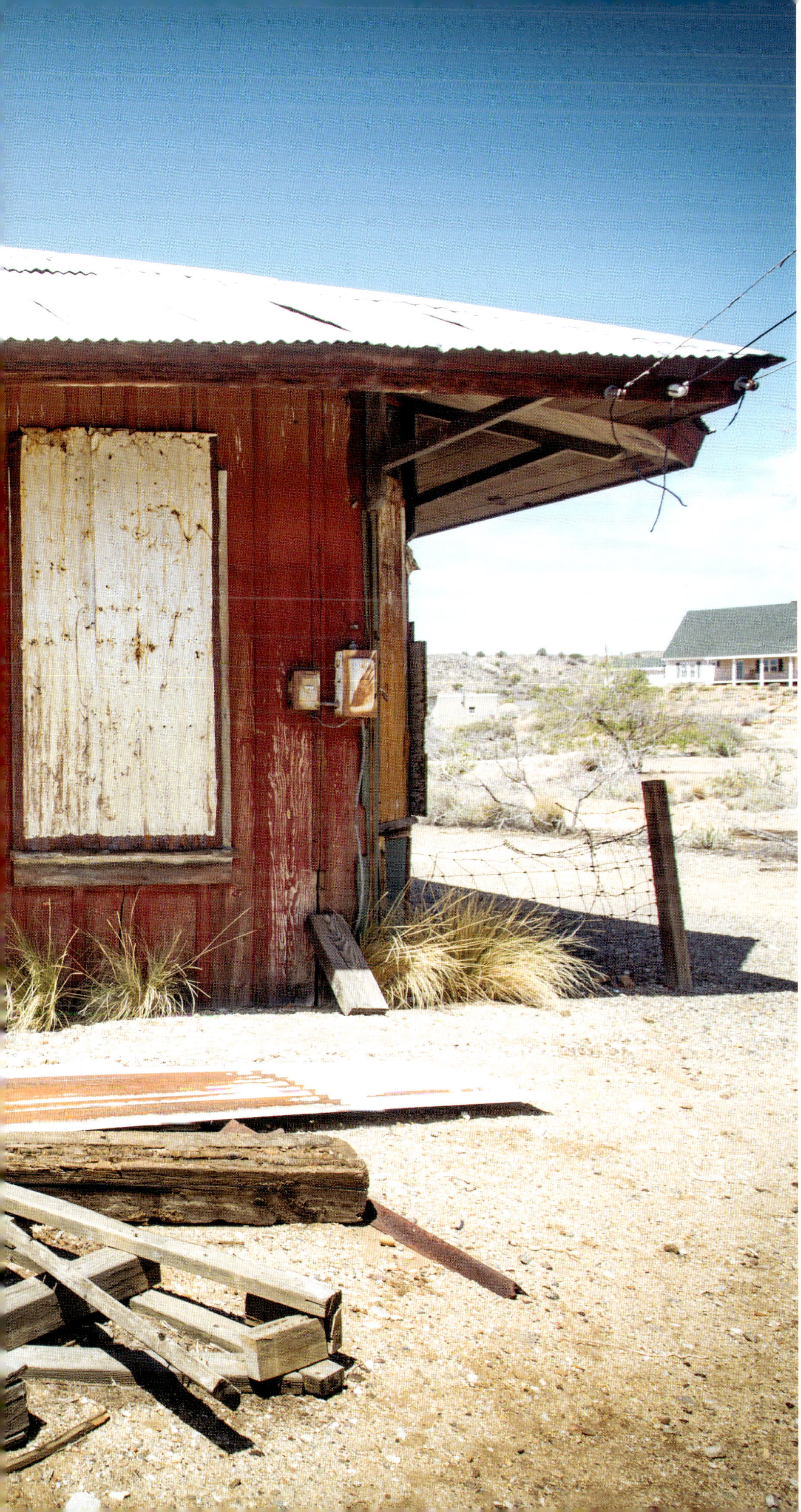

Chloride, Mohave County, Arizona, USA
Named for the local silver chloride mines, this is the oldest inhabited mining centre in the state. The Santa Fe Railroad built the line from Kingman to here in 1898, and trains ran until 1935 when the decline of the mines and the Great Depression brought about closure. The tongue-and-groove wood-built station remains intact, boarded up and with a corrugated iron roof.

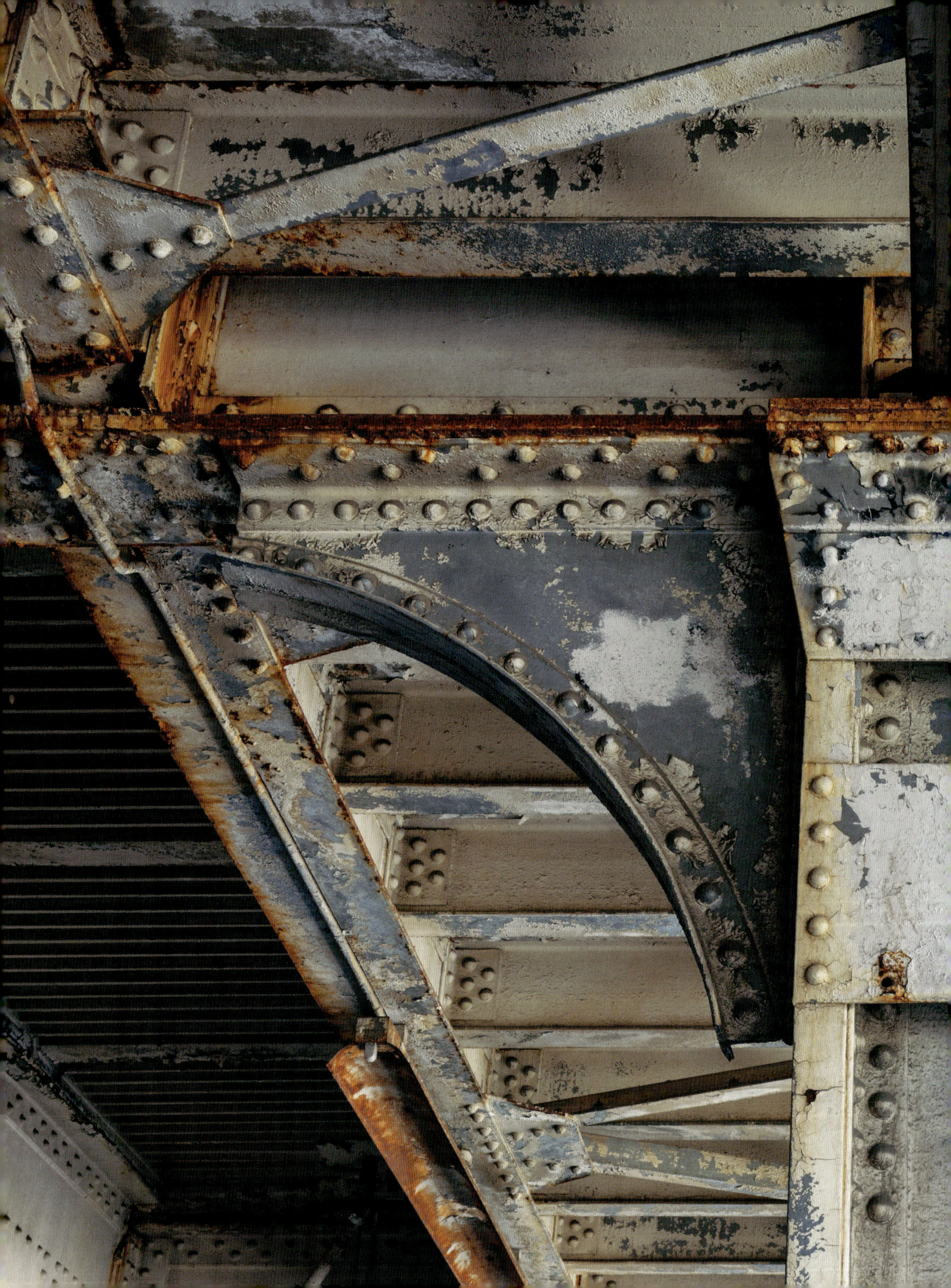

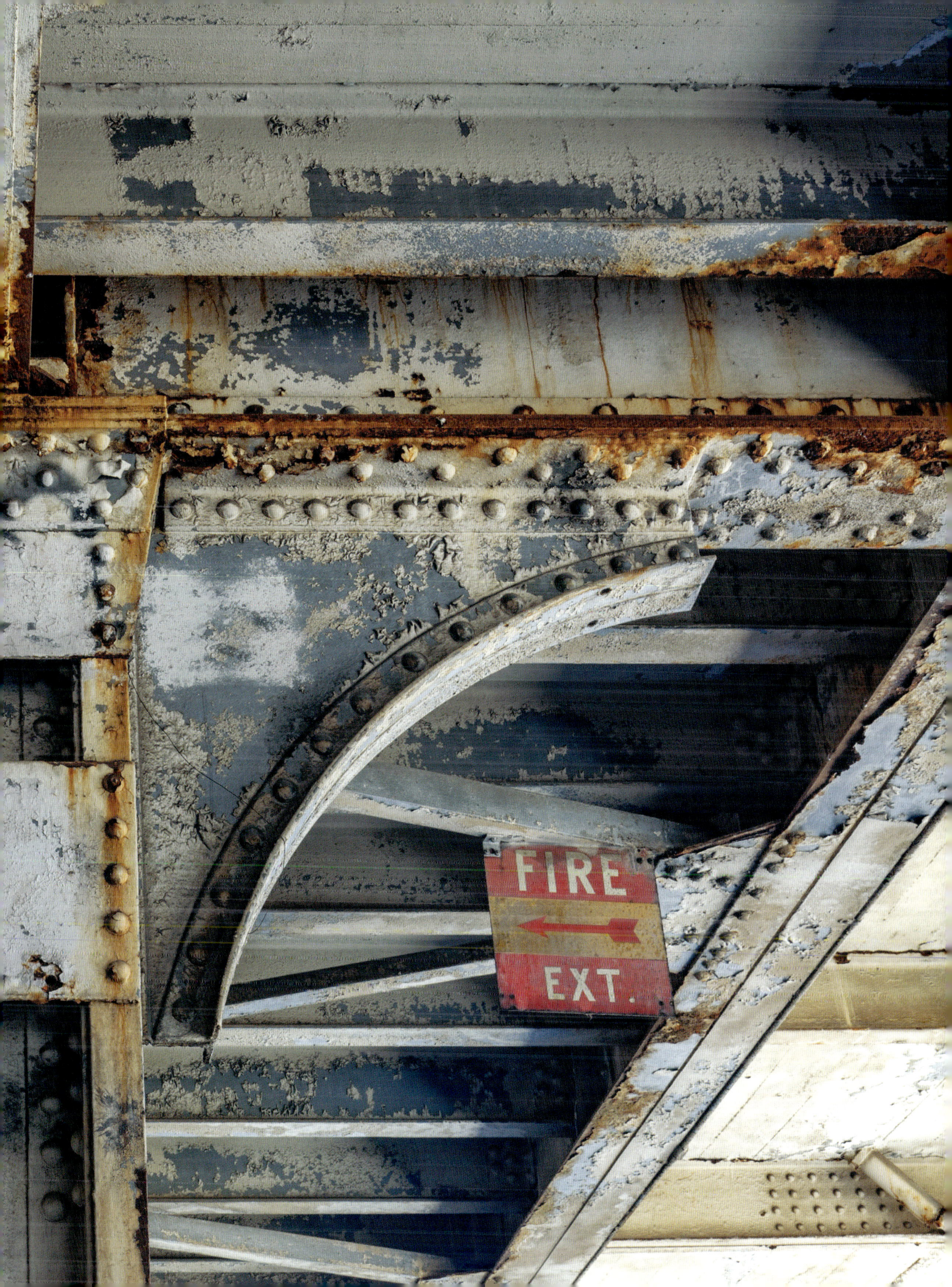
FIRE
EXT.

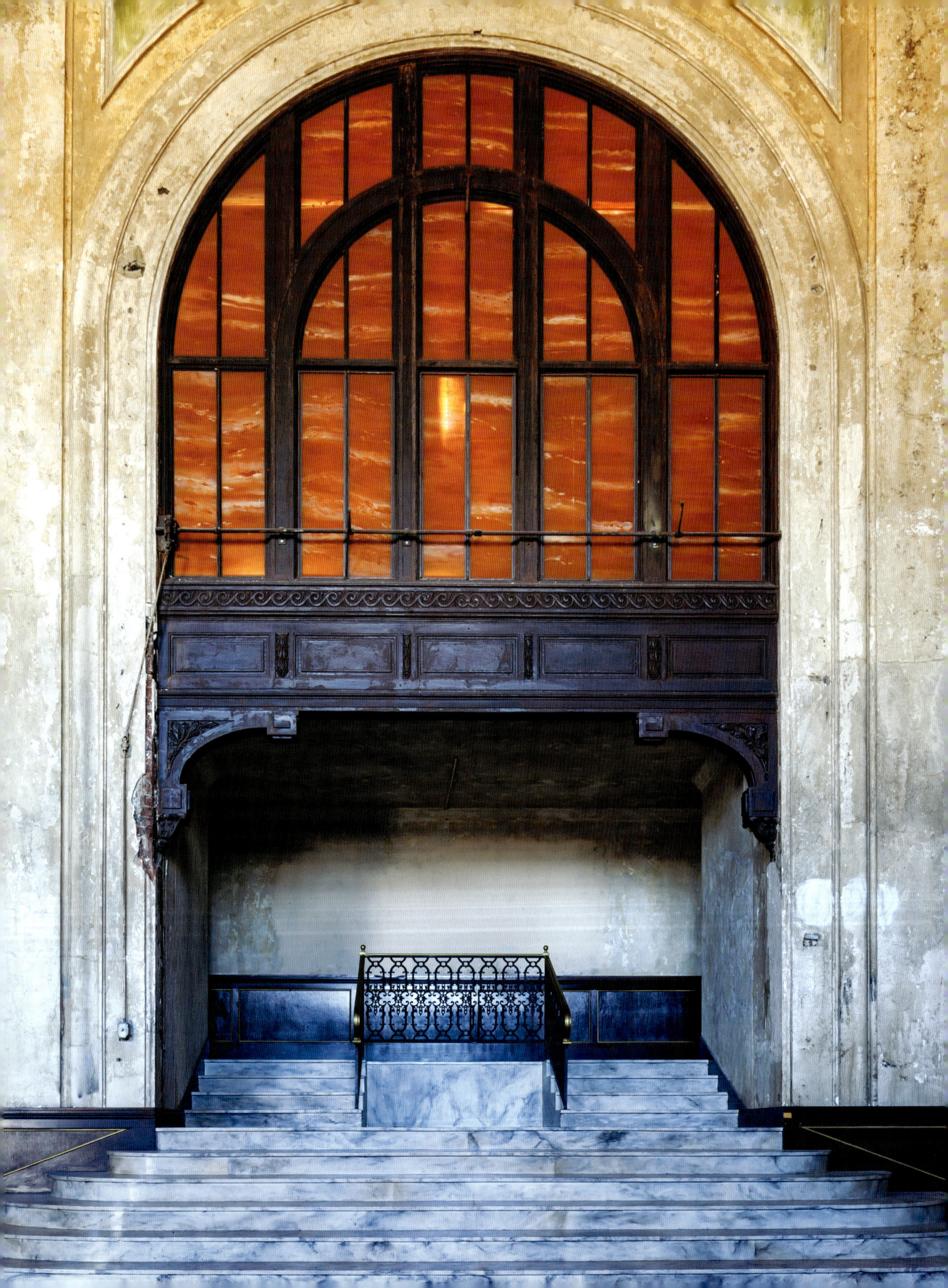

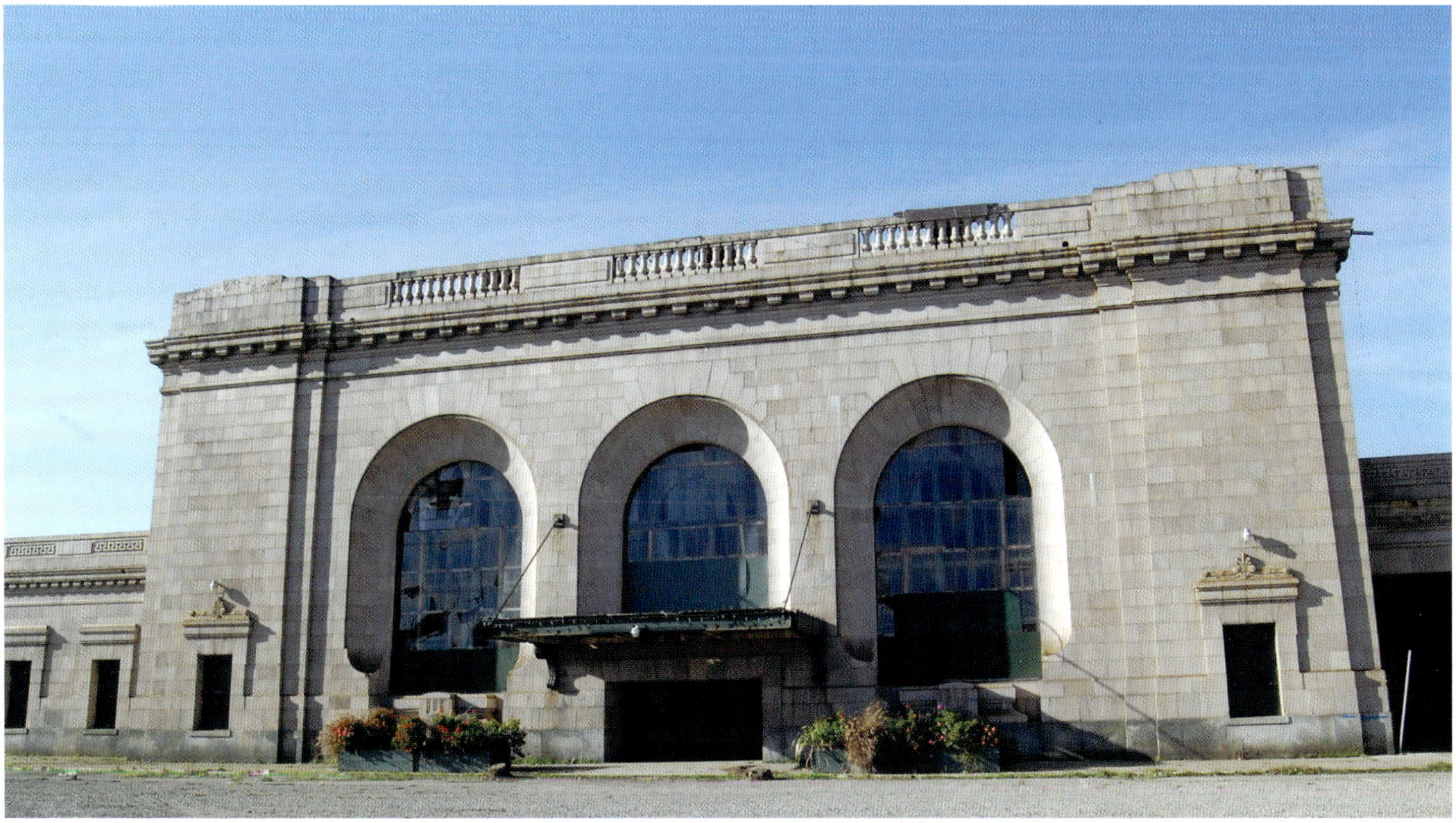

PREVIOUS PAGES AND ALL PHOTOGRAPHS HERE:

16th Street Station, Oakland, California, USA

Opened in 1912, this handsome Southern Pacific station was the western terminus of a transcontinental line. From 1916 to 1939 its elevated platforms, seen here from beneath, were used by the SP's East Bay Electric Lines interurban services, but this ended with the introduction of a rail service over the Bay Bridge.

OPPOSITE:

Unlike some abandoned city termini, the interior hall has been cleaned up and is used for private events. The Beaux-Arts ironwork and plasterwork are notable features.

ABOVE:

Façade of the station. Damaged in the 1989 Loma Prieta earthquake, it closed to all train services in 1994. Since 2005 the station has been in the care of a development project formed by local companies and action groups.

RIGHT:

The signal tower, which controlled the access tracks and sidings, has a strong concrete base. Glamorous trains like the California Zephyr and Coast Starlight expresses rolled past here during the station's glory days.

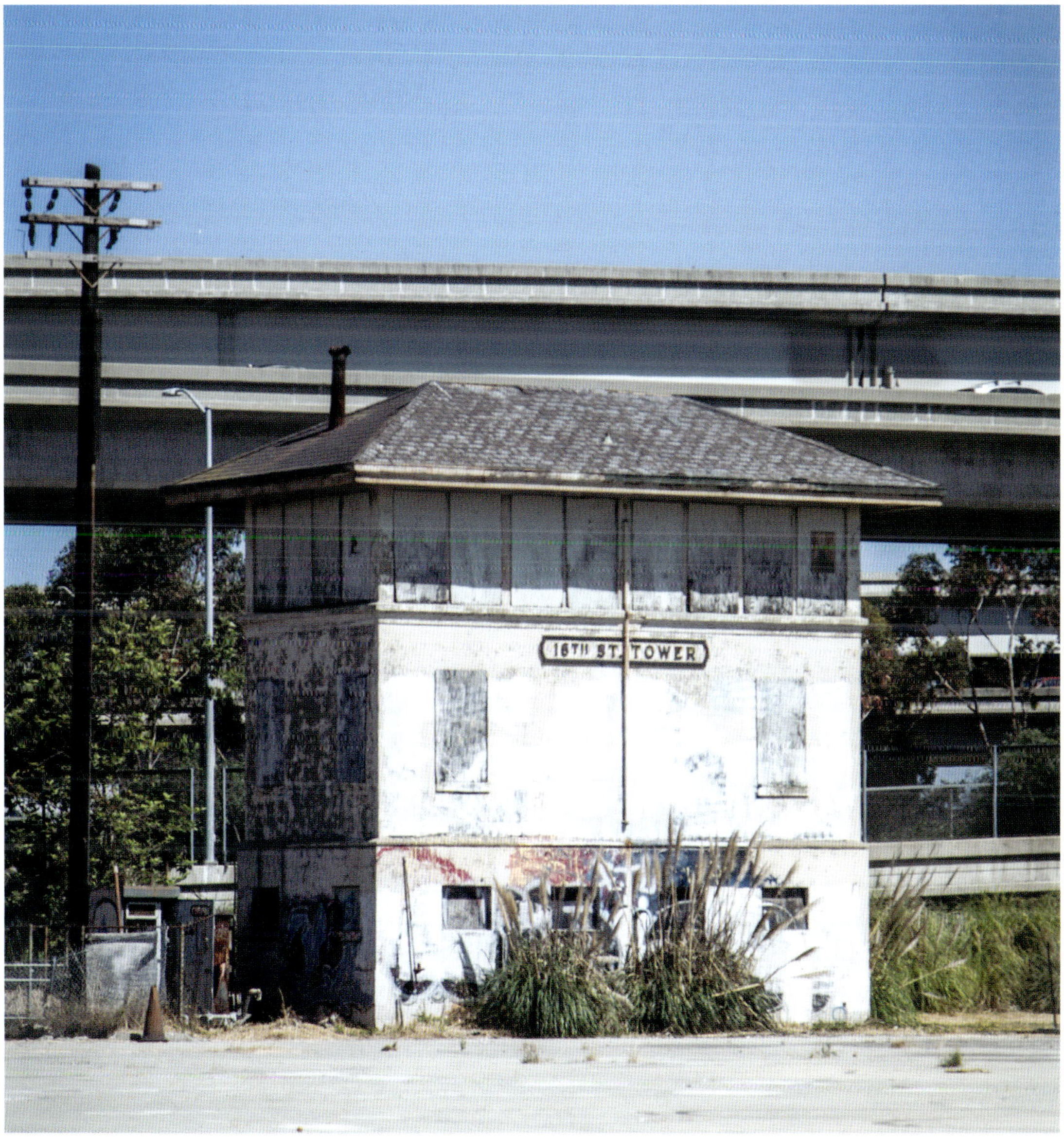

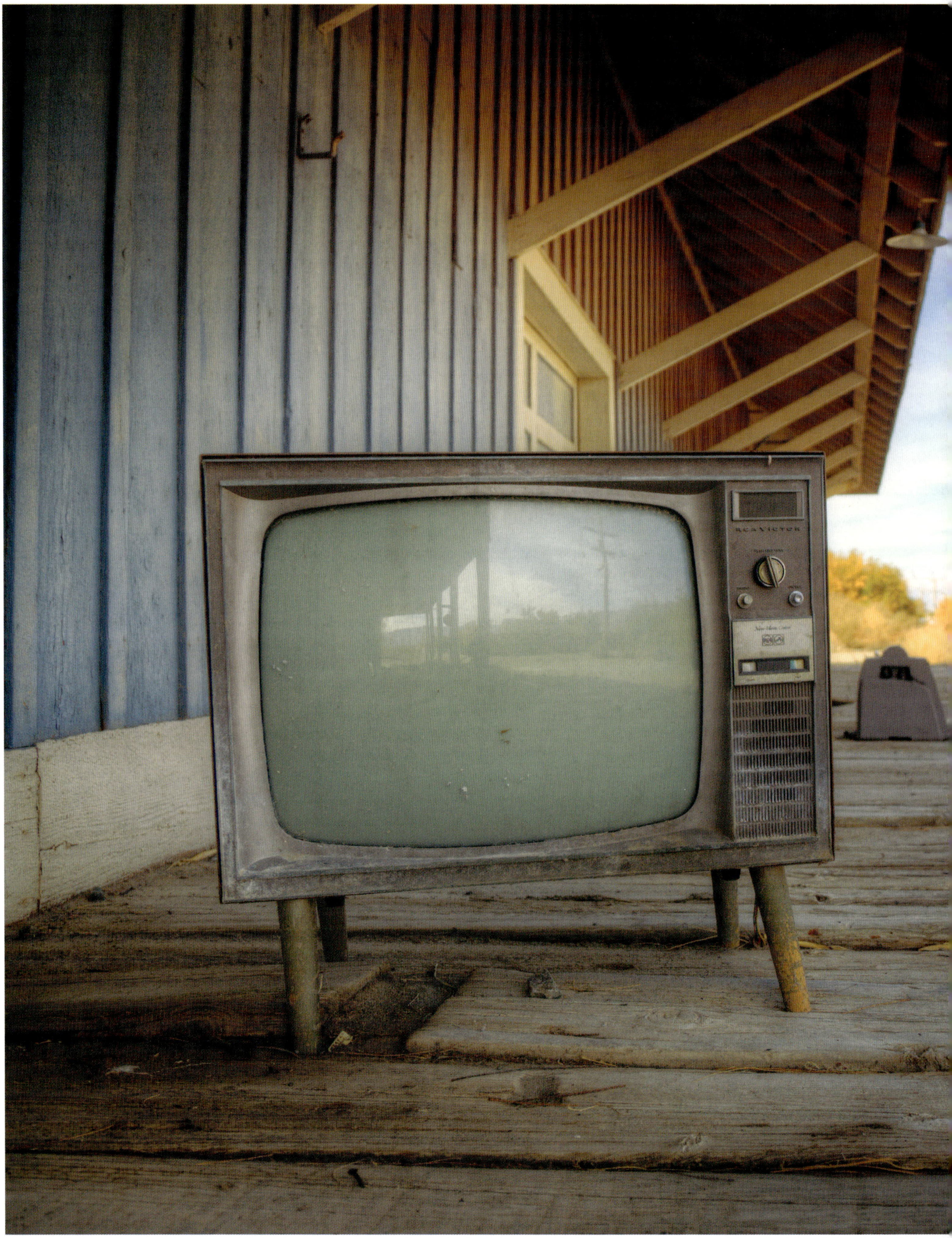
RCA VICTOR

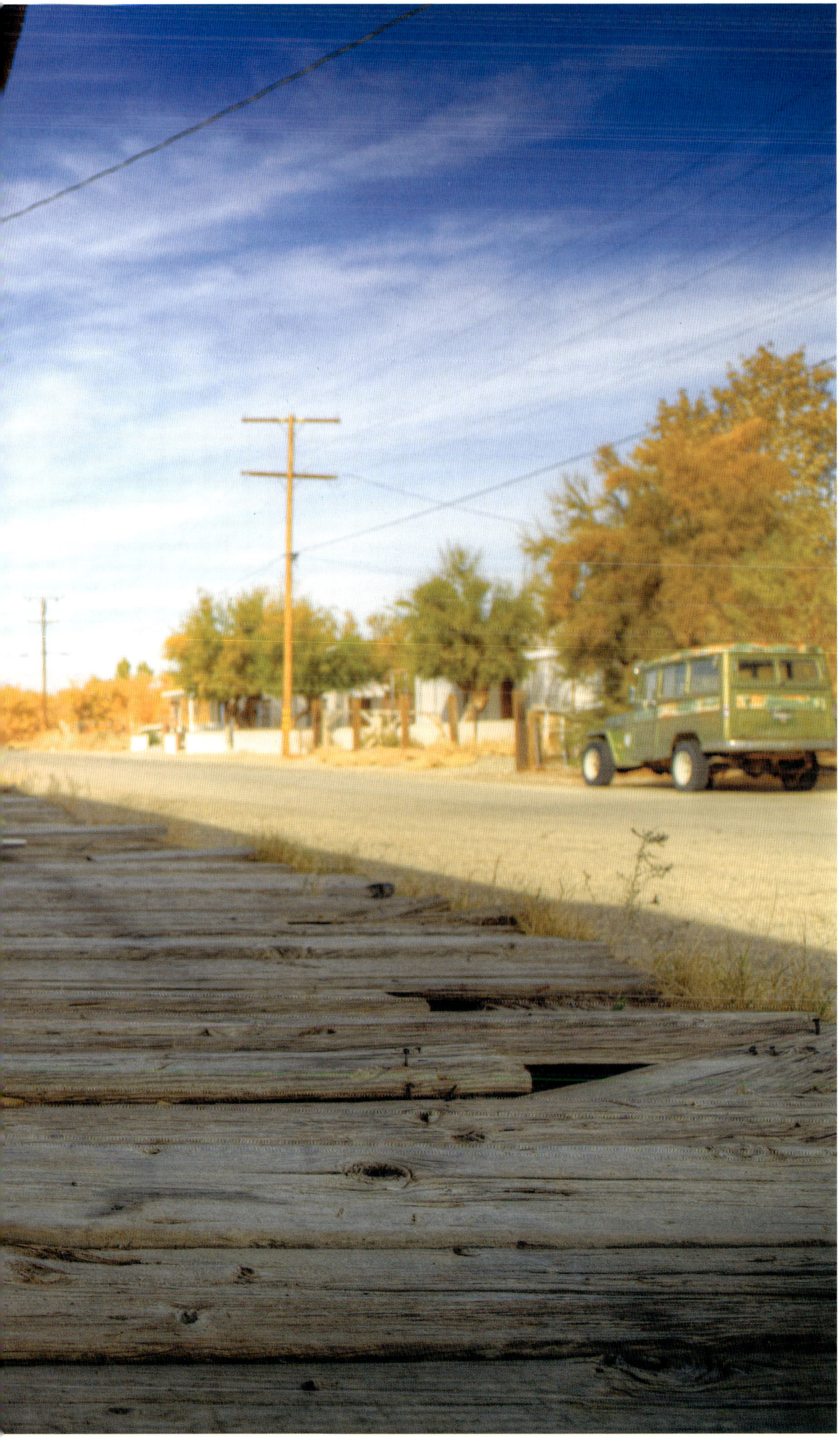

LEFT:

Keeler, Inyo County, California, USA

Keeler was the southern terminus of the 914mm (36in) gauge Carson & Colorado Railroad, which opened in 1883 and closed in 1960. Primarily a mineral-carrying line, serving silver and zinc mines in the area, its passenger cars were attached to the goods wagons. Several of the cars have been converted to dwellings in the township. The tracks were lifted in 1961 but the wood-built station survives.

OVERLEAF:

La Quemada, Zacatecas State, Mexico

Little remains of the station adjacent to the great Aztec archeological site of Chicomóztoc, on the Southern Pacific Railroad of Mexico's line between Guadalajara and Nogales, which became the Ferrocarril Sud-Pacifico de Mexico in 1951. Opened in 1927, its closure date is not recorded, but regular passenger services on the line ended in the mid-20th century.

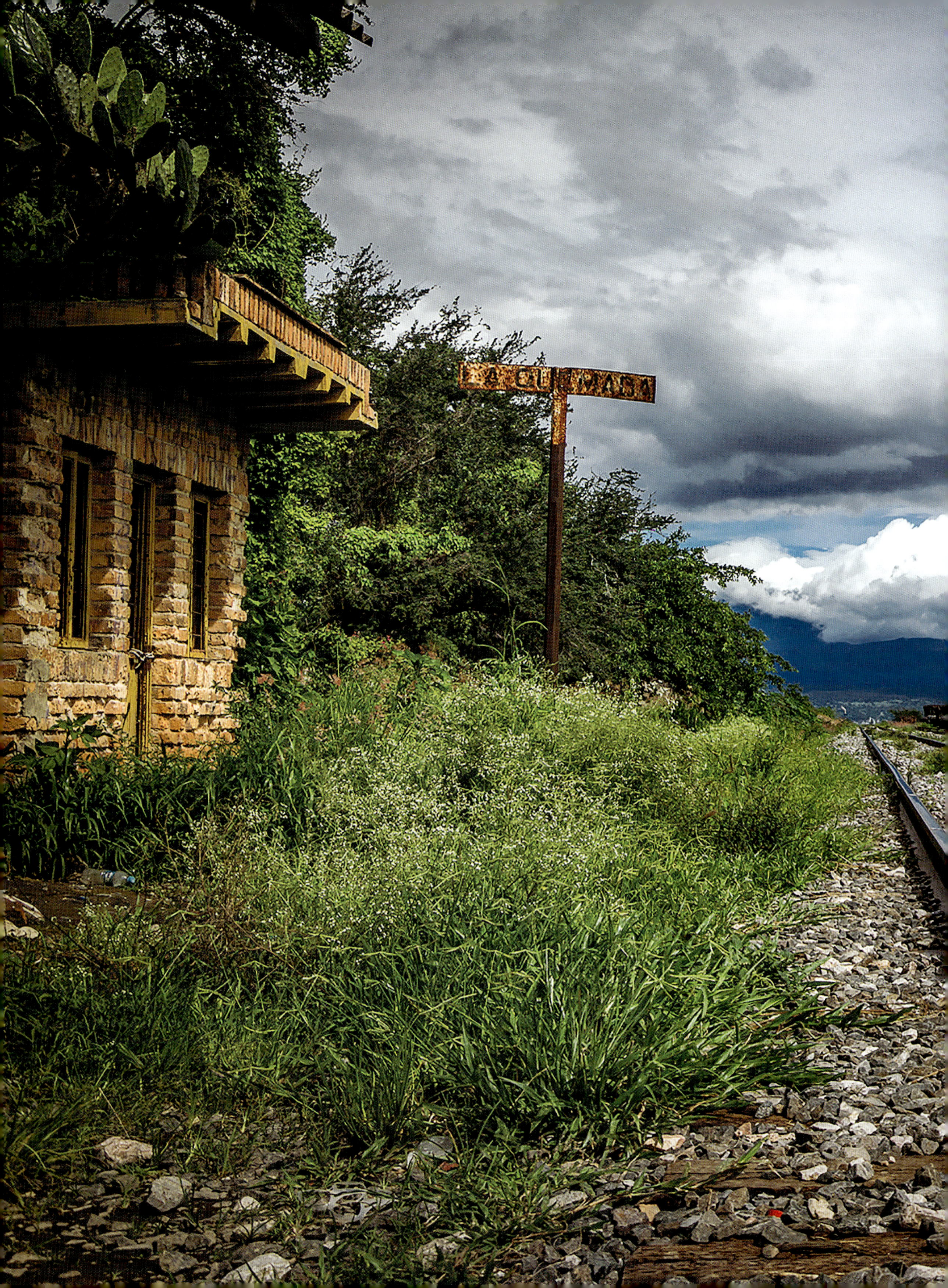

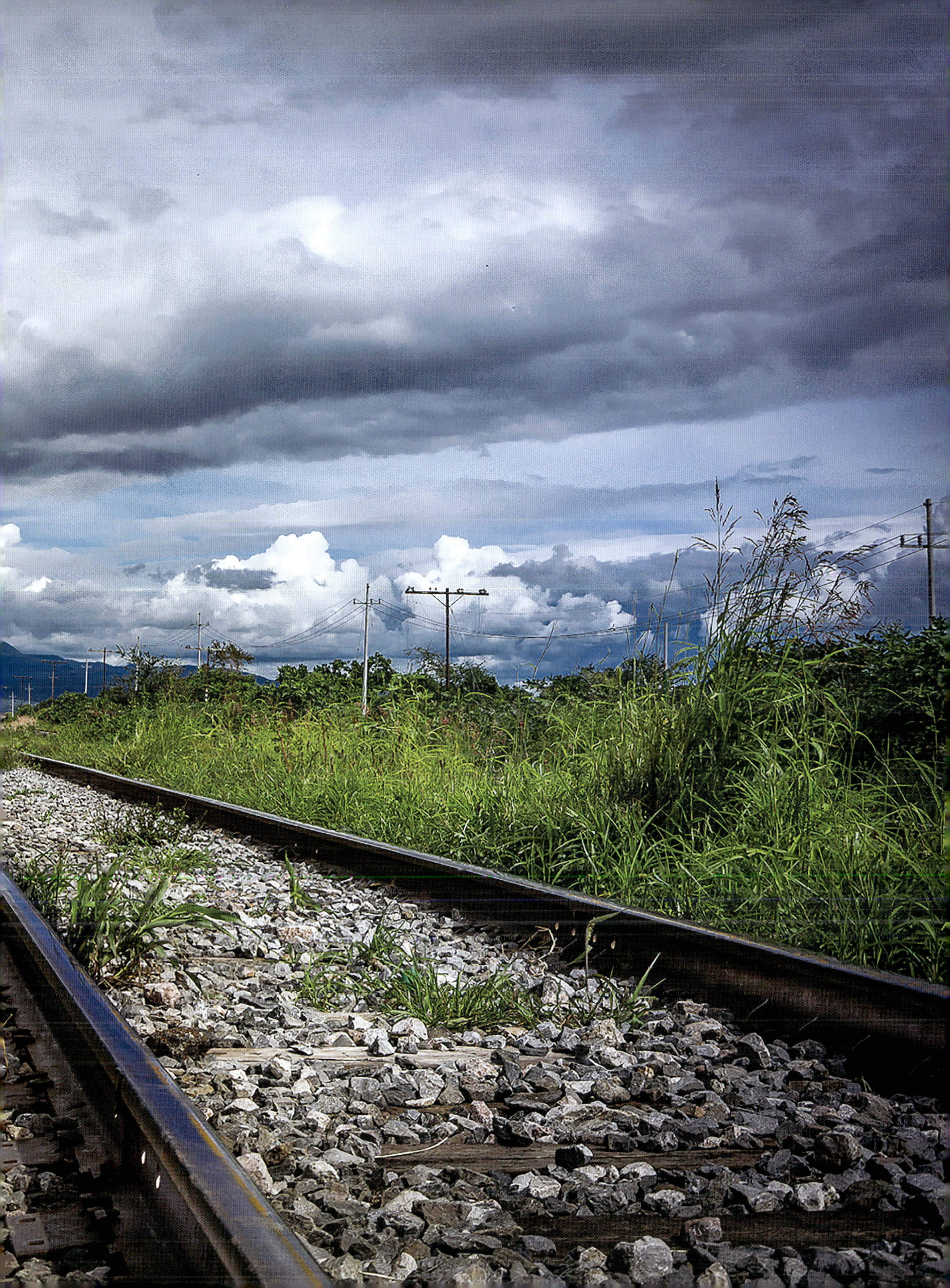

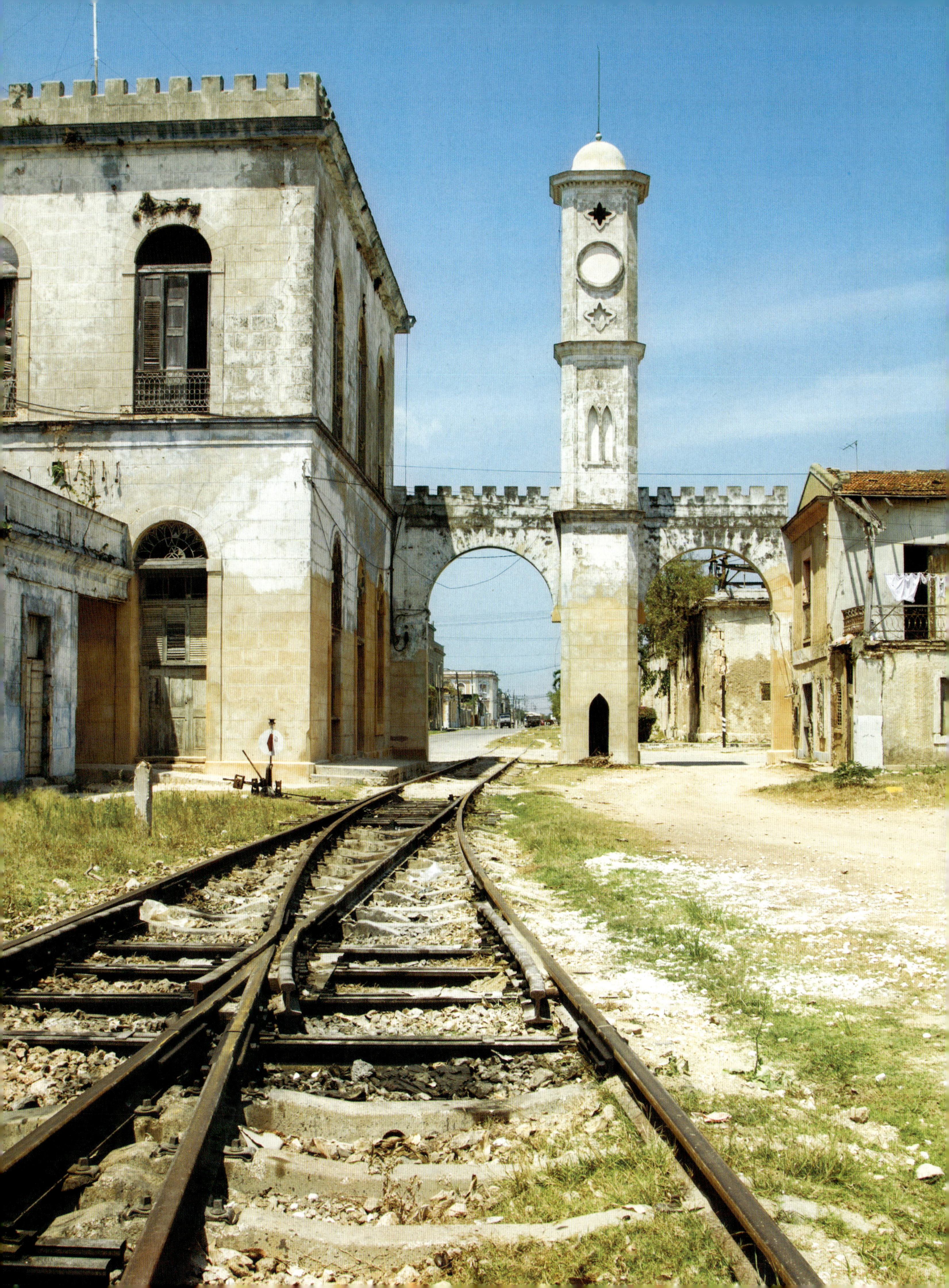

South America and the Caribbean

In Assis, Brazil, a proud new city terminus opened in 1875 had become an unused, decrepit ruin a hundred years later, its graffiti-daubed walls wired off as unsafe. By 2022 it was being reconstructed as a shiny new urban terminal for a new generation of fast, efficient trains.

The same thing is happening in many other places across the continent. In the 19th and early 20th centuries, railways were laid to open up tracts of country. That function was taken over by roads, and by the late 20th century railways generally were in a poor state across all of South America: unmodernized and suffering from years of neglect.

Now railways are once again seen as valuable assets, partly because of social changes, with more people living in large urban zones and requiring intercity transport; partly because the modern railway is seen as energy-efficient and environmentally friendly compared to overcrowded roads.

Among the most interesting developments are those in Ecuador, where modernization and restoration go hand in hand, to provide steam-hauled heritage trains that attract tourism as well as enhancing the local infrastructure. Still there are many abandoned lines and stations across the continent, offering food for thought to the traveller, as well as exciting evidence for historians and industrial archeologists.

To leave a disused railway line in a largely intact state is typical of South America; it seems to be done almost in preparation for possible re-use, though also there is not the same range of scrapping facilities as exists in Europe.

OPPOSITE:
Cardenas, Matanzas Province, Cuba
The tower between the archways looks like something borrowed from a church, but it once had clock faces for the benefit of travellers. Cuban railways have a few very modern (Chinese-built) lines and trains, and a decrepit though often picturesque network of other lines, mostly originating in the 19th century. Some are out of use, while others provide a daily or twice-weekly service.

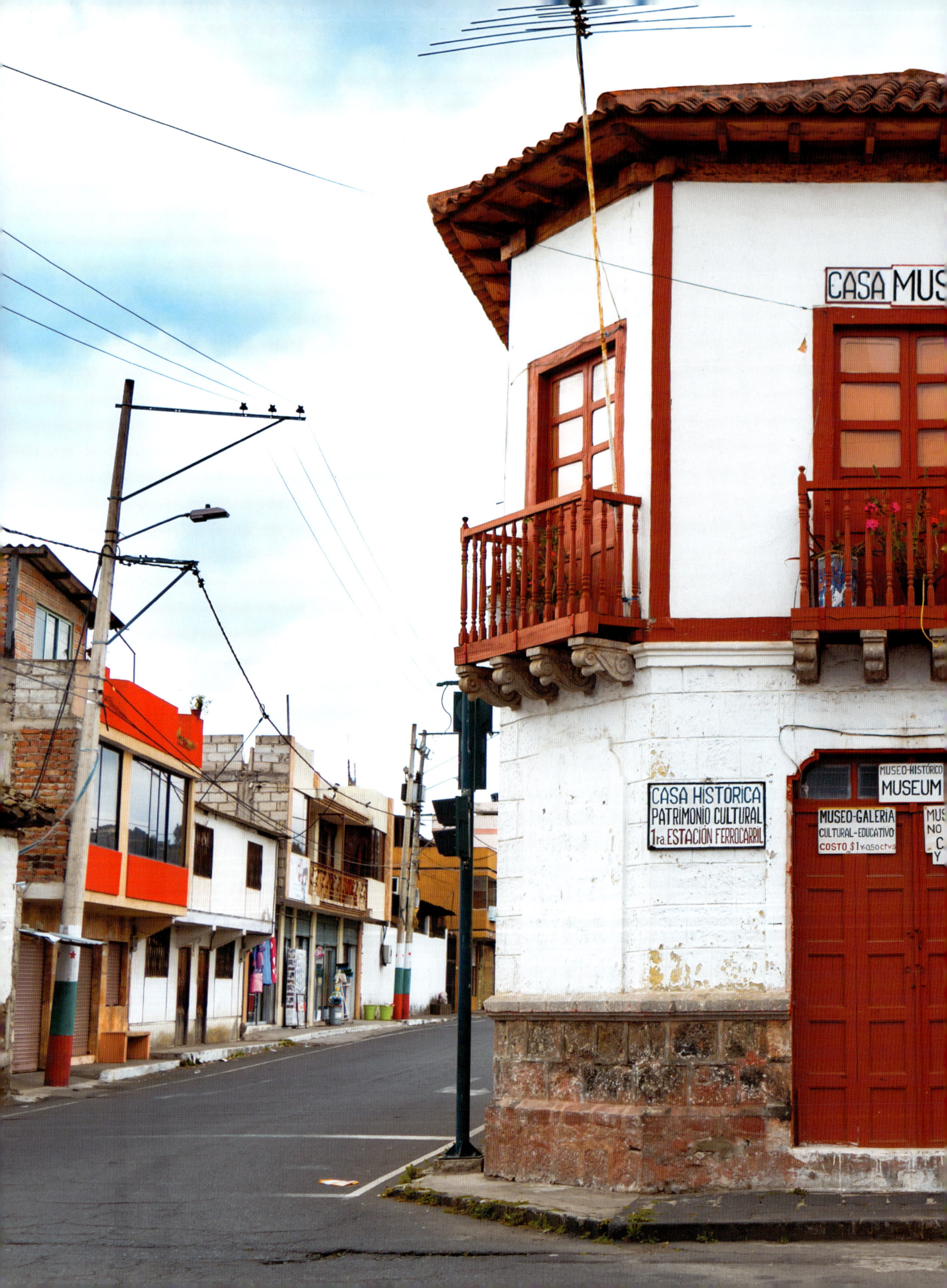
CASA MUS
CASA HISTÓRICA
PATRIMONIO CULTURAL
1ra. ESTACIÓN FERROCARRIL
MUSEO-HISTÓRICO
MUSEUM
MUSEO-GALERÍA
CULTURAL-EDUCATIVO
COSTO $1 y 0,50 ctvs

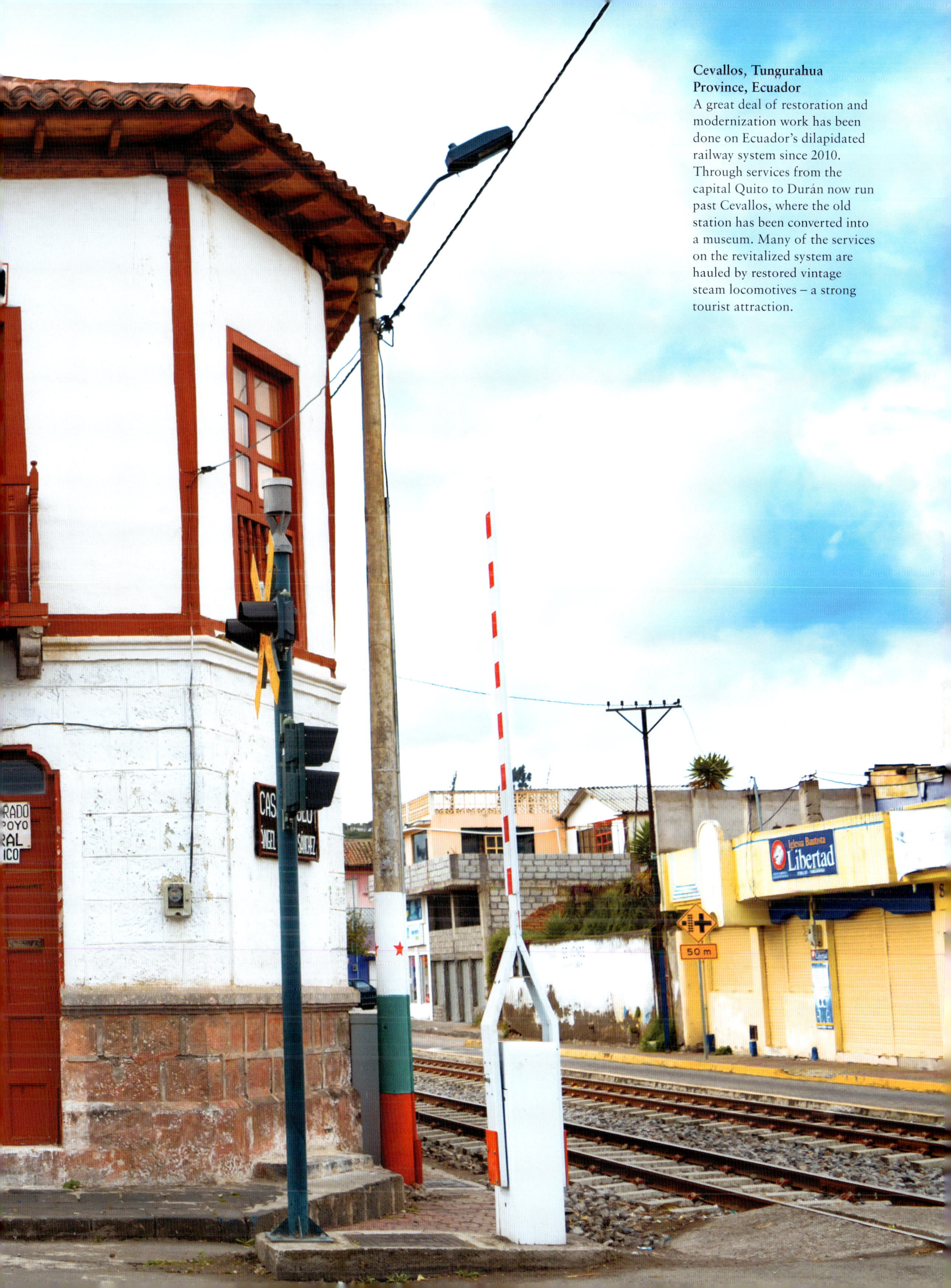

Cevallos, Tungurahua Province, Ecuador
A great deal of restoration and modernization work has been done on Ecuador's dilapidated railway system since 2010. Through services from the capital Quito to Durán now run past Cevallos, where the old station has been converted into a museum. Many of the services on the revitalized system are hauled by restored vintage steam locomotives – a strong tourist attraction.

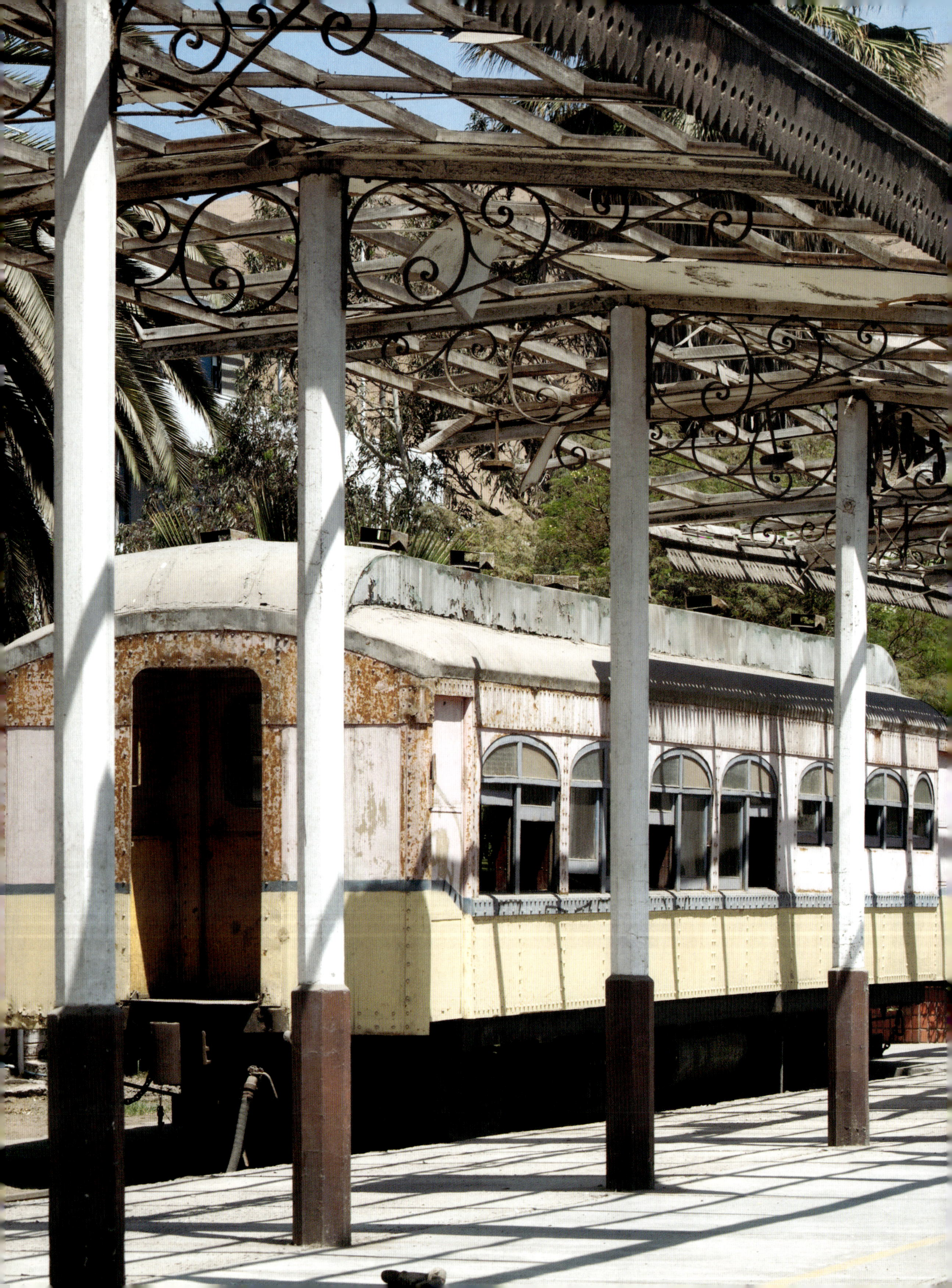

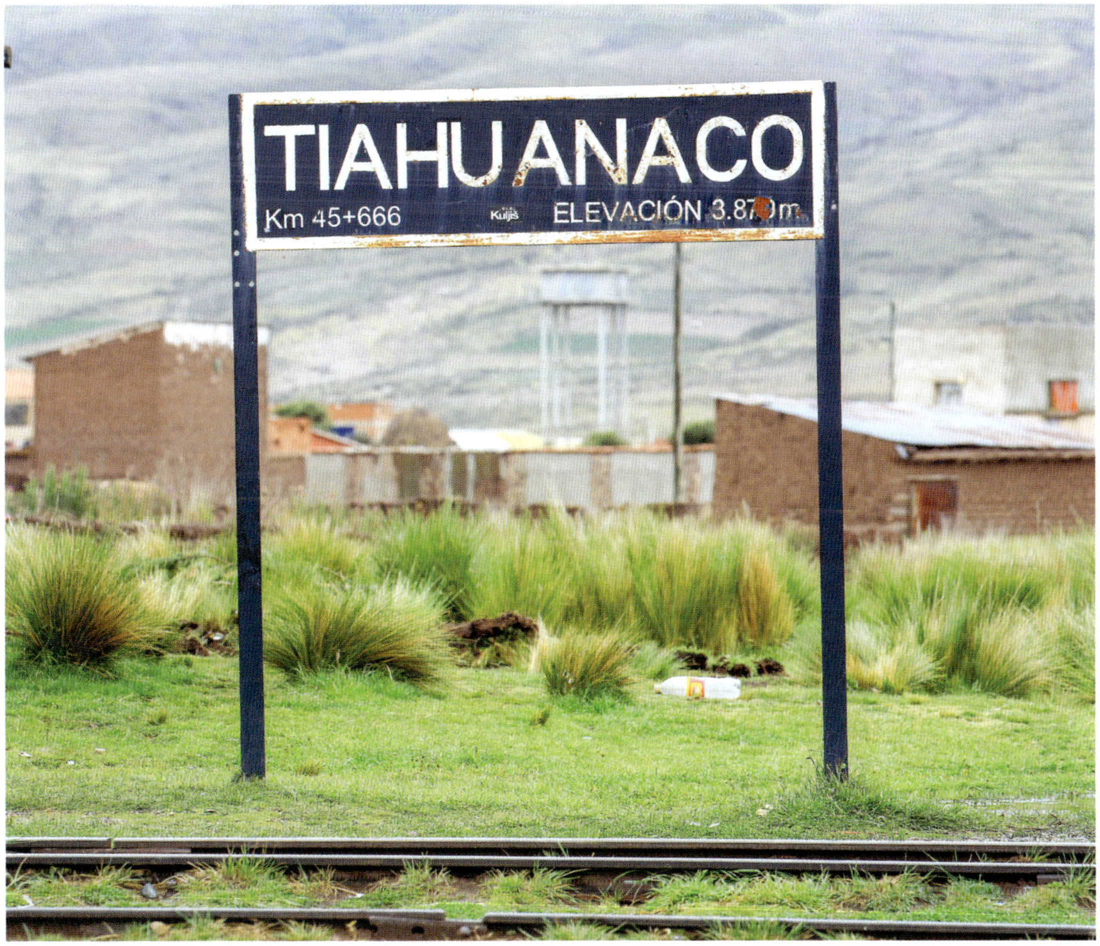

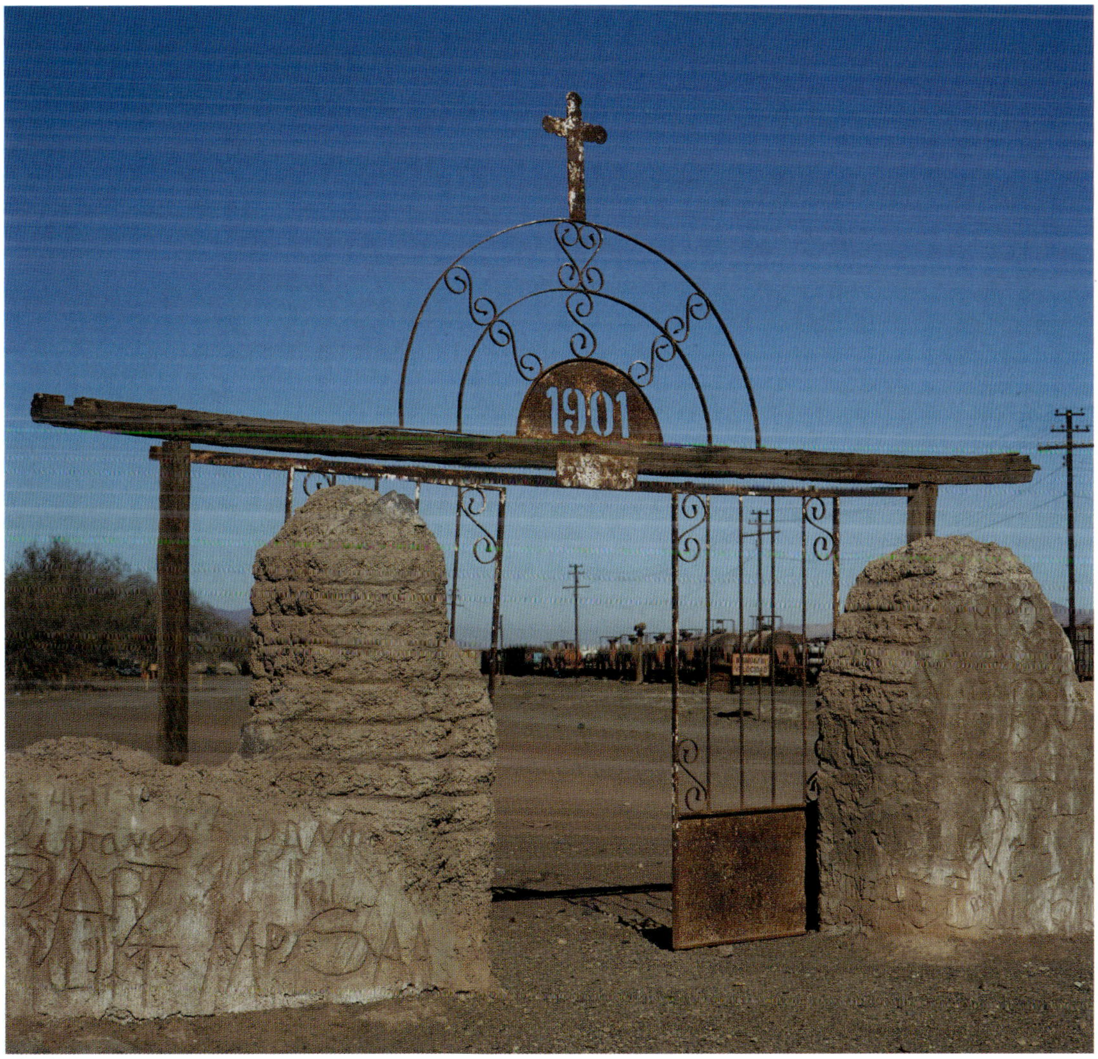

OPPOSITE:

Arica, Chile

Built in 1904, the 440km (273 mile) railway from the Chilean port of Arica to the Bolivian capital of La Paz has been mothballed for almost two decades. Reaching a height of 4257m (13,963ft), it is one of the world's highest railways. Work is now in progress to restore and upgrade the metre-gauge line. Perhaps one day, modern successors to the old sleeping car, seen here at the Arica terminus, will roll over the mountains.

LEFT:

Tiahuanaco, Bolivia

Close to Lake Titicaca, at a height of 3870m (12,673ft) this is the site of an Inca city, with some of the finest pre-Columbian period ruins. The station building remains intact with a corrugated-iron roof and though it is closed, a tourist train each month makes the trip from El Alto, past the archeological site at Tiahuanaco, and on to Giaqui on the shore of the lake.

BELOW LEFT AND OVERLEAF:

Baquedano, Antofagasta, Chile

Named for a famous general, Baquedano was a station long before this gate was erected in 1901. But by then it had become a busy junction for lines coming from the South American interior. Its station and locomotive depot, 75km (47 miles) from Antofagasta, still show the scale of railway operations. Here as elsewhere, some lines are being brought back into use.

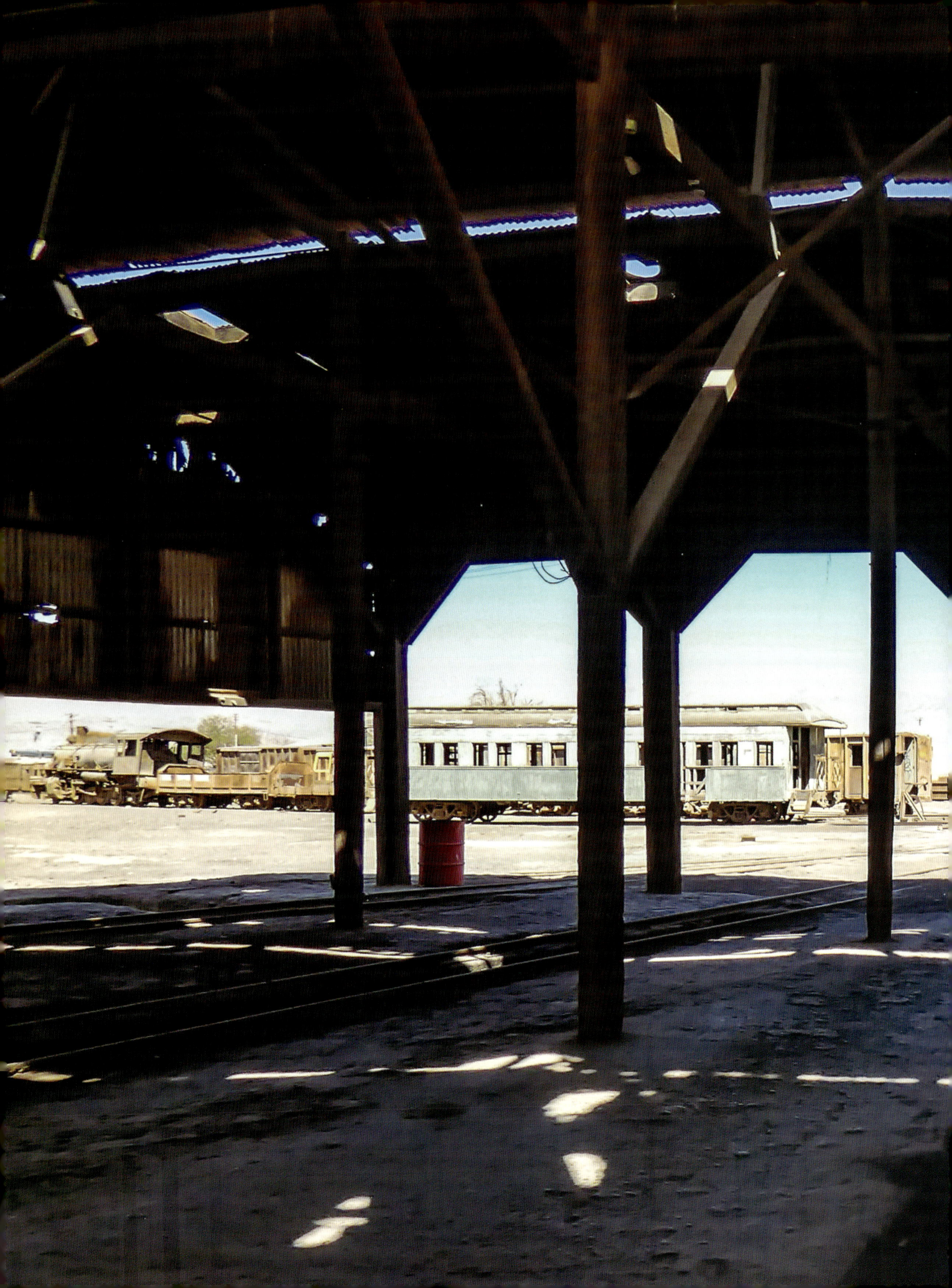

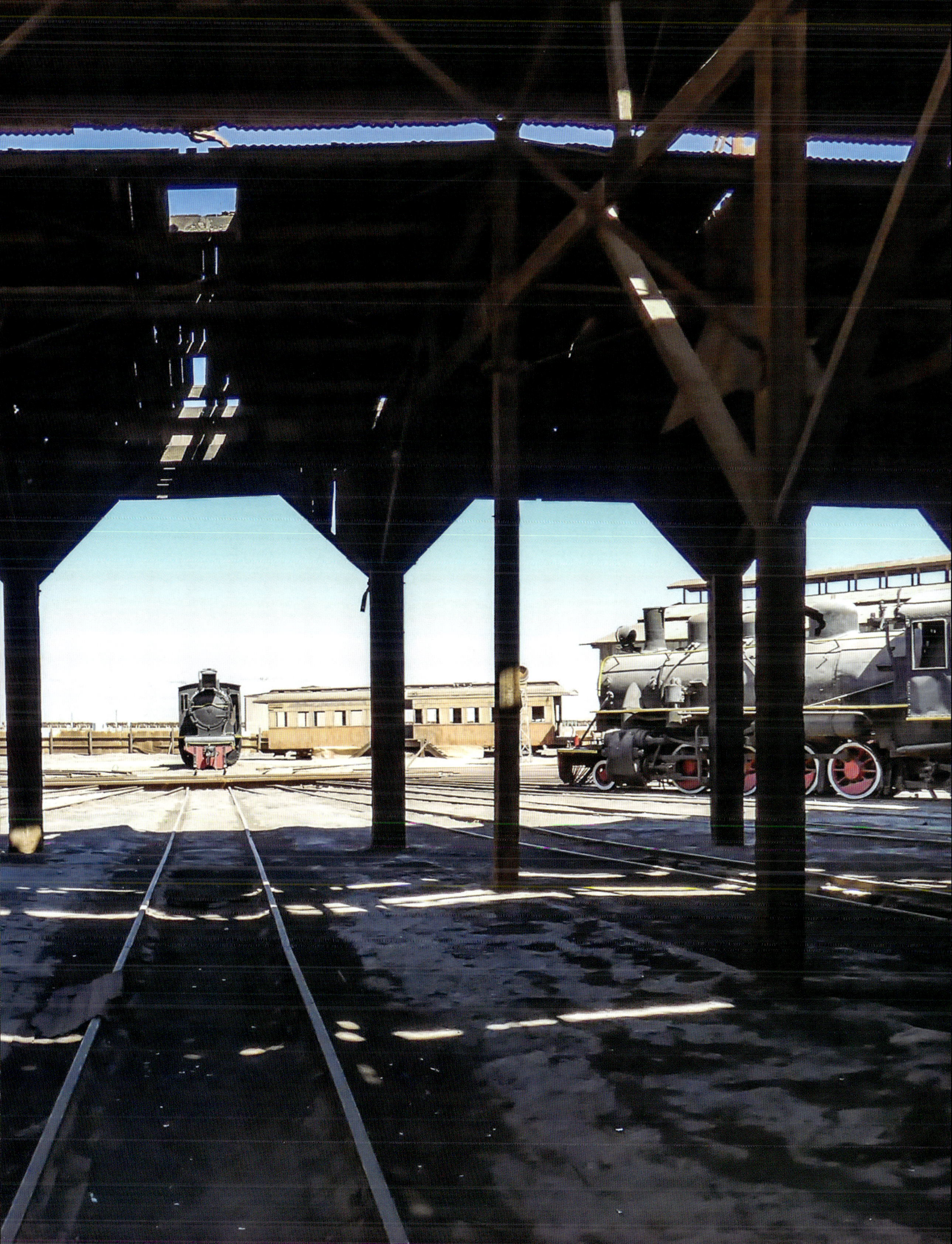

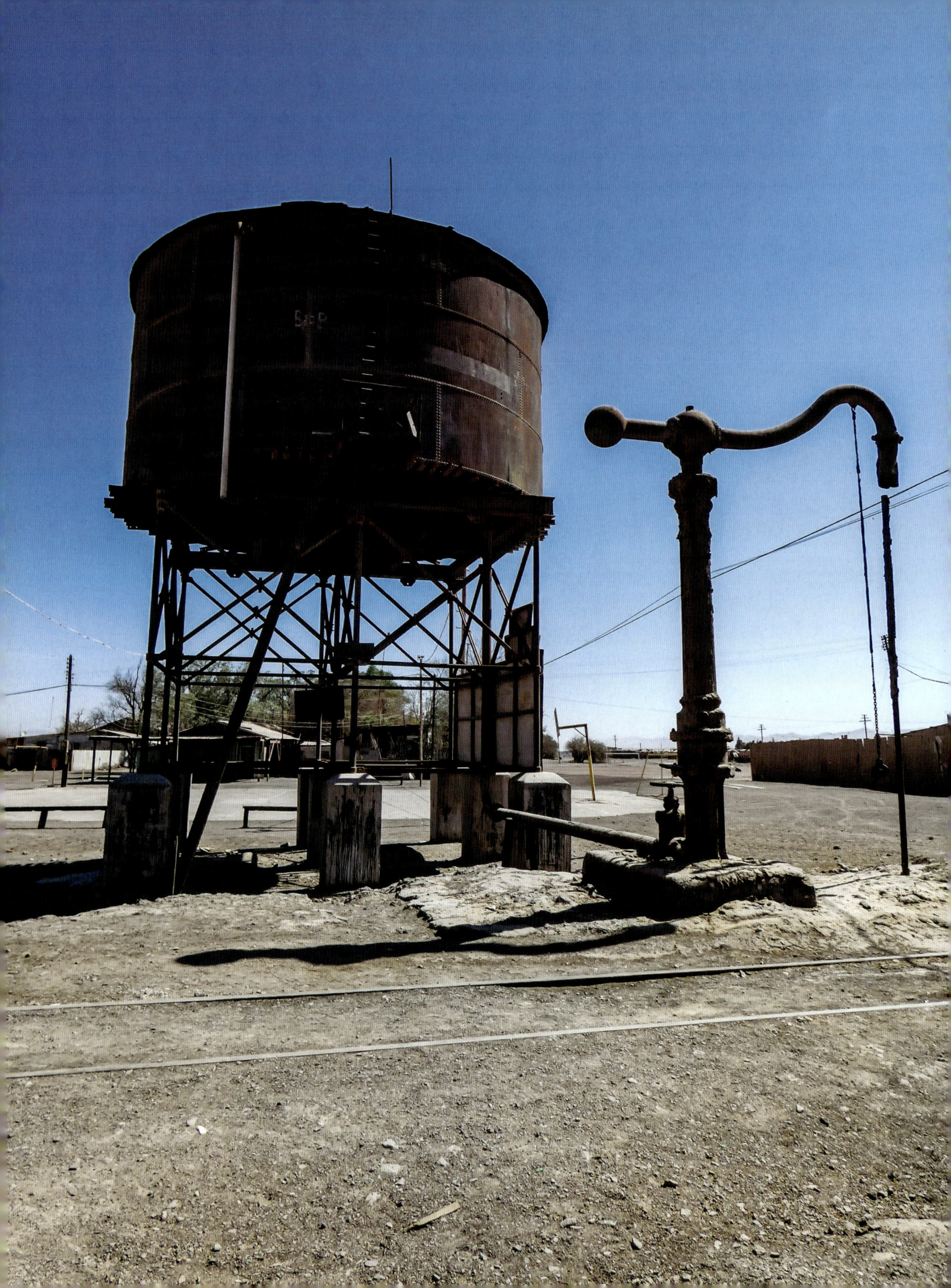

OPPOSITE:

Baquedano, Antofagasta, Chile

For the steam-powered railway, a water supply was essential, preferably pure, without salts and chemicals which could shorten the life of an expensive boiler. European manufacturers sent out catalogues showing a range of different styles for water columns, often in classical style, as here. The locally-built tank looks sadly unrefined by comparison.

BELOW:

Vinhedo, São Paulo Province, Brazil

The *Cia. Paulista de Estradas de Ferro* (Paulist Railway Co.) was a prime force from 1872 in the development of the Vinhedo region, even electrifying the line. It later fell on hard times, chiefly as the result of road-building, and the line was abandoned, leaving fragmentary infrastructure and the ruins of quite extensive station buildings.

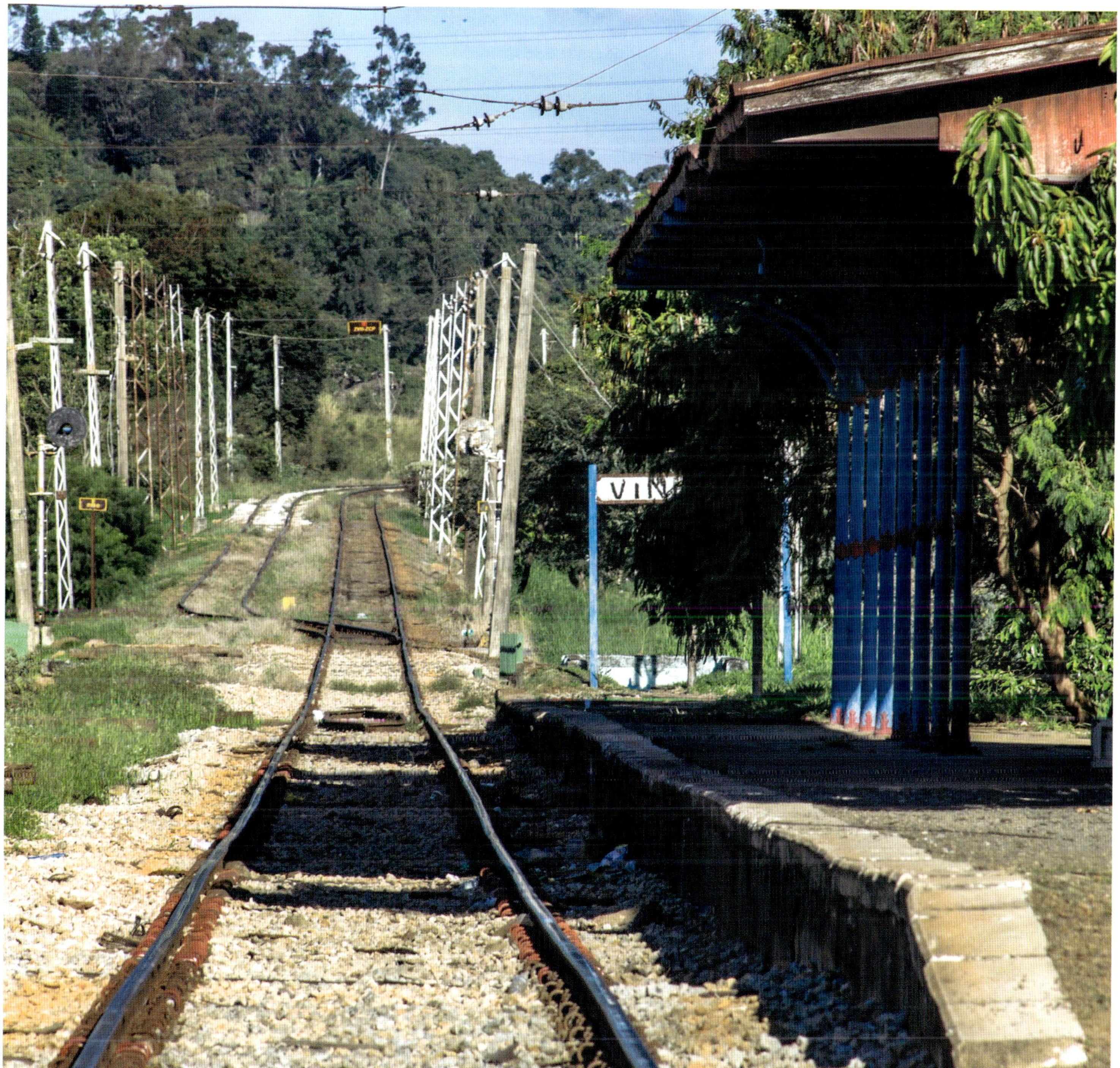

Quipapá Station, Pernambuco State, Brazil
Opened in 1885 when hopes of developing northeast Brazil were high, the station, at 427m (1400ft), was on the southern branch of the line from Recife on the coast to the inland centre of Maceió. As a long cross-country branch line, the railway could not survive and closed in 2020.

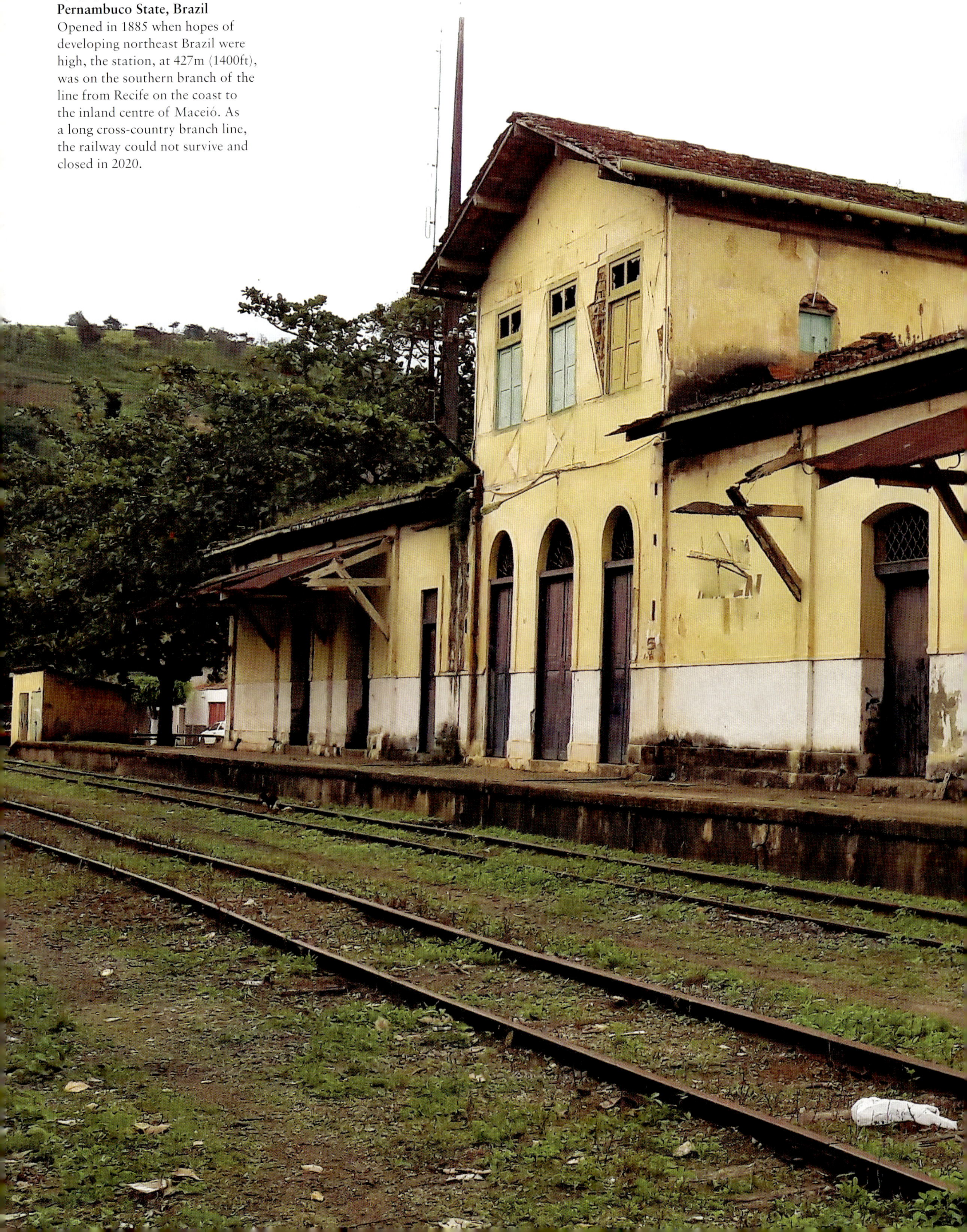

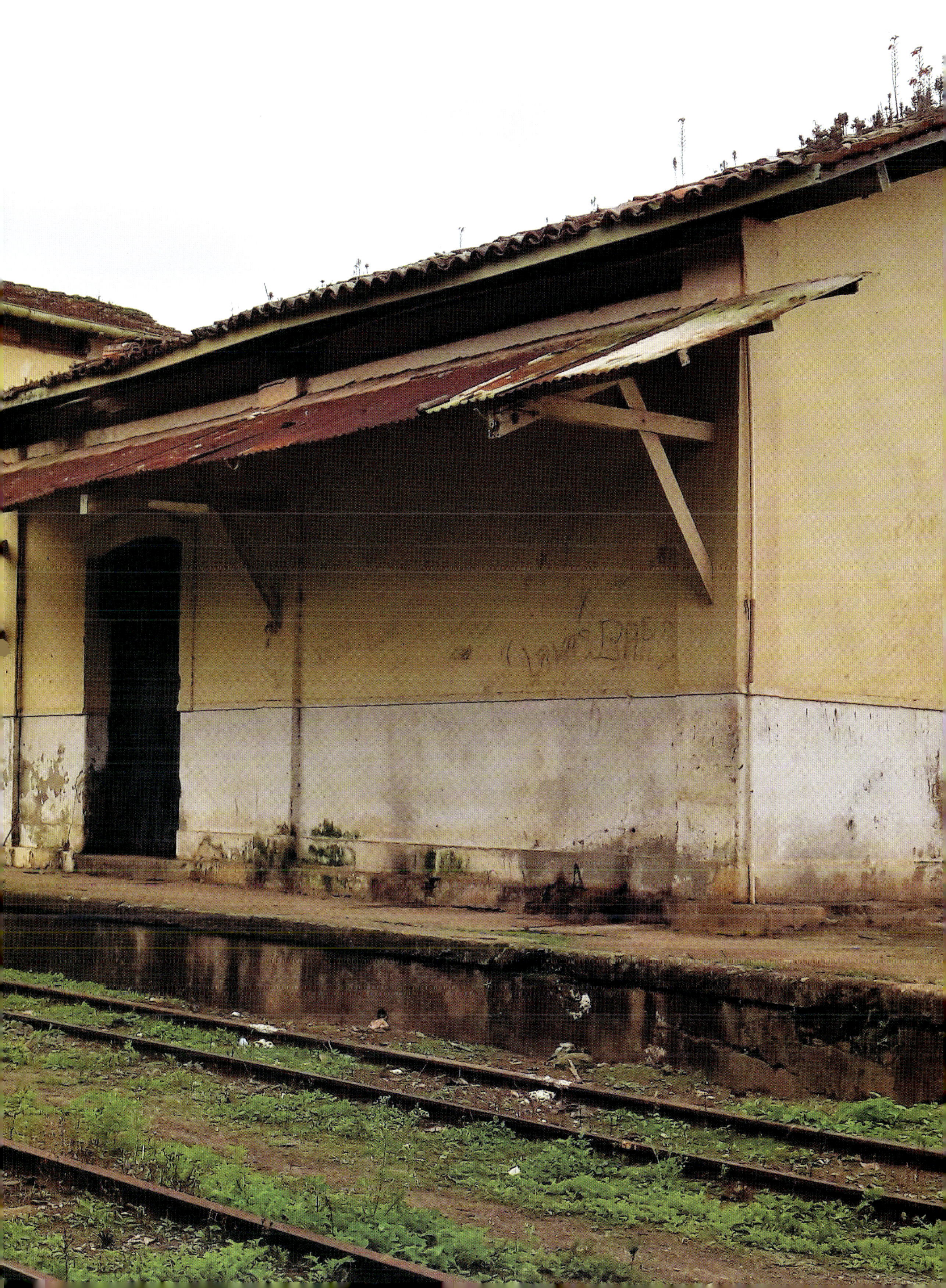

Old Railway Station, Assis, São Paulo Province, Brazil
This rather evocative image is already out of date. The old rails have been lifted, the utility awning (old rails used as pillars) will be just a memory. Assis' long-closed original terminus is about to experience a new lease of life as the *Terminal Urbano*, for modern fast trains to and from the city. With proper regard for the past, as much as possible of the original buildings will be integrated into the new terminal.

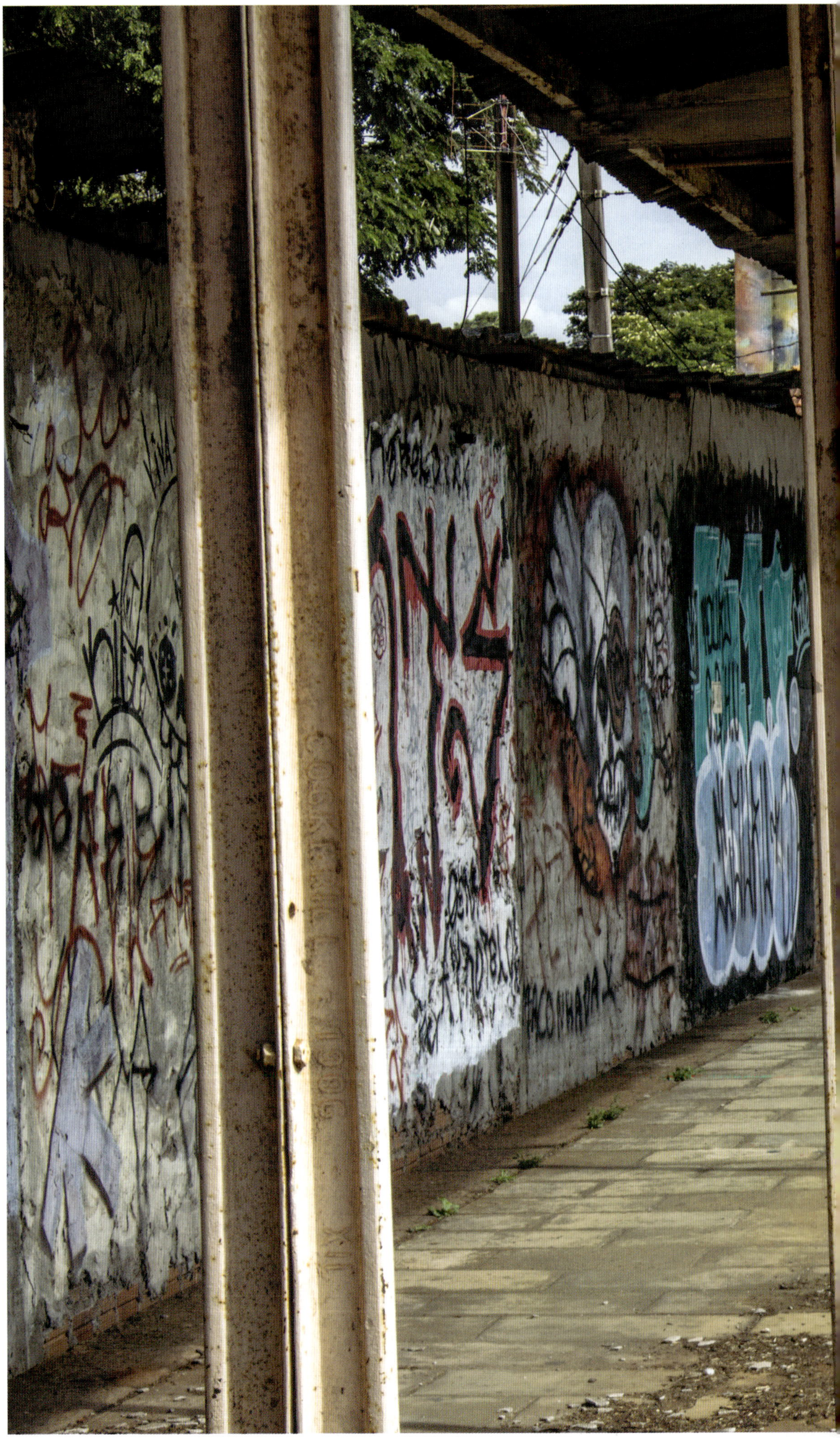

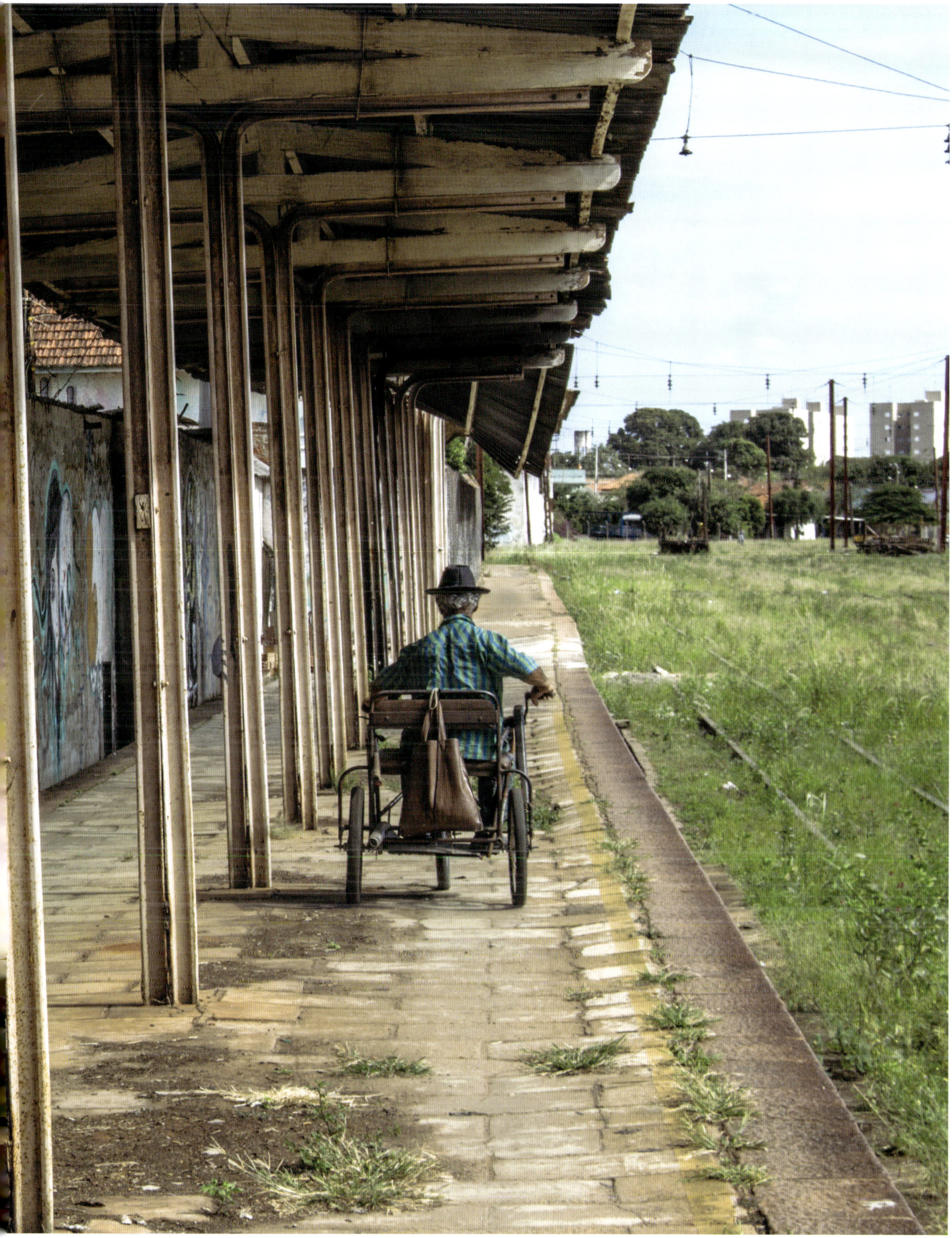

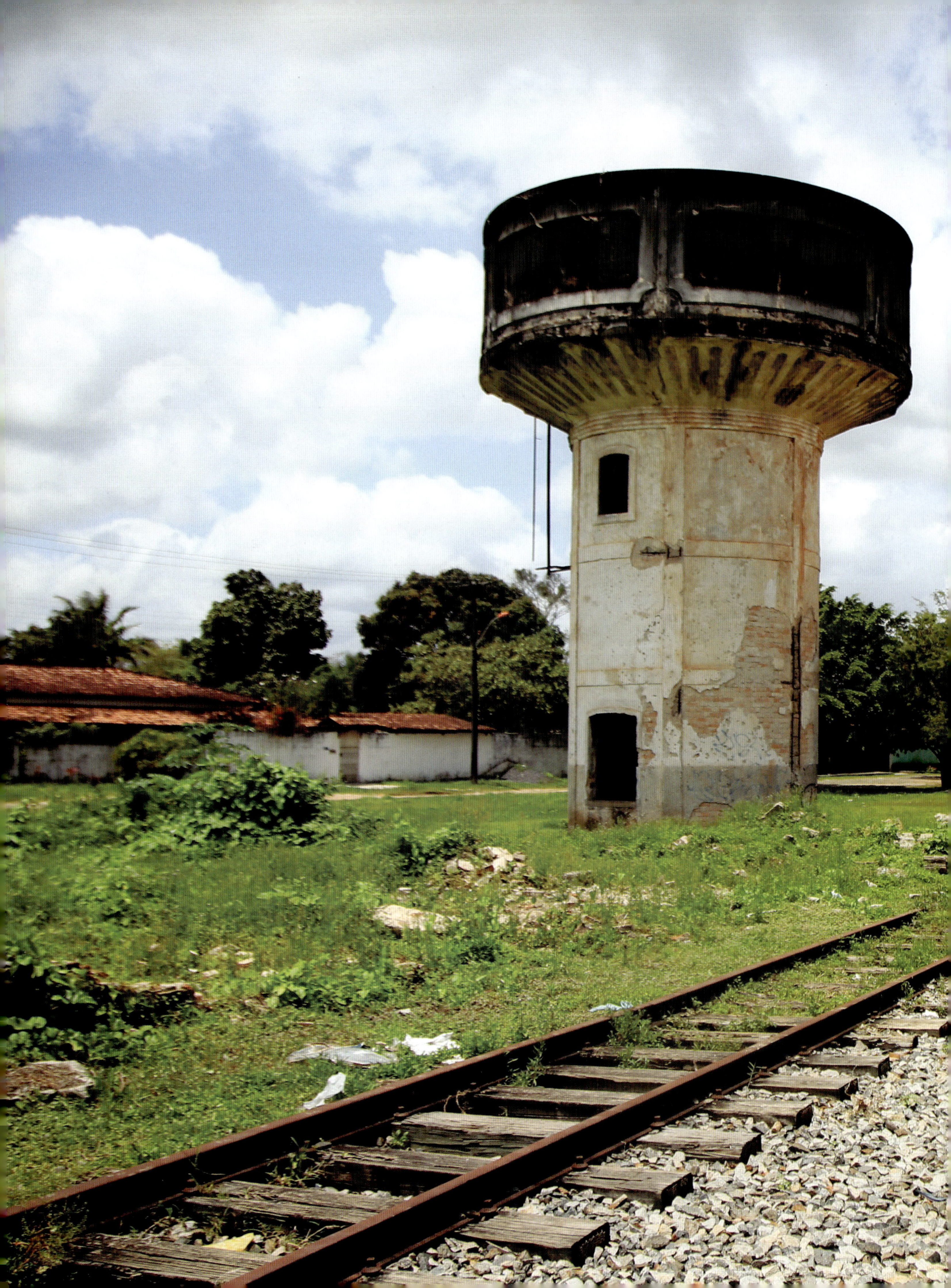

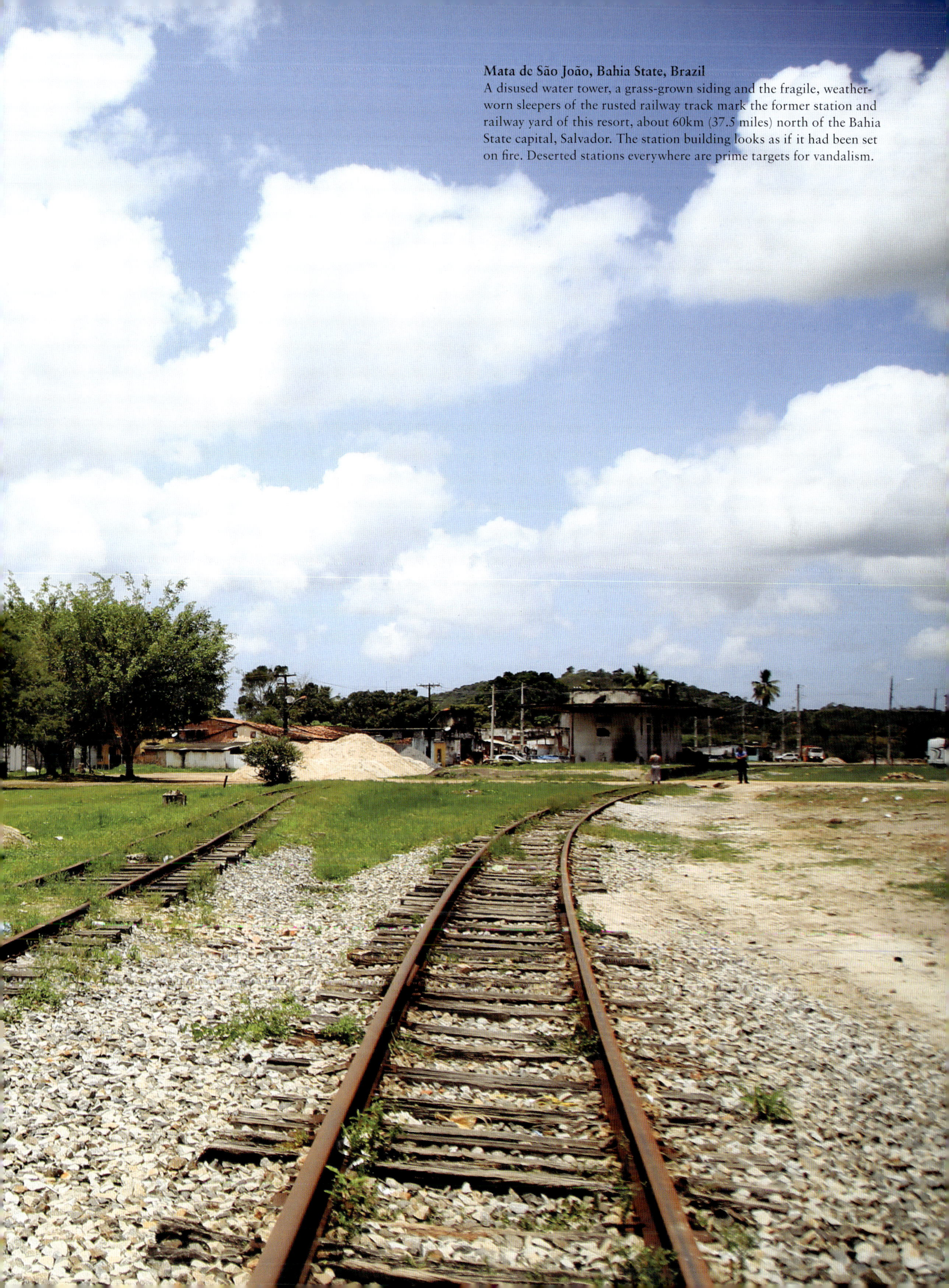

Mata de São João, Bahia State, Brazil
A disused water tower, a grass-grown siding and the fragile, weather-worn sleepers of the rusted railway track mark the former station and railway yard of this resort, about 60km (37.5 miles) north of the Bahia State capital, Salvador. The station building looks as if it had been set on fire. Deserted stations everywhere are prime targets for vandalism.

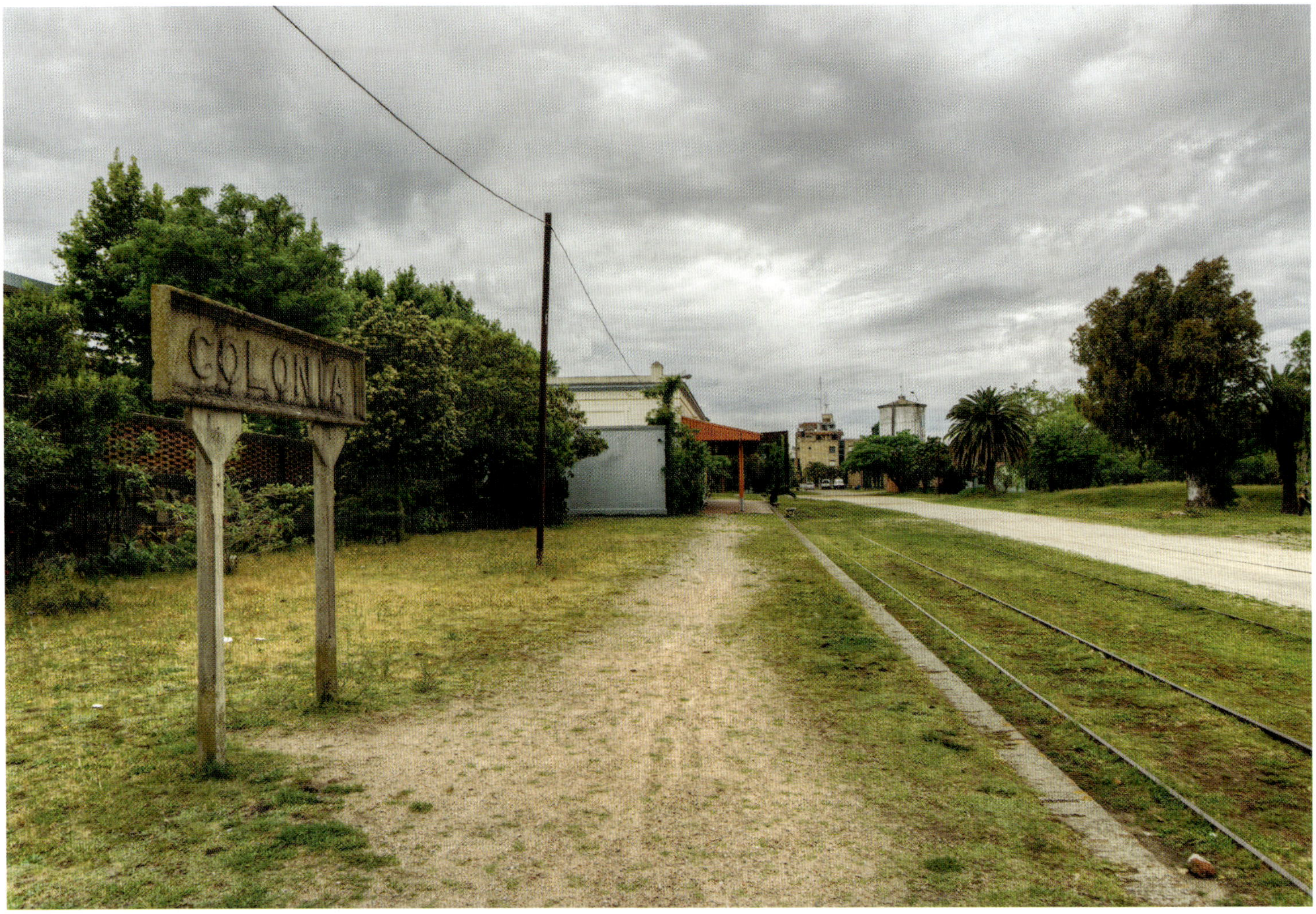
COLONIA

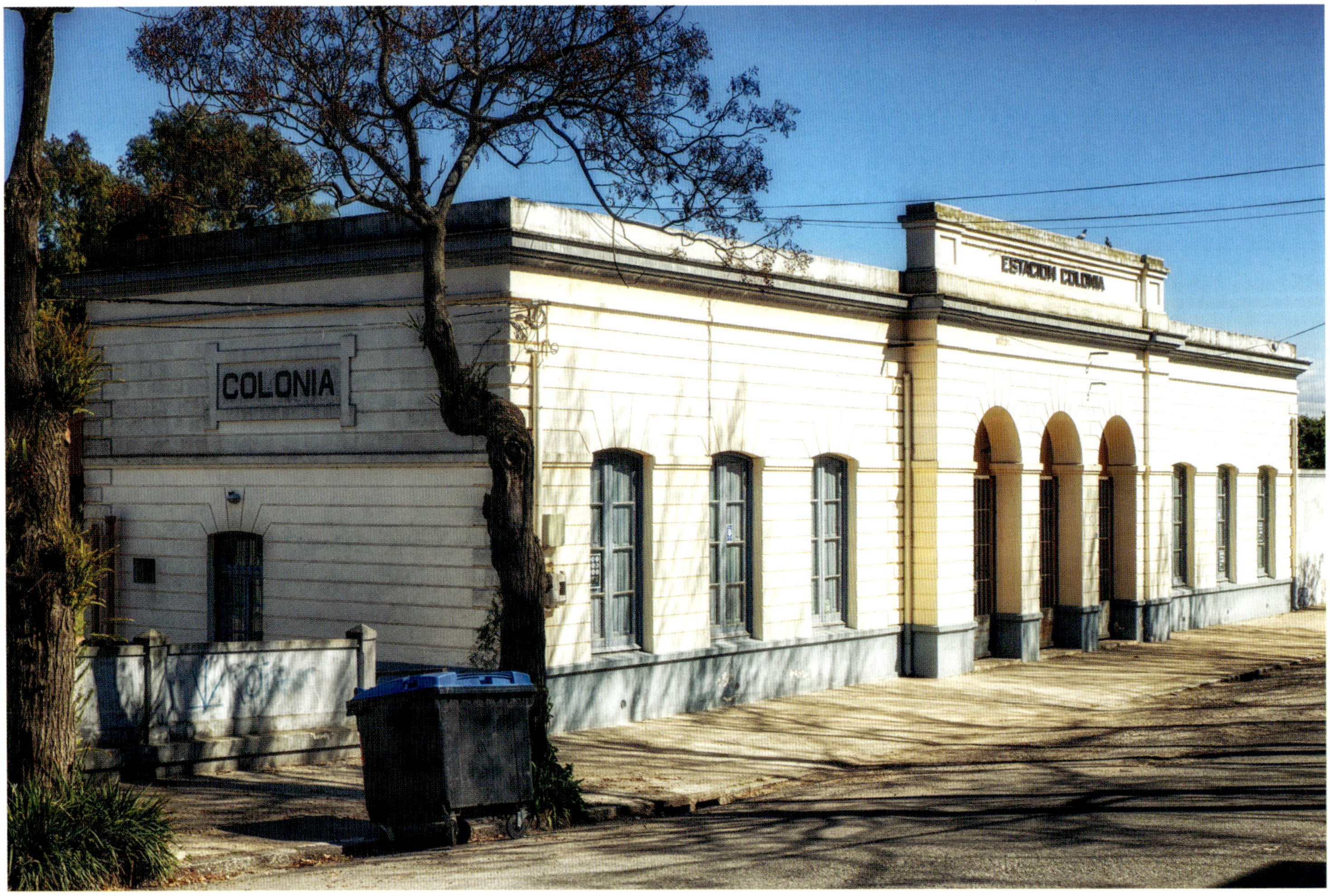
ESTACION COLONIA
COLONIA

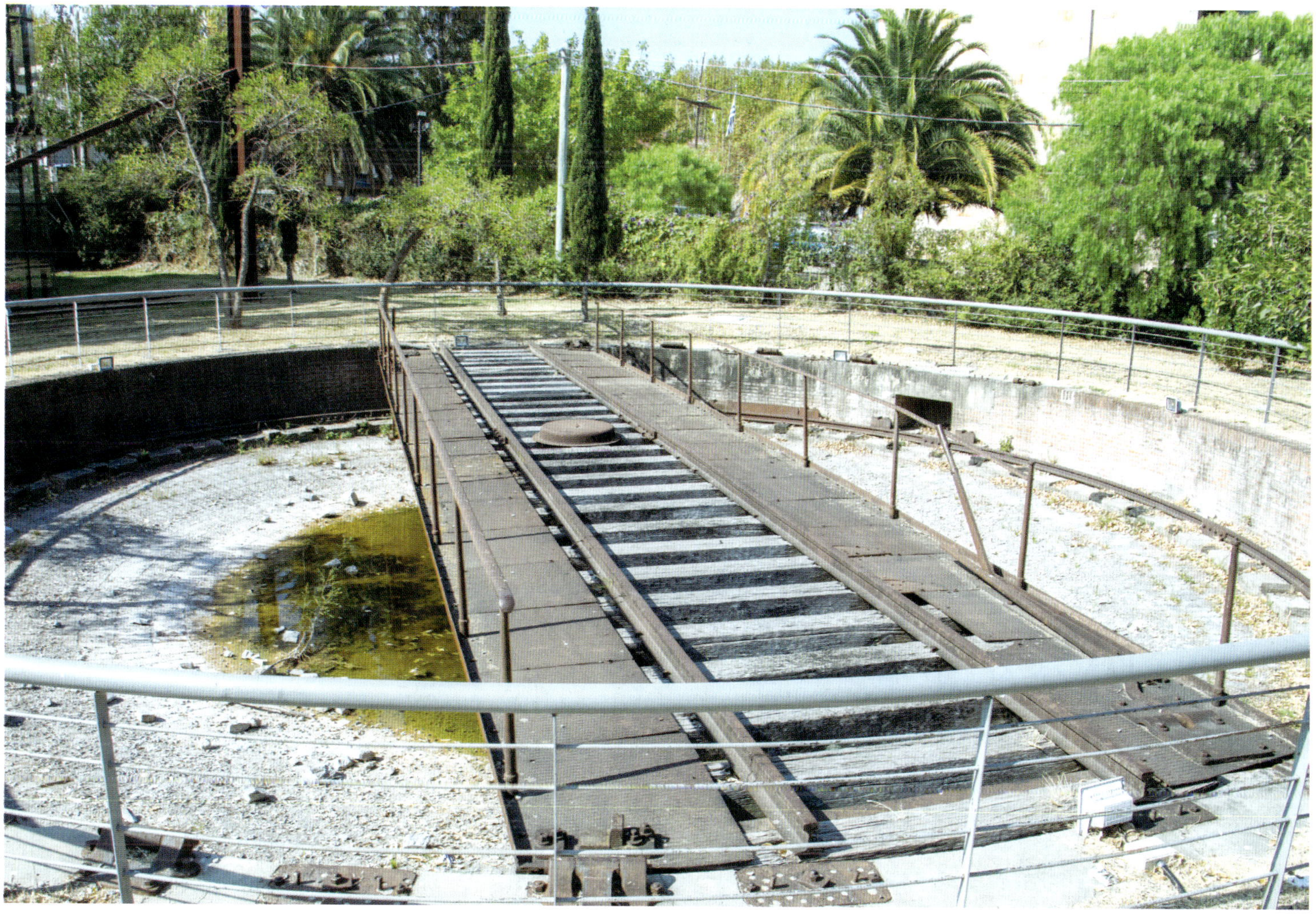

ALL PHOTOGRAPHS:

Colonia de Sacramento, Uruguay

Terminus of Uruguay's Montevideo–Colonia Railway, Colonia opened in 1901, linked with a ferry service to Buenos Aires in Argentina. Closed in 1985, it is now a railway museum, preserving features including the station building and some of its signage, goods warehouse, signal cabin and the locomotive turntable. A selection of vintage carriages can also be inspected.

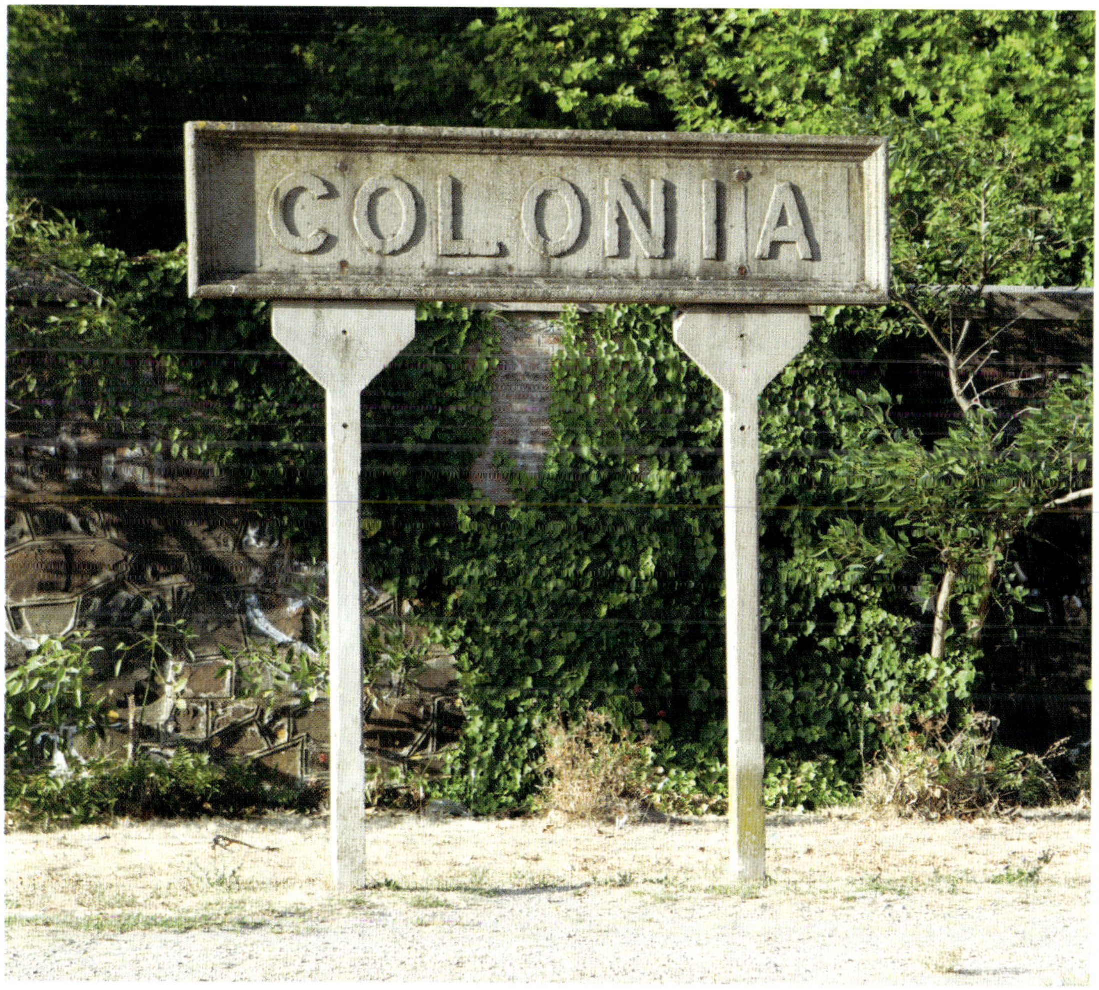

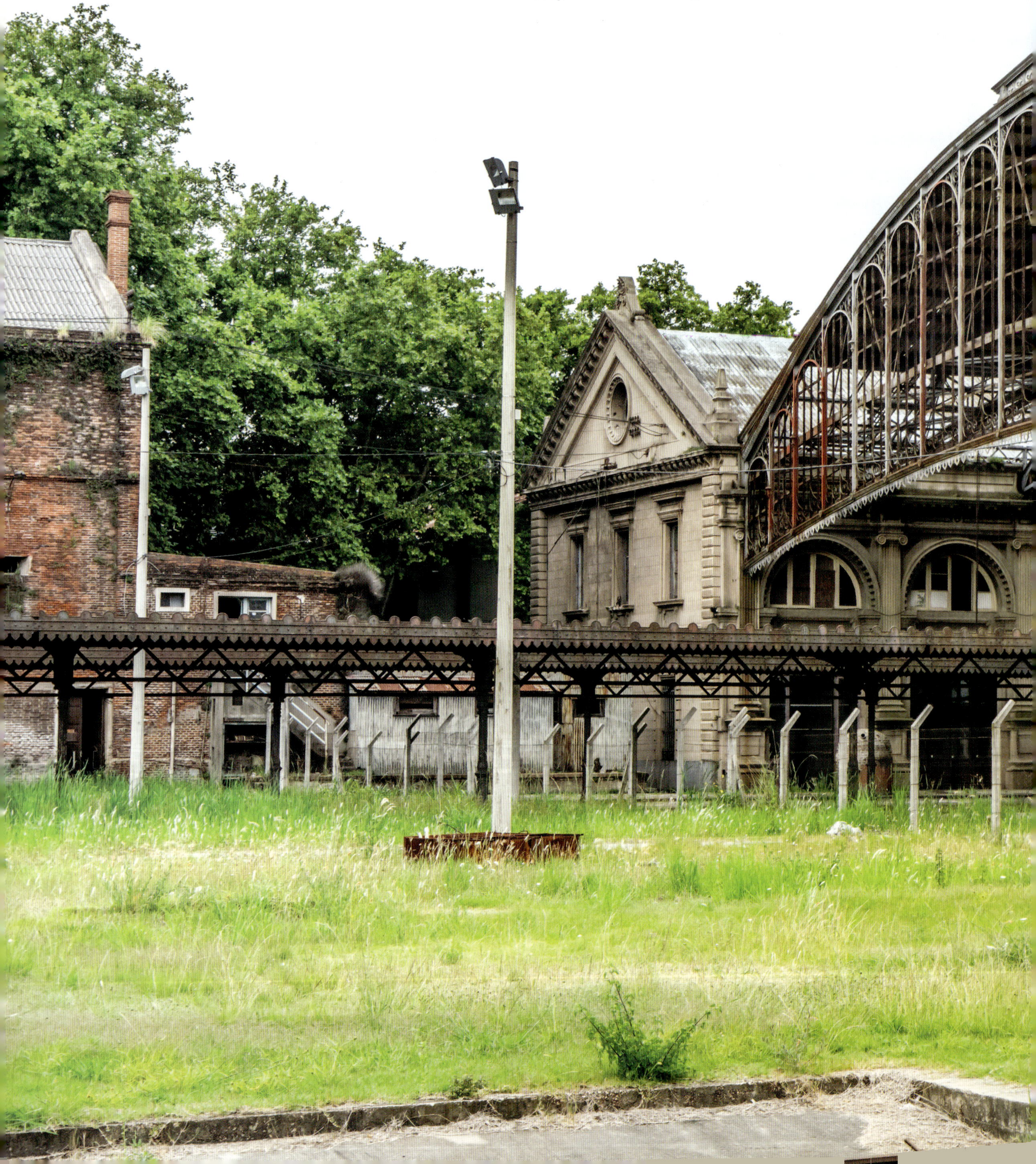

General Artigas Station, Montevideo, Uruguay
This was the Uruguyan capital's main railway station in the days when the country had an extensive passenger system. Opened in 1897, it was closed in 2003. The train shed looks French in style but its designer was an Italian engineer.

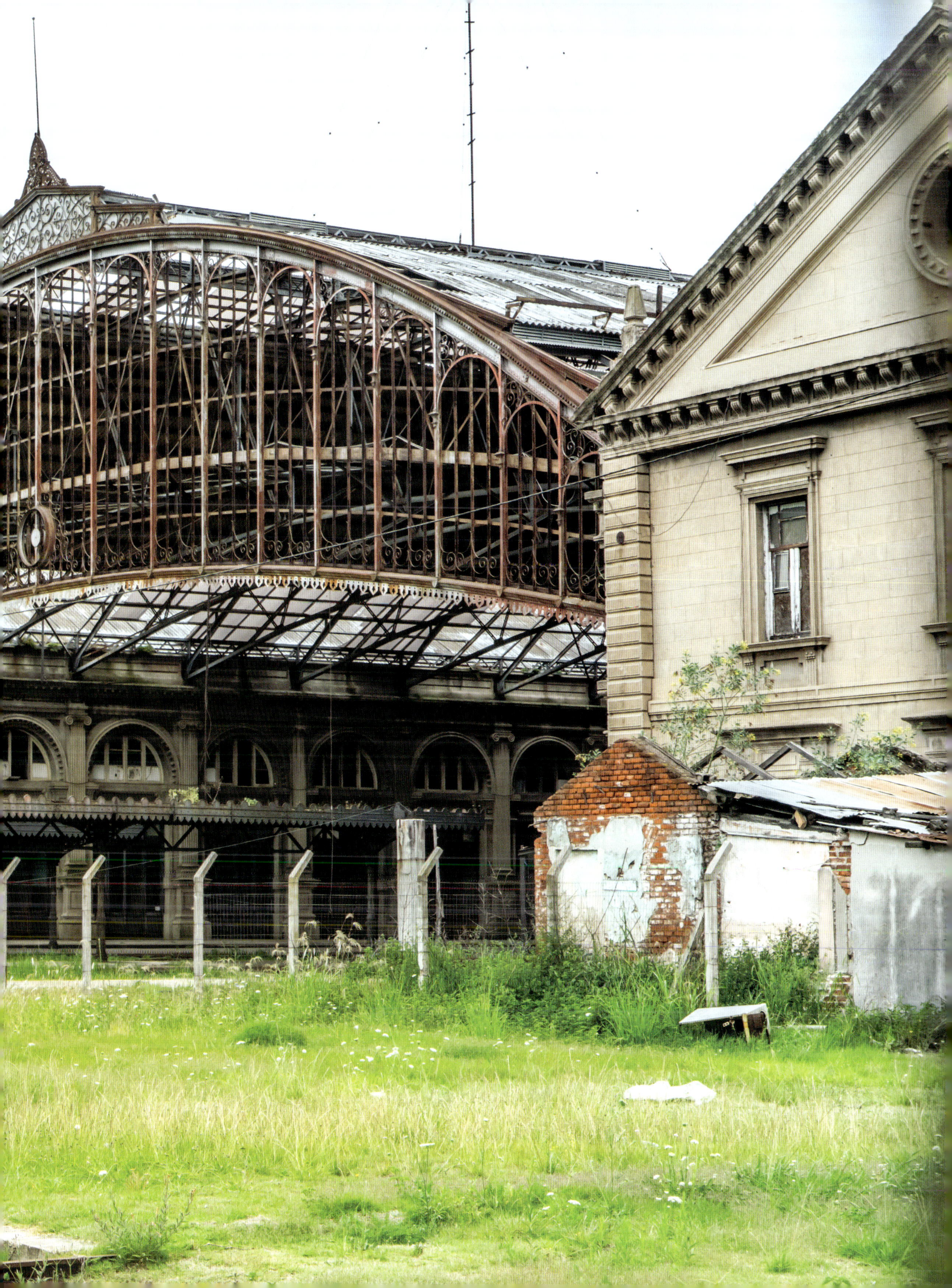

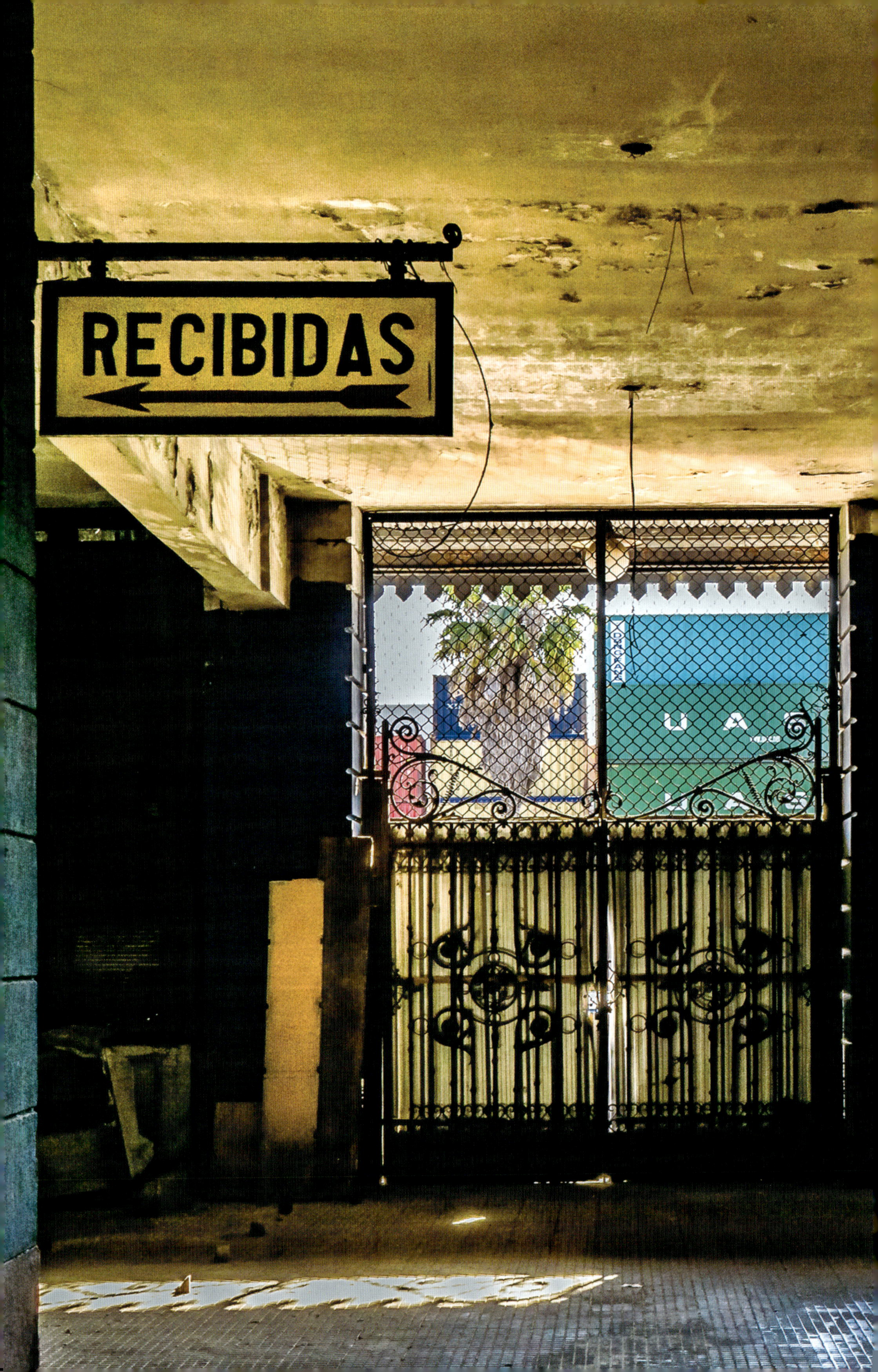
RECIBIDAS

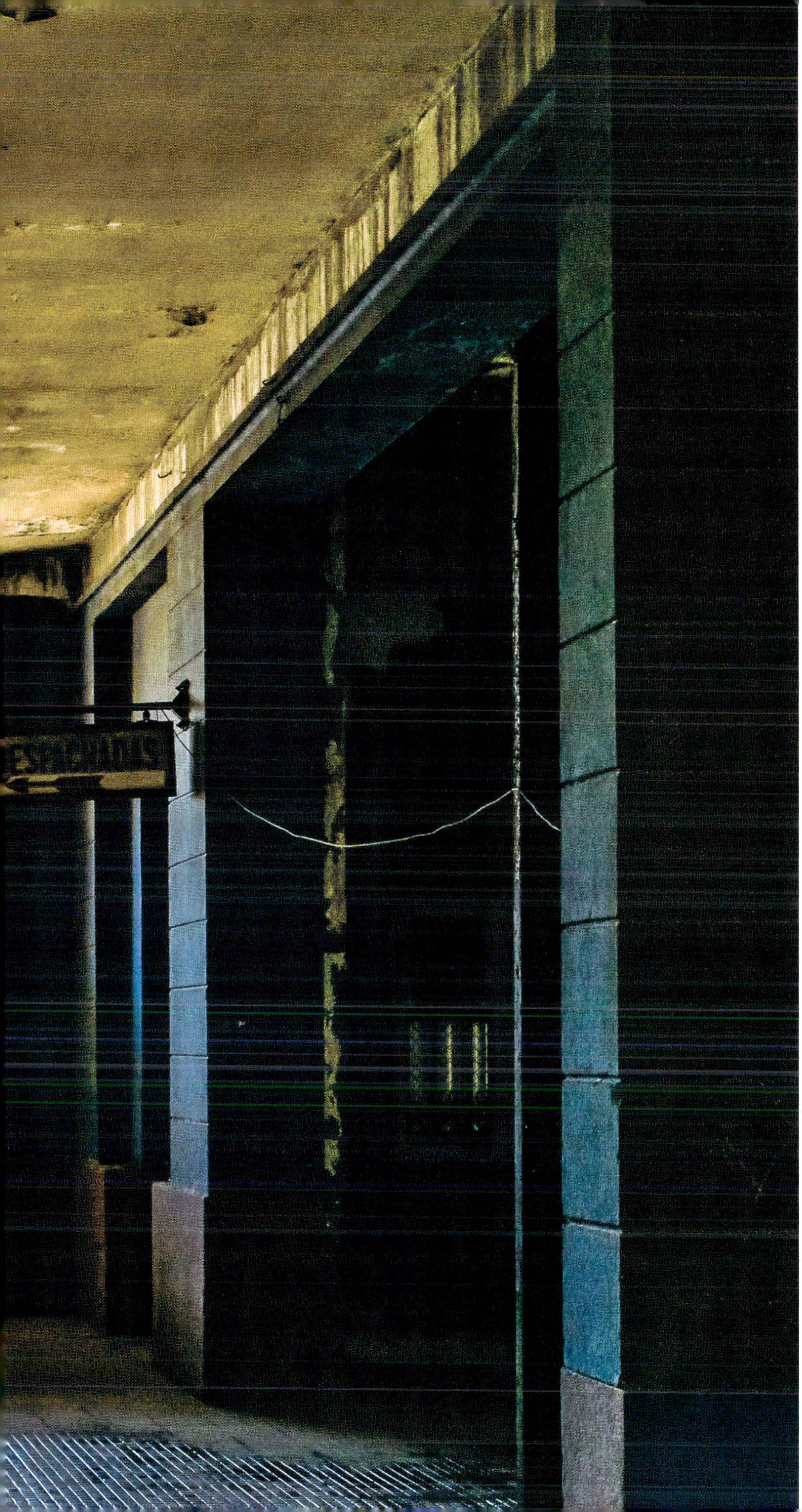

LEFT & OVERLEAF:

General Artigas Station, Montevideo, Uruguay

The arrivals and departures hall of the city's old central station still shows many attractive late-19th century features (Uruguayan railway construction was British-managed and financed) but is now in a state of decay. A much smaller, new station in the city is the terminal for some commuter trains.

The departures board (overleaf) shows that the Uruguayan rail system was focused on the capital, with lines spreading fanwise across the country, and international connections with Brazilian and Argentinian lines.

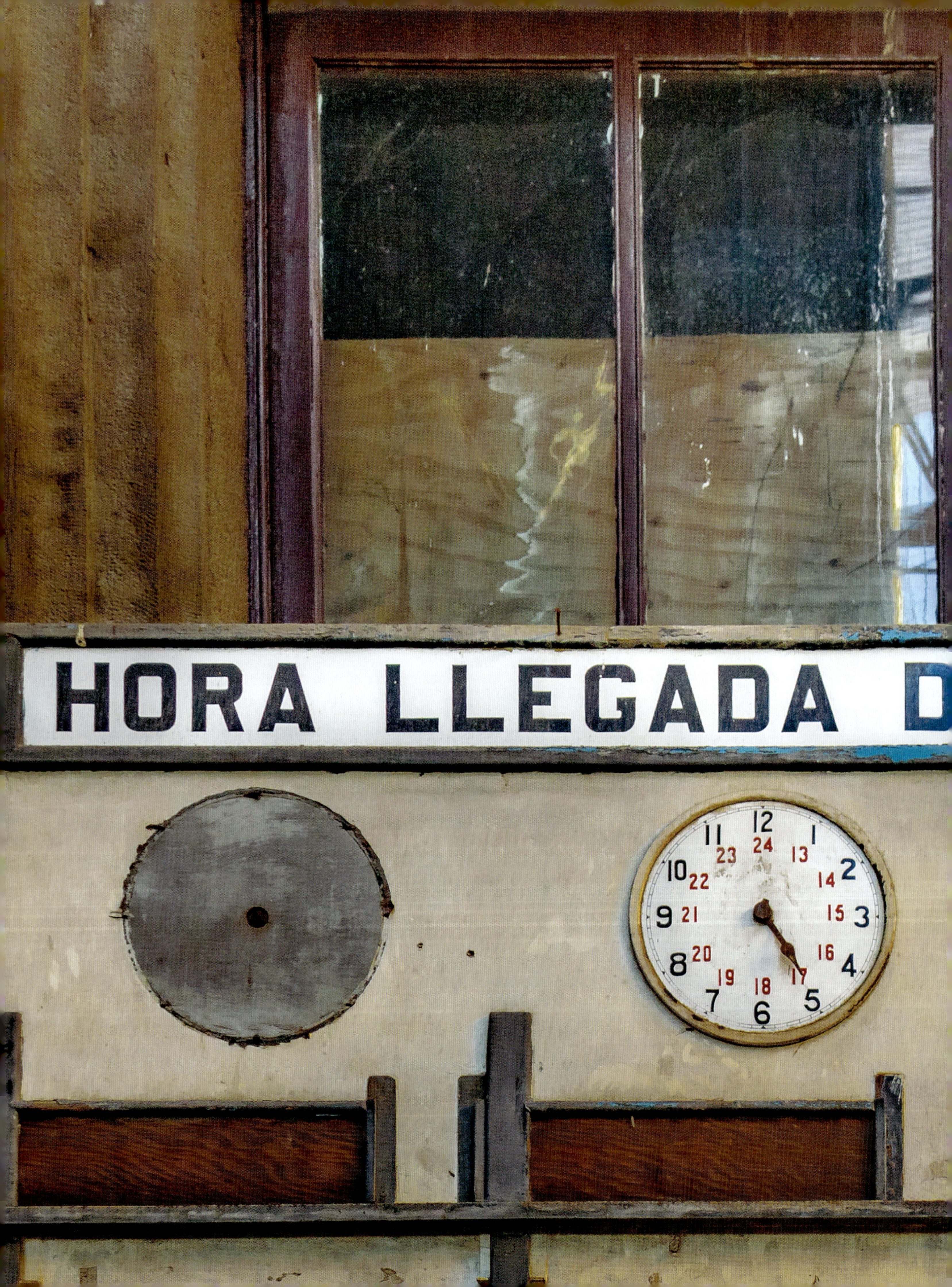
HORA LLEGADA D

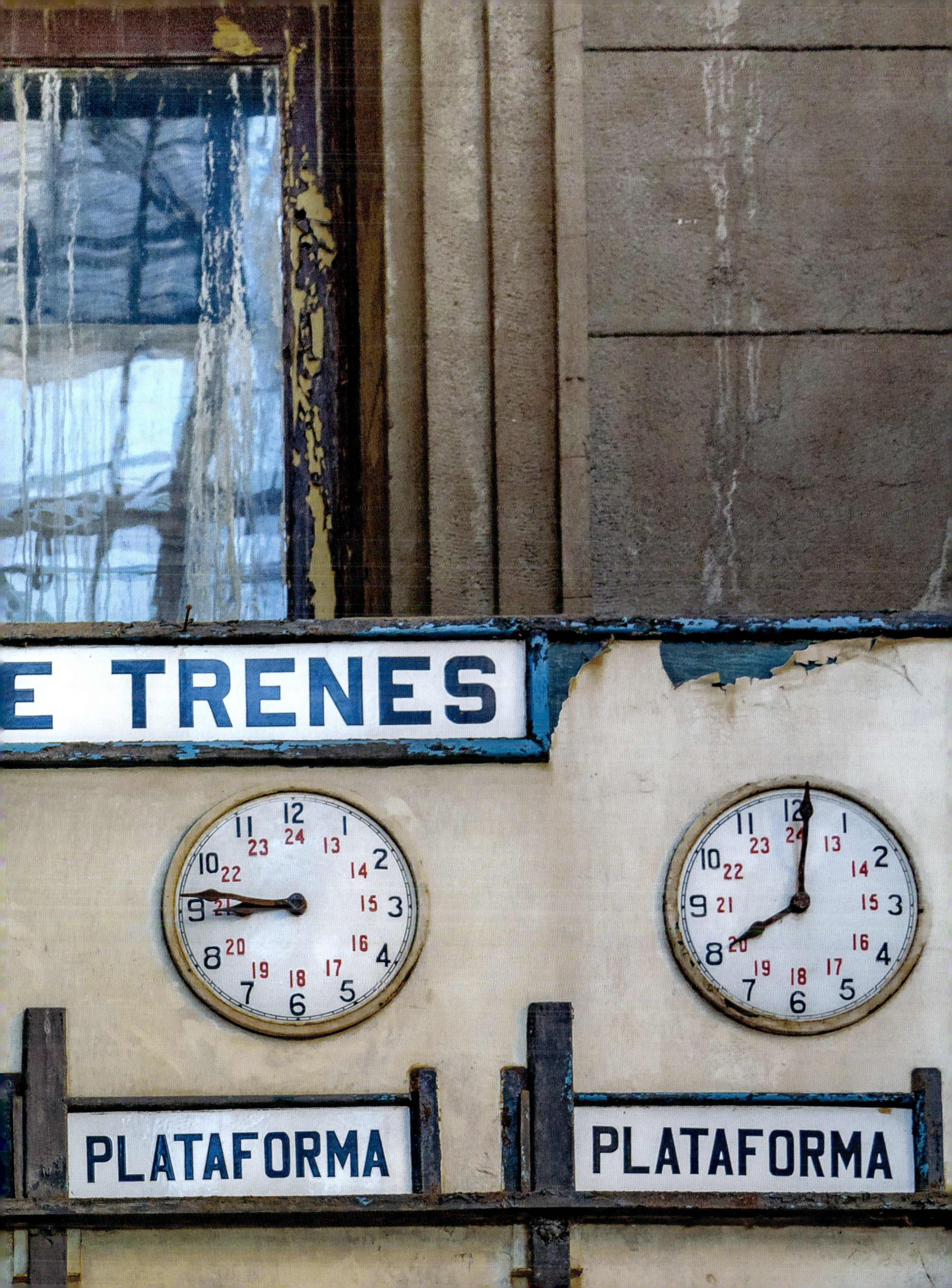
E TRENES
PLATAFORMA
PLATAFORMA

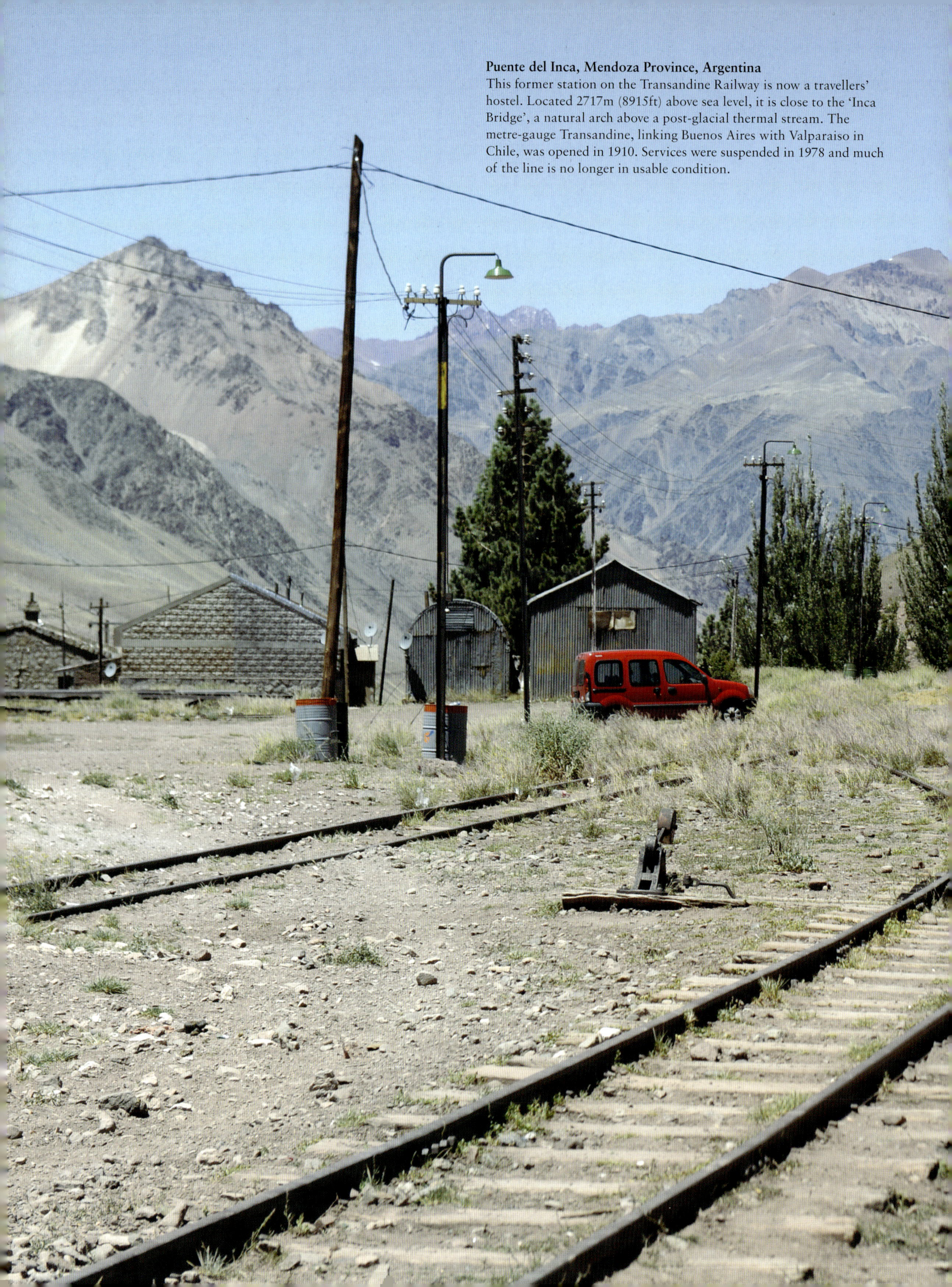

Puente del Inca, Mendoza Province, Argentina
This former station on the Transandine Railway is now a travellers' hostel. Located 2717m (8915ft) above sea level, it is close to the 'Inca Bridge', a natural arch above a post-glacial thermal stream. The metre-gauge Transandine, linking Buenos Aires with Valparaiso in Chile, was opened in 1910. Services were suspended in 1978 and much of the line is no longer in usable condition.

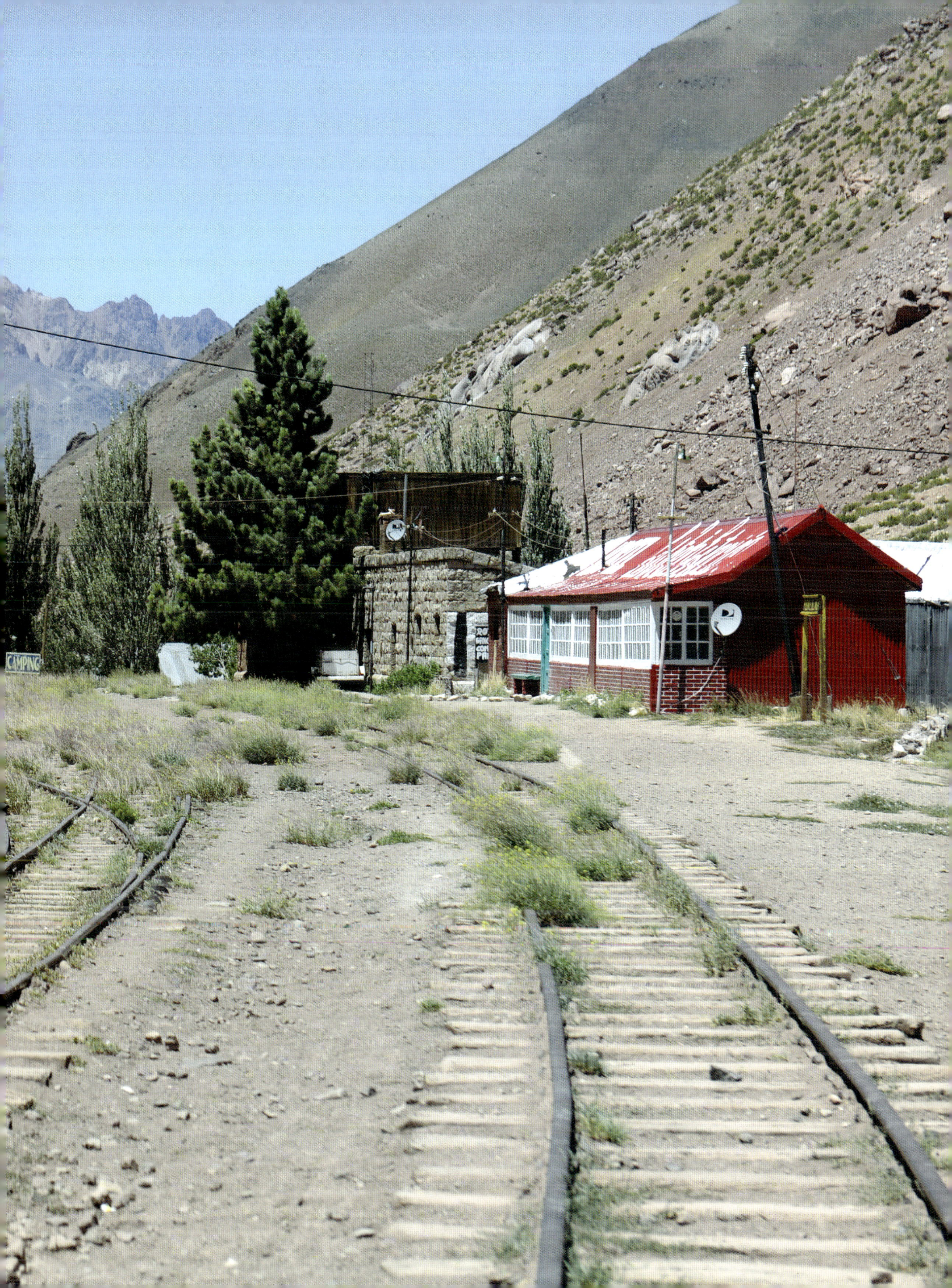
CAMPING

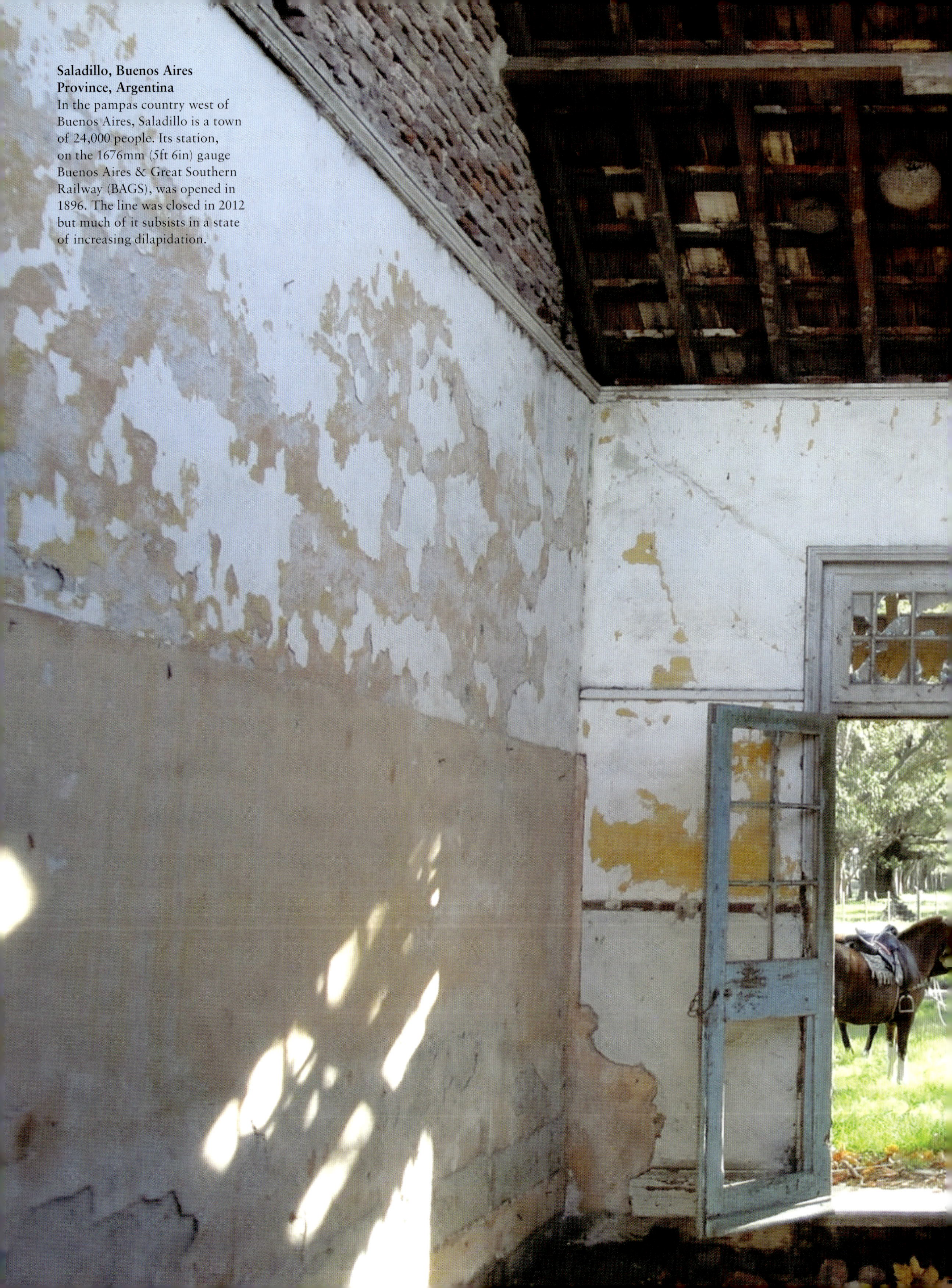

Saladillo, Buenos Aires Province, Argentina
In the pampas country west of Buenos Aires, Saladillo is a town of 24,000 people. Its station, on the 1676mm (5ft 6in) gauge Buenos Aires & Great Southern Railway (BAGS), was opened in 1896. The line was closed in 2012 but much of it subsists in a state of increasing dilapidation.

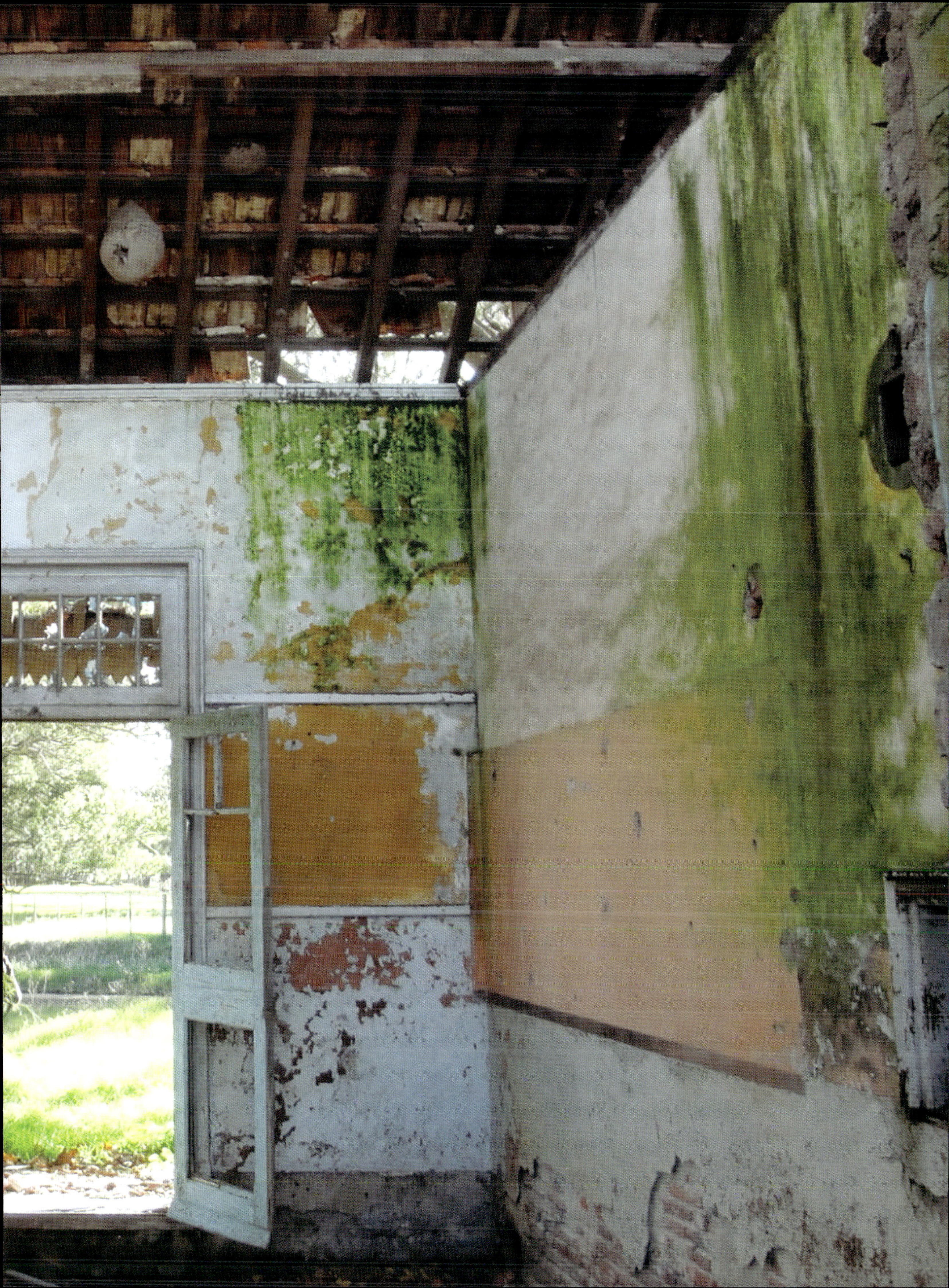

Arribeños, Buenos Aires Province, Argentina
The station opened in 1902 on the Ferrocarril Buenos Aires al Pacifico (BAP), a typical wayside station on what was intended to be an international main line. Passenger services continued into the 1990s, when many other local stations across the country were also shut down. Occasional freight services still use the line, and there are hopes that passenger trains may be reinstated.

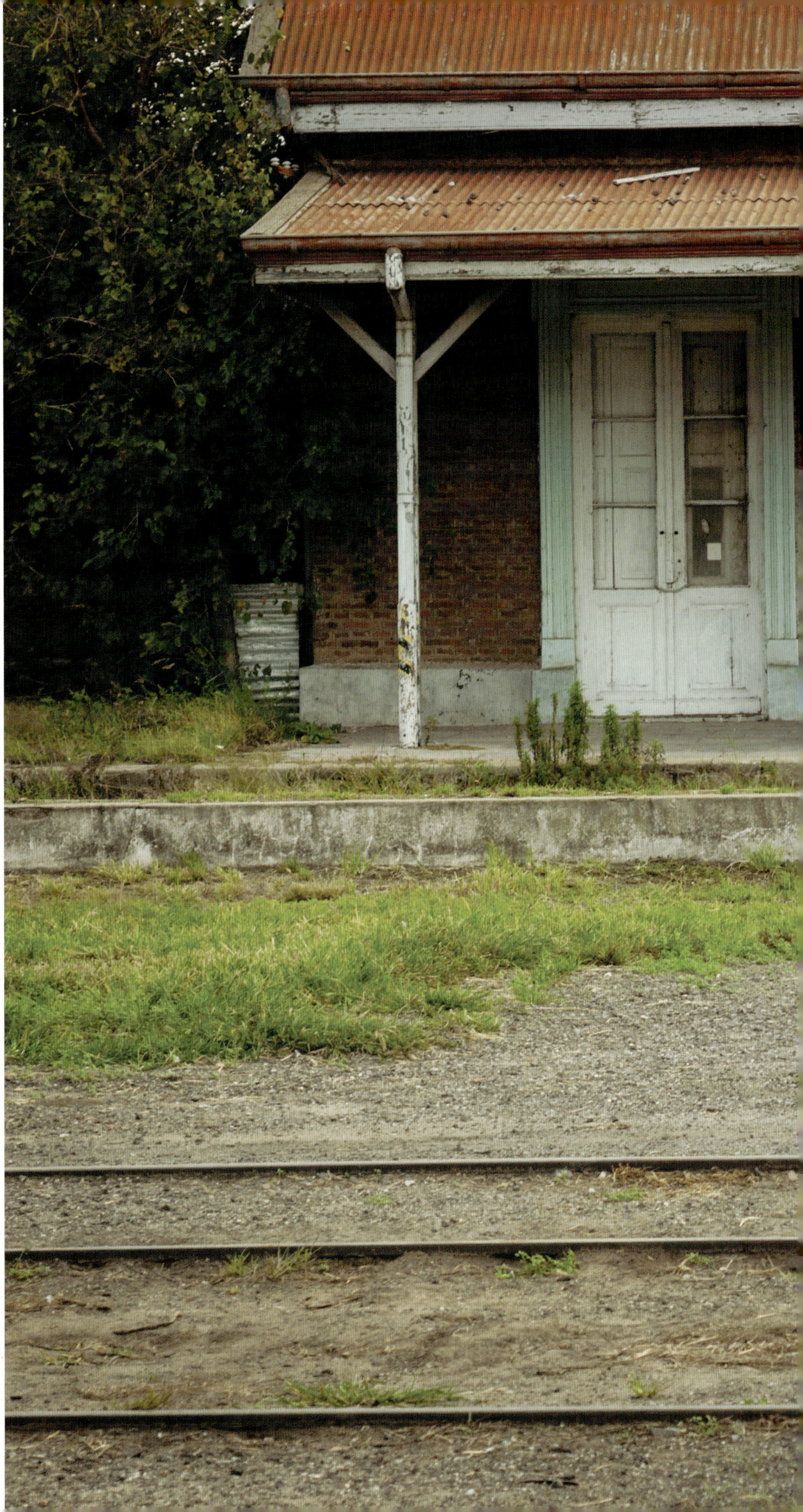

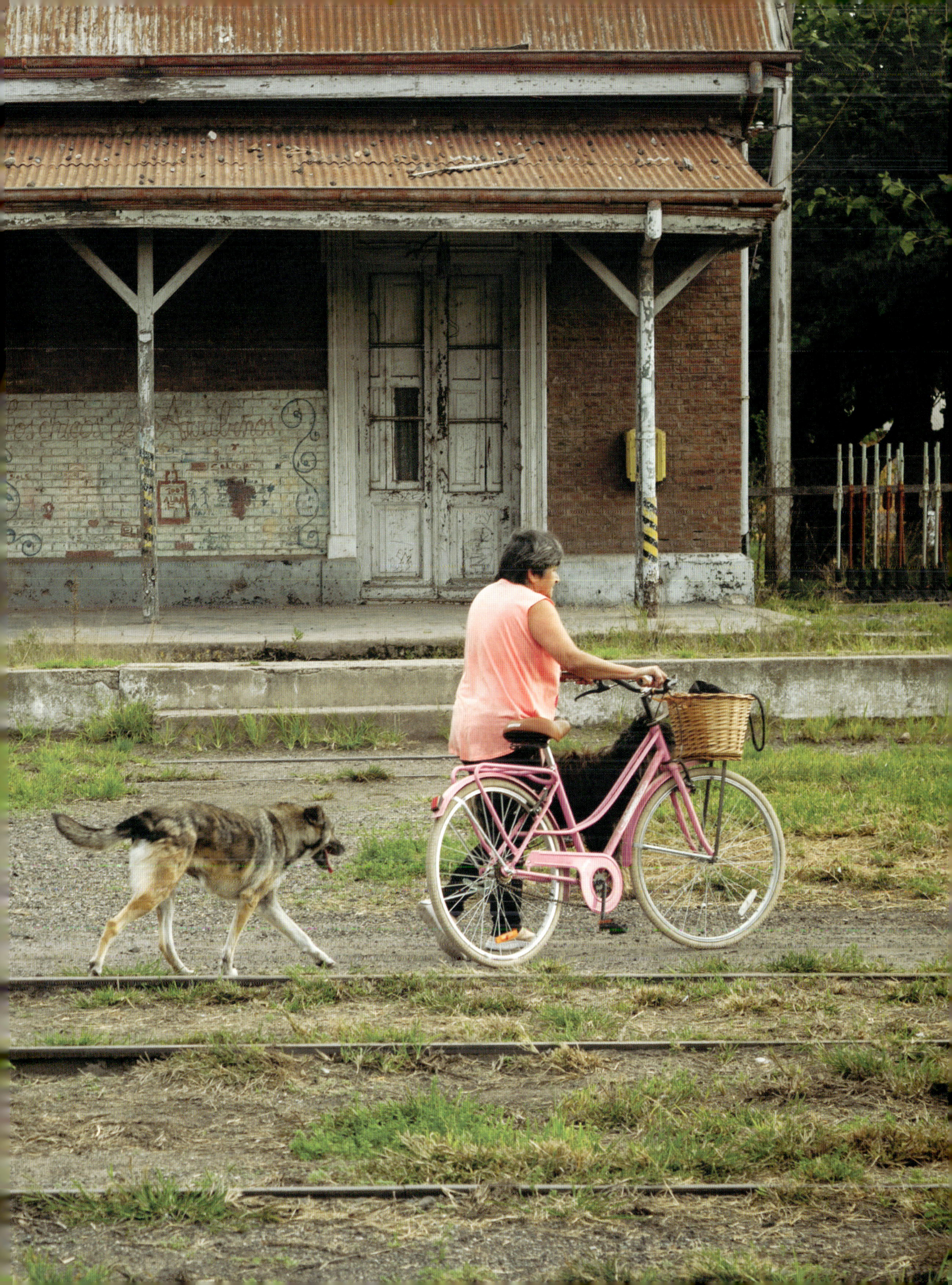

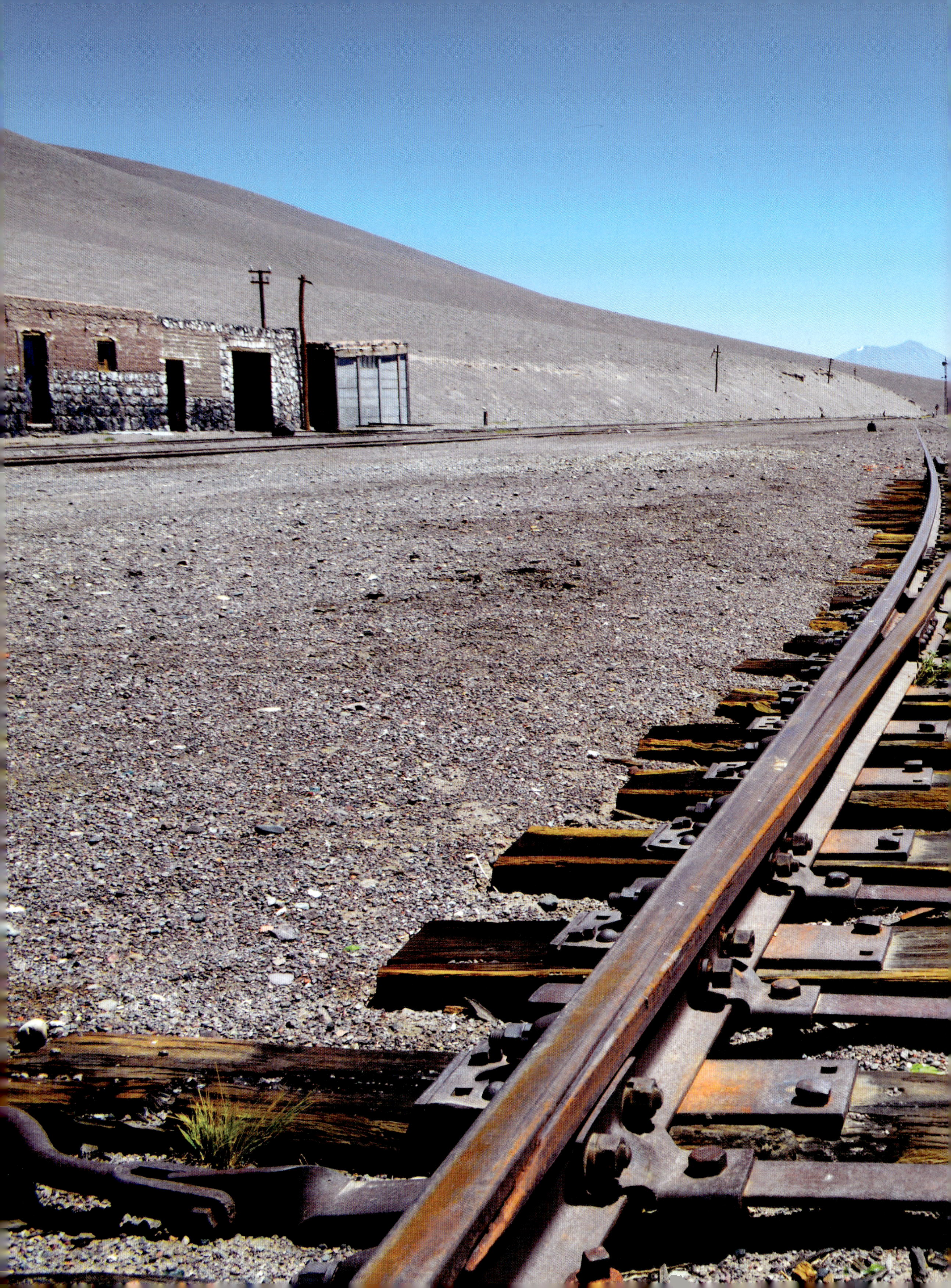

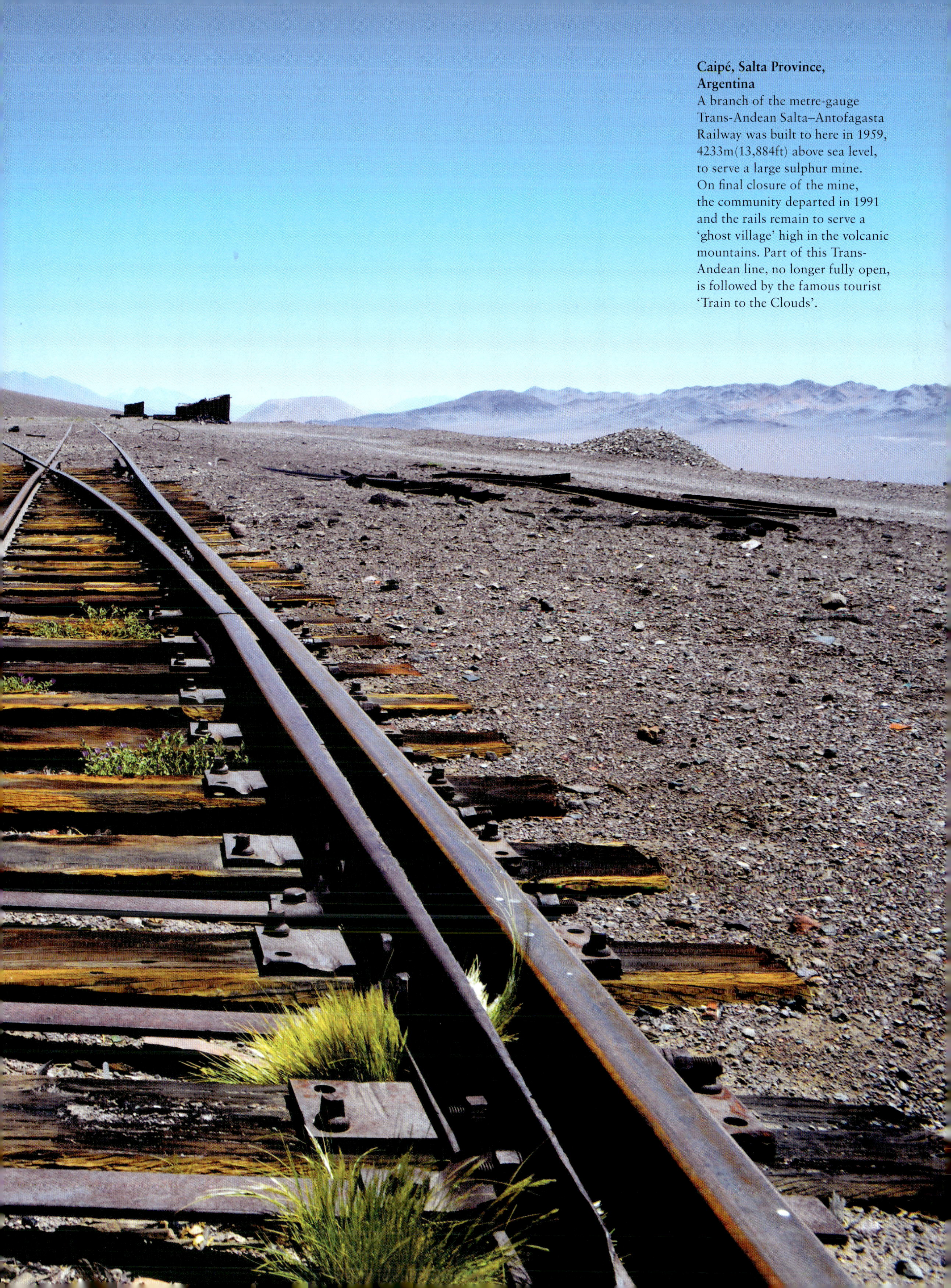

Caipé, Salta Province, Argentina
A branch of the metre-gauge Trans-Andean Salta–Antofagasta Railway was built to here in 1959, 4233m(13,884ft) above sea level, to serve a large sulphur mine. On final closure of the mine, the community departed in 1991 and the rails remain to serve a 'ghost village' high in the volcanic mountains. Part of this Trans-Andean line, no longer fully open, is followed by the famous tourist 'Train to the Clouds'.

Picture Credits

Ajznponar: 66 top (Creative Commons Attribution-Share Alike 3.0 Unported License)

Alamy: 7 (Rick Pisio/RWP Photography), 10 & 11 all (GP Essex), 14/15 (markferguson2), 17 (Stephen Knox), 18 (David South), 19 both (Jim Laws), 20 top (Kumar Sriskandan), 20 bottom (Dominic Jones), 21 (wyrdlight), 22/23 (Richard Naude), 24/25 (Avalon/Construction Photography), 26/27 (Cofiant Images), 28/29 (John Bentley), 32 & 33 all (Eden Breitz), 40 top (Photo12), 40 bottom (Raquet Photography), 41 (agefotostock), 43 bottom (Meritxell Cortes Francisco), 50 (Rob Cousins), 55 bottom (Stephen Birch), 76 bottom (Chris Bloom), 77 (Keith Parsons), 78/79 (Eric Nathan), 80 & 81 top (Edwin Remsberg), 81 bottom (Robert Harding), 90 & 91 all (Berengere Cavalier), 92/93 (Ian Fleming), 94/95 (Shay Levy), 96/97 (Eddie Gerald), 105 top (Graham Prentice), 106/107 & 108 bottom (Sipa US), 108 top & 109 bottom (Imaginechina Limited), 109 top (Andrew Brenton), 112 top (Joonso Kim), 113 & 114/115 (Trevor Mogg), 118/119 (Hideo Kurihara), 124 & 125 all (REY Pictures), 131 bottom (Stephen Dwyer), 149 (CNmages), 174 (Ei Katsumata), 175 bottom (Zuma Press), 176 & 177 (ML Pearson), 184/185 & 186 (Rick Pisio\RWP Photography), 187 top (Jonathon ORourke), 187 bottom (Joanna Kalafatis), 188/189 (Rod Edwards), 196 (Philip Game), 197 bottom (Jeremy Richards), 208 top (Jan Jerman), 209 top (Jeremy Graham), 209 bottom (Photononstop)

Dreamstime: 6 (Mulderphoto), 30/31 (Ivan Vander Biesen), 34/35 (Steven Hatton), 48/49 (Fernando Garcia), 52/53 (Rubengalvezfoto), 56/57 (Claudio Testa), 58 top (Nenad Nedomacki), 58 bottom (Dragoncello), 59 top (Slobodan Makic), 60/61 (Emmanouil Pavlis), 71 (Nadzeya Dzehtsiarova), 76 top (Hongqi Zhang), 82/83 (Javarman), 84 top & 85 (DiversitySignatures), 99 bottom (Kateryna Levchenko), 122/123 (Lowlihjeng), 128/129 (Miroslav Liska), 130 top (Filedimage), 131 top (Kim Britten), 134/135 (Brian Scantlebury), 136 (Intoit), 140 bottom (David Butler), 141 bottom (Rlesyk), 148 (Thomas Maroulis), 152/153 (Demerzel21), 154 & 155 all (Shermancahal), 160/161 (Kenishirotie), 172/173 (Alisha Hicks), 182/183 (Littleny), 201 & 204/205 (Alf Ribeiro), 206/207 (Joa Souza), 210/211 (Ivo De), 212/213 & 214/215 (Danflcreativo), 218/219 (Alex Irzyk), 222/223 (Adrian Gallardo)

Getty Images: 8 (Ciaran Murphy/500px), 12/13 (2c image), 16 top (McLean Photography), 36/37 (Cartsen Koall), 38/39 (Andreas Mohaupt), 51 top (Westend61), 59 bottom (Gabrielle Therin-Weise), 63 & 64/65 (Julian Ward), 70 both (LeliaSpb), 72/73 (Franz Aberham), 74 (Zelda Gardner), 84 bottom (Seyllou), 86/87 (Pierre-Yves Babelon), 98, 99 top & 100/101 all (Joseph Eid), 102 & 104 all (Deshakalyan Chowdhury), 110/111 (Lingjing Liu/EyeEm), 130 bottom (Southern Lightscapes-Australia), 138/139 (Insight Imaging), 140 top (Richard F Wintle), 141 top (Dougall_Photography), 156/157 (MyLoupe), 162/163 (Timothy Fadek), 164/165 & 166/167 (Raymond Boyd), 175 top (GeoStock), 178/179 (David Kozlowski), 190/191 (Cavan Images), 216/217 (Daniel Garcia)

iStock: 62 bottom (gionnixxx), 88/89 (mtcurado), 158/159 (Nicola Patterson), 192 (JulianneBirch), 202/203 (Cidney Photos)

Shutterstock: 16 bottom (Joe Dunckley), 42 (bellena), 43 top (EQRoy), 44/45 (Obatala-photography), 46/47 (Rjc 9666), 51 bottom (Xesproductions), 54 top (Juanjo-Almeria1), 54 bottom (Philip Openshaw), 55 top (Shanti Hesse), 62 top (Mehmeto), 66 bottom (Nightman1965), 67 top (Fotokon), 67 bottom (Maxim Mitsun), 68/69 (Zhukov Oleg), 105 bottom (Sabih Ullah), 112 bottom (Yeongsik Im), 116/117 (Shaun in Japan), 120 & 121 all (Dave Siva), 126/127 (Husni), 132/133 (Gareth Lowndes), 142/143 (James McSweeney), 144/145 (Brian Logan Photography), 146/147 (Joseph Sohm), 150/151 (kropic1), 168/169 (BD Images), 170/171 (Sabrina Janelle Gordon), 180/181 (Rexjaymes), 194/195 (Jorge Burneo Celi), 197 top (JordiStock), 198/199 & 200 (Enfocale), 208 bottom (Atosan), 220/221 (Lucas_¬creativepics)
Ministry of Land, Infrastructure, Transport and Tourism of Japan: 178/179